Henry Stephens Randall, George Wilkins Kendall

Sheep Husbandry

With an Account of Different Breeds, and General Directions in Regard to Summer

and Winter Management, Breeding and the Treatment of Diseases

Henry Stephens Randall, George Wilkins Kendall

Sheep Husbandry
With an Account of Different Breeds, and General Directions in Regard to Summer and Winter Management, Breeding and the Treatment of Diseases

ISBN/EAN: 9783337247744

Printed in Europe, USA, Canada, Australia, Japan

Cover: Foto ©Lupo / pixelio.de

More available books at **www.hansebooks.com**

AN ACCOUNT OF THE DIFFERENT BREEDS,

AND GENERAL

DIRECTIONS IN REGARD TO SUMMER AND WINTER MANAGEMENT, BREED-
ING, AND THE TREATMENT OF DISEASES.

𝔚𝔦𝔱𝔥 𝔓𝔬𝔯𝔱𝔯𝔞𝔦𝔱𝔰 𝔞𝔫𝔡 𝔬𝔱𝔥𝔢𝔯 𝔈𝔫𝔤𝔯𝔞𝔟𝔦𝔫𝔤𝔰.

BY HENRY S. RANDALL, LL. D.,

LATE SECRETARY OF STATE OF THE STATE OF NEW YORK.

WITH HIS LETTER TO THE TEXAS ALMANAC ON

SHEEP HUSBANDRY IN TEXAS,

AND

GEO. W. KENDALL'S ON SHEEP RAISING IN TEXAS.

New-York:

ORANGE JUDD COMPANY,

245 BROADWAY.

1880.

Cork.

PREFACE.

So full and complete is the exposition of the subject as discussed in the following pages, and so clearly are the scope of the work and the circumstances which prevailed with its accomplished author, to pass it through our hands to the public, explained in his own "INTRODUCTION," that only in compliance with a common custom in book-making, might any thing have been deemed necessary in the way of preface; were it not to acquit ourselves of the obligation to tender thus publicly to Col. RANDALL, not our own thanks merely, but those of the agricultural community, for the great benefit which must ensue to it, in the proportion that this instructive contribution to the stock of our agricultural knowledge and literature may command the attention to which it is, on every account, so well entitled.

An agricultural correspondence, reaching far back, and spreading widely over the Southern States, to which has been more recently added considerable extent of personal observation, had with us, already established the conviction, that in no other part of our country, perhaps, does there exist a resource at once so fruitful, and so little availed of, as that which is possessed in that region, for the prosecution of this—one of the most interesting and important branches of Husbandry that any country can enjoy.

But while it has been easy to perceive this defect so apparent in their agricultural economy, amounting in the aggregate to a national loss of no inconsiderable magnitude; it was not so easy to expose, as Col. Randall has done, the fallacy of the difficulties that were supposed to stand in the way, or to indicate how the real impediments which do exist may be overcome, or materially mitigated.

Something of these imaginary difficulties, for successful Sheep Husbandry, may, as we believe, be assumed to have their origin in the prejudices engendered in the minds of Southern agriculturists, by the sweeping condemnation of it to be found in the celebrated and deservedly popular essays of ARATOR, by Col. John Taylor—*clarum venerabile nomen!* and it may be that these prejudices are referable in a degree, also, to the concurrent opinions of the no less celebrated John Randolph, "of Roanoke," who, even on the floor of Congress, gave them utterance in vehement and bitter denunciation against the harm-

less animal itself—going so far in his animosity to it, and to all encouragement of the great industry which it was formed to subserve,
as to declare, that he would at any time go out of his way *" to kick a
sheep!"* Nor would it be unreasonable to apprehend that these impressions against the policy and profit of sheep-breeding, as an important object of attention for the Southern land-holder, have taken root
the more kindly in the minds of a people unaccustomed, if not naturally averse to that careful and minute attention which the successful
prosecution of this business demands—a people whose sons, it may be
feared, still find it easier, if not more commendable, to follow in the
venerated footsteps of their sires, than to encounter for themselves the
labor of investigation, and the trouble (together with some expense)
of new arrangements incident to every new employment of labor and
capital. The general impression, in fact, is, (the reader will judge
how far it is just,) that cultivators of the soil everywhere are, of all
classes, the least apt to embark in any new enterprise, however promising. They *talk* and *talk* about it, but rarely go about; and perhaps it may be better that it should be so; yet it is well to remember
that precipitancy is one thing, and *torpor* quite another! We once
knew a farmer (so called) in Calvert county, who, being told, as he sat
toasting himself in the chimney corner on a cold winter's night, that
the house was on fire! without moving from his seat, answered, *"call
the people!"*

In opposition to all that has been urged or imagined against Sheep
Husbandry in the South, on the score either of ill-adapted climate,
deficiency of suitable forage, want of adequate demand for wool, or
other obstacles, the whole subject has been so admirably and thoroughly
canvassed in the work here offered, that further argument would be
superfluous; otherwise we might oppose to the hitherto prevailing belief, if not prejudice, the experience of some, on a limited scale, and
the well-settled opinion of yet many more among the most enlightened
of our acquaintances in that region—gentlemen uniting ample oppor-
'unities with close habits of observation on all questions of rural
economy, and who have not hesitated to express the confident belief,
that profitable and interesting as has been the growing of cattle in
western Virginia, an equal amount of capital and attention, devoted
to sheep and *wool growing* in the same section of country, would be
yet more remunerating. Looking for reliable information yet further
south, and back to a period more remote, even anterior to our *declaration of independence*, it may not be out of place to quote an evidently careful and intelligent author of a work on the climate and
products of each of the then English colonies. Speaking of *Georgia*,
and her well-ascertained adaptation to the growth of silk, the vine,
the olive, madder and *wool*, he remarks: " Wool, we [England] take

in large quantities from abroad, because it is of a kind we 'cannot produce in England : our colonies on the continent of North America, *South of New York*, produce a wool *entirely similar to the Spanish.* No staple they could produce would, therefore, be more advantageous to Great Britain. It is well known that a piece of fine broadcloth cannot be made without Spanish wool; it is also known that the Spaniards have of late years made great efforts to work up their own wool; if they should succeed, or if they should by any other means prevent the export of it, our woollen fabrics, though they might not be stopped, would at least be burdened with a fresh expense and a new trouble; all which would be prevented by *encouraging* the *import* of wool from America : and at the same time that this good effect was wrought, another would be brought about, *in cramping the manufactures of the colonies.*"

Unfortunately for the agricultural interest of our country particularly, the desire to " cramp the manufactures of the colonies," here so candidly avowed as the settled policy of England, not only survived the Revolution, but has been so well fostered by our own subserviency to it, as to render our independence, in respect of this and other no less important industrial pursuits, rather nominal and fictitious than substantial and true; nevertheless, with the odds of pauper labor and immense capital against us, thanks to the ingenuity and enterprise of our people, we need not despair of final success with any thing like fair consideration on the part of our own government. For this opinion we need have no better authority than that of SAMUEL LAWRENCE, the enlightened and liberal proprietor of the Middlesex Mills, at Lowell, who says, " the business of manufacturing wool in this country is on a better basis than ever before, inasmuch as the character, skill, and capital engaged in it are such as to defy foreign competition." Occasional revulsions, such as the present, will occur from causes abroad over which we have no control, but let not the wool grower relax in the care of his flock, for the same far-seeing manufacturer has declared that he could point to articles of wool now imported, that will require thirty millions of pounds of medium and fine quality to supply the demand.

After all, then, on viewing the importance of the inquiry to numerous friends for whose welfare we profess to entertain unaffected concern, and the great extent of the district which seemed to us to be so well adapted to the growth of sheep and wool—the magnitude of the interests involved swelled upon the contemplation, begetting a conviction that as a question of practical agriculture, it was not to be worthily and well treated by a few hasty and superficial essays, or by more elaborate compilations in relation to the oft-repeated natural history of the animal, its prominence in scriptural annals, &c., unsustained by that laborious and discriminating comparison of facts and authorities to illustrate its uses and its value, and by that fulness of personal experience in the breeding and management of the various races, "in sickness and in health," which constitute the excellence of these letters to Col. Allston.

Under all these circumstances, the reader of the work here presented may well judge how fortunate that it should have been undertaken, *con amore*, by a gentleman so well prepared by general scholarship, by exact practical knowledge, and by extensive inquiry into the mercantile and manufacturing, as well as the agricultural bearings of the question.

It is due, however, no less in justice to ourselves than to truth, to add, that in urging him to undertake it, we had no idea of committing the author to such an amount of labor, even had we foreseen that being, as he says, a "labor of love," it would have thus ended in producing, as in our judgment it has, decidedly, the best work on the subject of She   .   .s at any time appeared in our country.

May we not refer, for the soundness of this opinion, as well to its originality and strictly American character, as to the comprehensiveness with which it presents the subject in its various relations, instructive alike to the merchant, the manufacturer, the political inquirer, and the legislator; as to the practical farmer ?   Nor is it to be characterized alone by its utility in these respects; for the reader will agree with us that its entire fairness and freedom from narrow views and local prejudices, much enlarge, in a moral and instructive view, its title to general confidence and favor.

Finally, as far as the public judgment may be anticipated in reference to a production originally appearing disadvantageously, in detached parts, and not until now finished and embodied; if we may conclude from the favorable manner in which such portions have been reviewed and recommended, by some of the leading journals of the country, the writer may well felicitate himself on having rendered a most acceptable service not only to his brother farmers, but to his countrymen generally; while we may unaffectedly, and, as we think, confidently add, it establishes for Col. Randall himself a claim to stand in front of those whose pens, some of them under high motives of patriotism, have been engaged in illustrating one of the most important of all our industrial pursuits; nay, one which may be considered essential, in an eminent degree, to our national independence.

J. S SKINNER.

INTRODUCTION.

Tʜᴇ subject of Sheep Husbandry has recently attracted more attention in our Southern and South-western States, than at any previous period. The want of a staple or product, the cultivation of which should render productive the capital invested in millions of acres of mountain and other lands, which do not now yield a farthing of income, and which, from their soils, situation, or other circumstances, are unadapted to the growth of any of the present Southern staples, has struck every Southern man, as well as every traveller of ordinary intelligence, who has passed through the regions indicated. The want, too, of some class of domestic animals to constitute the basis, or *pivot* as it were, of a system of convertible husbandry on the tillage lands of the South, to take the place of the present imperfect rotations of crops, and new and old field-system, has become apparent to many of her more investigating agriculturists.

The fact that the mountain and other unproductive lands alluded to cannot be made to profitably yield any vegetable products but pasturage ; that for the present, and for a long time to come, at least, the bulk of them will not afford a pasturage adapted to the support of large animals ; could not but suggest the growing of wool, as their best, if not their only available staple. The similarity of their general climate, too, with that where wool is most cheaply grown on the Eastern Continent, was a consideration promising favorably to this husbandry. And, finally, it had not failed to strike men of ordinary commercial intelligence, that of those animal staples, to the production of which a Southern climate is adapted, the Sheep furnishes a vastly more marketable one than any of the larger grazing animals.

The superiority of the Sheep over other animals for supporting the fertility of tillage lands, by converting a portion of their products into manure, was not so apparent. But the well-known fact that they receive the preference for this purpose, in some of the best agricultural countries of the world, made it sufficiently probable to demand a full investigation, before adopting an adverse conclusion, especially as what has been said in relation to climate and the marketableness of animal staples, was as applicable to these lands, in the South, as to those adapted only to grass.

But Sheep Husbandry as a system, and especially a system tested by experience, was scarcely known in any of the Southern States excepting in western Virginia. Whether the theoretical considerations and natural circumstances which apparently favored its introduction would be met, in practice, with unforeseen obstacles, was a matter calling for grave circumspection. The Southern agriculturist is ever wary of innovation, and very properly averse to rash experiment. He knew, it is true, that his roving and untended "native" sheep obtained subsistence, and found no

7

enemies to their health but the wolf and cur, on all the Southern zones. But
whether the local climate and herbage of those different zones—the low, level,
Tertiary sands of the Atlantic plain—the granite hills of the middle, and the elevated Paleozoic or Transition regions of the mountain zone—would be found to
agree with the more valuable breeds of sheep; whether their wool would retain its
qualities or degenerate in these several localities; whether a greatly increased sup
ply of wool would find a remunerating price in market; whether the mountain
could be converted into sheep-pastures, and wool produced on them without an
expense which would absorb all the profits; whether Sheep Husbandry could be
made a substitute for "resting," or expensive artificial manures, in restoring to the
cotton, tobacco, and grain lands of the middle and tide-water zones the fertility
withdrawn by tillage; and various other important correlative questions were all
problems to him. And to add to the difficulties of forming a correct opinion, and
especially of instituting safe and satisfactory experiments, he was ignorant of all
the practical details and manipulations of Sheep Husbandry : he knew little of the
various breeds, and their respective adaptation to his wants.

For information on the subject of practical Sheep Husbandry and breeds of sheep,
there are a multitude of European, and several American works, of great value.
But for the answers to the questions in the preceding paragraph, which involve the
particular bearings and adaptation of this husbandry, of the different breeds, etc.,
to the agricultural circumstances and wants of the various regions of the South—
where was the inquirer to find the desired information ? Some well-written letters,
embracing portions of these topics, have appeared from time to time in our agricultural journals. They have been of great value in drawing attention to the subject.
But they have not usually occupied limits sufficient for the examination of more
than a single phase of the general subject, or they have been mere *coup d'œils* of
that subject, omitting all but a few important facts and considerations of a general
character. They have, too, usually been replied to, or published contemporaneously
in the same or other agricultural journals, with contradictory statements—sometimes with crude and erroneous speculations—calculated to confuse or mislead the
inexperienced inquirer. Beyond these occasional Letters in the agricultural journals, nothing, so far as I am aware, has appeared on this subject.

A practical farmer, I have bred nearly all the approved varieties of almost every
kind of domestic stock—of *every* kind commonly kept on Northern farms—and
have been familiar with the details of their management and husbandry. I have
owned flocks of sheep, and been more or less familiar with them, from my childhood ; and for the last fifteen years have made their economy, their habits, their
comparative profitableness with other kinds of stock, and the comparative value of
their breeds, matters of careful and constant observation and experiment.

When Corresponding Secretary of the New York State Agricultural Society, a
few years since, the facts drawn out by me in an extensive correspondence with
eminent Southern agriculturists, united to what knowledge I had previously obtained by reading and personal observation of the Southern States, led me to th
impression that there were numerous considerations and natural circumstances
strongly indicating the expediency of introducing wool-growing extensively into
those States. But at that time, my attention, in common with that of many if not
most, of the Northern flock-masters, was turned towards the prairies of the North-
west, as a region capable of sweeping away all American competition in this branch
of husbandry. Glowing estimates and calculations had been predicated on very
partial experiments. The value of the natural grasses, the character of the winters
and general climate, and the general facilities of the prairies for wool-growing, were
then little understood here, and had been made the subjects of much favorable exag-

geration. Facts subsequently ascertained, have, it cannot be denied, materially changed the impressions of our flock-masters on this subject. Whether correctly or incorrectly, they no longer fear Western competition in growing *fine* wool. My own coincides with the popular impression on this topic, if we consider that competition in its relations to a period not far distant in the future.

The adoption of these views led me to again turn my attention, never entirely withdrawn, more particularly to the capabilities of the South for this branch of husbandry. My conclusions and the reasons for them will be found in the following Letters. In a letter to Hon. Robert J. Walker, Secretary of the Treasury, published in his Treasury Report of 1845, and in a series of letters published in the Virginia " Valley Farmer," the same year, I stated some of the general conclusions I had then arrived at on this topic. These publications were followed by letters from gentlemen residing in Virginia, North and South Carolina, Tennessee, Mississippi, Louisiana, and Arkansas, making farther inquiries, and usually imparting more or less local information on the subject. Some of these were practical men, only seeking information on practical points ; others, eminent for intelligence and legislative experience, embraced a more comprehensive field of investigation, and sought from me, as probably from other sources, to ascertain by a wide range of general facts and statistics, the probable bearing, now and in future, of an extensive system of wool-growing on the Agriculture, Commerce, Manufactures, domestic consumption—in short, the whole domestic economy of our Southern States.

Answers to these questions demanded careful investigation, and involved a great variety and complexity of details in the practical department of the subject, rendered far more numerous by the wide differences existing between the soils, established husbandry, and even the climates, of the three distinct and well-defined zones already alluded to. The location of some of my correspondents was on the mountains of Virginia, the Carolinas, and Tennessee—others on the hilly zone of the same States—others on the Tertiary sands of the tide-water zone, and the Cretaceous plains of the Mississippi and Arkansas. To give opinions on all the topics referred to, and in reference to natural circumstances so various, supported by even a respectable show of corroborating facts, was an undertaking requiring considerable time and labor : to repeat them separately to each correspondent, was wholly out of the question.

Requested by Mr. Skinner, a little more than a year since, to prepare a series of Letters on Sheep Husbandry, and especially on Sheep Husbandry in the South, for The Farmers' Library, it occurred to me that a compliance with his request would enable me to answer each of my correspondents by once writing; and moreover, I could feel, under such circumstances, that I could properly afford to bestow an amount of time and elaboration on my communications which I should otherwise find impracticable. And I confess, I also thought if the information I could impart would prove of value to my personal correspondents, it might also prove so to many others among the numerous readers of a popular agricultural magazine. The liberal offer of the Publishers to provide all such cuts as I should choose to direct, was an additional inducement to adopt this medium of communication. I have often felt the want of these in agricultural letters of my own, and in reading the works of others. In describing a breed of sheep, for example, to a person who has never seen them, the best chosen words convey but a vague impression. In many other cases also, cuts exhibit at a glance what it would require much circumlocution to describe ; and they in many instances convey ideas to the mind with a definiteness, correctness, and exemption from possibility of misunderstanding, which words alone never could. The cuts include portraits of all the breeds which I supposed could of possibility possess, or *claim* to possess, superior value, for any region o

B

locality within the United States; all the necessary anatomical figures, with those of the less known insect and parasitic enemies of the sheep; and finally, representations of every implement, fixture, or process employed in Sheep Husbandry where I thought they would convey important information—and particularly *new* information—more clearly than it could be done by words. Many of the latter class of illustrations have never before been, so far as I am aware, attempted; and, representing as they do the results of years of inquiry and experiment, I trust they may prove of service to *beginners*—particularly in regions where Sheep Husbandry has been hitherto little known.

The Letters were begun and concluded exclusively as a "labor of love." To possess the consciousness that even a limited portion of my fellow-men have been benefited by my labors, would be all, and the noblest recompense to which I could aspire. Nor do I feel, that in attempting to benefit the agriculturists of one section of our country, by urging them to appropriate a branch of industry now giving subsistence to those of another section, I am seeking the good of the former at the expense of the latter. Every region has natural advantages, or those resulting from the natural course of events, for different branches of industry. A right to these advantages enures from a right to the soil; and the former is just as natural and sacred a right as the latter. To attempt to wrest them from the holder by legislation, is oppressive; to withhold from him any knowledge necessary to the enjoyment of them, is unfraternal and unmanly. If Virginia can grow wool, or any other staple, more cheaply than New York, let her do it. She will only force New York to fall back on the production of some other staple, or to adopt some other branch of industry. And why not? Why should there not be a division of production, where it is called for by natural circumstances, at least within the limits of a common nation? It is doubtless well for every region, whether extensive or limited, to produce its own necessaries of life to the greatest *economical* extent. But an attempt to force Nature against her manifest capabilities, for the sake of attaining a fancied local independence, is to inflict a real evil, in the hope of attaining an imaginary good. History is full of instances where the prosperity of large masses of individuals, and even whole nations, has been crippled, in futile efforts to upbuild this or that branch of industry, in spite of natural obstacles, or against the competition of regions possessing greater natural advantages. Among the foolish, selfish, and even iniquitous legislation of past ages, there has been none perhaps productive of more real mischief to human industry than the intermeddling enactments of governments, ostensibly designed for its benefit. Masses of men, because divided by a rivulet, speaking a different language, or owning the sway of different potentates, have aspired to that physical independence of each other, and of the whole world, which the God of nature rendered economically, if not absolutely impossible. The vexatious restrictions on trade and commerce imposed in pursuit of this object by one government, were met by retaliatory ones by others, until international commercial legislation became a confused labyrinth of enactments—their absurdity only equaled by their mischievousness. And like the elephants formerly used by barbarian nations in battle, they nearly as often trampled down their friends as their enemies. The era of these things is rapidly passing away. That patriotism which includes only a province or State, among one common people, is beginning to be recognised as narrow and sordid: nay, among intelligent men, that philanthropy is beginning to be thought meagre and unexpansive which stops even at the boundaries of Nations.

In preparing the following Letters, I have labored under disadvantages inseparable from the circumstances under which they have been prepared. I have written them from month to month, amid the hurry of other pursuits, with little idea of what would be their ultimate limits—usually with one or more of the immediately preceding numbers in the hands of the printer, and consequently not under my inspection. I have not therefore had that opportunity to proportion the space devoted to the several topics, avoid repetition, and correct errors, possessed by him who completes and revises, before any portion of his manuscript is rendered unalterable by stereotyping.

Reliance on insufficient authority has in a very few instances led me into errors, but fortunately, so far as discovered, they have been of trifling importance, and in relation to matters of no especial moment. Those thought worthy of notice have been corrected in subsequent parts of the body of the work. The causes I have named, therefore, affect rather the literary character of the Letters, than their general accuracy.

In stating important facts and conclusions, I have consulted such writers of reputation as were within my reach. Among the foreign ones who have prepared works on Sheep Husbandry, or expressed important opinions on some of its separate topics or facts, or who have alluded to the Sheep Husbandry of particular countries or nations, reference has been had to the following, either by consulting their works, as I have in most instances been able to do—or by quotations from them found in the works of other writers of reputation;—Anderson, Bakewell, Barnes, Barrow, Bischoff, Blacklock, Bourgoing, Bright, Carr, Coventry, Culley, Cunningham, D'Arboval, Darwin, Daubenton, Dick, Ellman, Gasparin, Gilbert, Goese, Harrison, Hogg, Hood, Howitt, Hubbard, Jacob, Lang, Lasteyrie, Leeuwenhoek, Lichstenstein, Linnæus, Low, Luccock, Maitland, Malte-Brun, McCulloch, Moffat, McKenzie, Paget, Parkinson, Parry, Petri, Pictet, Powell, Reaumur, Rodolphi, Sinclair, Slade, Southey, Spallanzani, Spooner, Stephens, Swaine, Trail, Trimmer, Valasnieri, Vanderdonk, Von Thaër, Walz, Western, Willmer & Smith, Youatt, Young, and some others. Of our domestic writers, I have aimed to consult all of the most prominent ones. It is not necessary to enumerate them, extending, as the list would, to hundreds.

The examination of these writers, foreign and domestic, has been no recent undertaking with me. For years, I have found it a source both of instruction and pleasure, to peruse their works. Where they have proposed any thing new to me, which I thought promised favorable results, I have usually sought the first opportunity to put their propositions to the *experimentum crucis* of actual trial. I have often thus learned valuable facts. But I have nearly or quite as often ascertained that what may be true of one breed, in one climate, or under one set of circumstances, is not true when all or a part of these conditions are changed. The English and German systems of management, for example, I regard as almost wholly inapplicable here, on account of the entire different relation which the prices of land and labor bear toward each other in those countries and our own. And I sometimes have had the conviction forced upon me, that writers even of reputation have assumed positions in relation to practical matters, which they must have derived from other sources than direct personal experience.

While I have carefully reviewed and collated the opinions of other writers on doubtful practical points, I have in all instances, as will be seen in the following pages, preferred the results of personal experience and observation, to adverse authority, however eminent. Compilations, it seems to me, are sufficiently abundant, and I have thought it better to give my own opinions, leaving them to stand or fall, as they shall be found accurate or inaccurate. Where I have found it necessary to rely on others for any fact, or have quoted their opinions, I have uniformly given them credit. To my kind correspondents, particularly my Southern correspondents —many of whose communications are not published on account of their reluctance to be cited as authority for facts, where their modesty leads them to underrate their own comparative knowledge and experience—I tender my thanks for their assistance.

I have addressed the Letters to Col. R. F. W. ALLSTON, of Waccamaco Beach, near Georgetown, South Carolina—a gentleman to whom I am indebted for much valuable information on the subject of Southern Agriculture, and who has ever evinced a most earnest desire to contribute to the improvement of that Agriculture.

HENRY S. RANDALL.

CONTENTS.

SHEEP HUSBANDRY

IN THE

UNITED STATES,

IN A SERIES OF LETTERS TO R. F. ALLSTON,

OF SOUTH CAROLINA.

LETTER I.

EFFECT OF CLIMATE ON THE HEALTH AND WOOL-PRODUCING QUALITIES OF SHEEP.

—

Introductory Remarks...Wool-Growing and Manufacturing Statistics of the Southern States compared with those of New-York...Effect of Warm Climates on the Health of Sheep...Sheep in the Southern States below latitude 32°...Effect of Climate on Wool-Producing Qualities of Sheep—on the Quantity of the Wool...Weight of Fleeces in the Southern States indicated by U. S. Census of 1840—Important Omissions in that Census—Other important Errors in it...Table of Weights of Fleeces in Four Counties where they average highest in each of the Southern States and in New-York—Latitude, Topography and Climate of those Counties...Warmth of Climate conducive to the Production of Wool—Reasons.

—

R. F. W. ALLSTON, Esq—

Dear Sir : That spirit which prompts communities and States to attempt to render themselves independent, so far as the supply of physical wants is concerned, of other communities and States, is an eminently proper one, up to certain limits. Beyond these, it degenerates into mere sectional selfishness, as deserving of reprobation in the community as in the individual—nay, more so, for it militates more widely against the interests and happiness of mankind. Agriculture supplies the most of our physical wants which are not administered to spontaneously by Nature. In this great department of human labor, it is not difficult to decide how far the inhabitants of each particular region are called upon to rear from the earth what their wants require. Nature herself has, in the distribution of soils and climates, both indicated and limited the production of many of the agricultural staples, by geographical boundaries, sometimes topically

15

and sometimes by whole regions. This compels those practicing Agriculture, both as individuals and masses, to make that "division of labor" which, as in the mechanic arts, gives a better knowledge of its principles and a greater expertness in its practical manipulations. It also creates the necessity of exchange. Exchange is commerce, and commerce begets and diffuses civilization.

Agricultural production, then, should be controlled by the demand or want, and by the adaptation of the country to such production. It would be absurd, for example, for New-York to attempt to raise its own rice and cotton, instead of exchanging the surplus of what it can most readily produce for that rice and cotton, or selling its surplus where it is wanted, and buying the rice and cotton with the proceeds. But, on the other hand, it would be equally absurd for New-York to be dependent on South America or Australia for her wools, when she can raise that staple just as well as those countries, and thus save paying for transportation and the hire or commission of the agents of exchange.

Though Nature both indicates and limits the production of staples by soils and climates, she too rigidly enforces the primal curse, or perhaps we should say blessing, of labor, to bring forth each, indigenously, in the regions adapted to it, or ever to place them there, unless transported by the enterprise and industry of man. The potato and maize were a recent gift from this continent to the eastern. The debt has been repaid by rice, the sugar cane, the horse, the cow, the sheep, and a multitude of other plants and animals. How singular is the history of some of their deportations! The sugar cane, now furnishing an important staple in some of our own Southern States, originated in the eastern confines of Asia; was not vouchsafed to the Greek and Roman; traveled into Arabia about the last of the thirteenth century; passed thence into Africa; was carried by the Moors into Spain; by the Spaniards and Portuguese into the West India Islands; and thence we received it. Rice, the great staple of your own State, sir, a plant of which it has been said that it "has altered the face of the globe and the destiny of nations," originated also in Asia, and has traveled by the same slow stages, until it has reached that low zone which skirts our south-eastern shores, to render its vast marshes, otherwise useless, as profitably productive as the best grain or cotton lands of the Southern States.

Here, sir, we find an instructive lesson. Other regions there are in our Southern States, now, nearly as useless as would be her "hammocks" without rice, inviting the introduction of some other great staple to supply, if feasible, a home demand, and a surplus for profitable exportation. If this great object can be achieved, and by the same means, the husbandry of the regions now under cultivation be made to assume that mixed and convertible character which will both add to their present proceeds, and better sustain their fertility, for future demands on them, a benefit will be conferred on the South the present and final results of which it would be difficult to overestimate. Repudiating theoretic speculation and vague conjecture—advancing just so far and no farther than we find our way illumined by the broad and certain light of facts, let us inquire what important staple there is, not now extensively produced at the South, which would come within and at the same time fill the requirements I have mentioned.

Woolen fabrics constitute an important item in the imports of the Southern States, and for these they exchange the proceeds of no inconsiderable proportion of their industry with the Northern States and with Europe.

The following table will exhibit the population, and the amount of home

production in these staples, acc)rding to the U. S. census of 1840, in the
States south of the Ohio and Potomac, and west of the Mississippi, (including Louisiana,) in 1839. To these are added, by way of comparison,
the statistics of the State of New-York, under the same heads, for the
same year:

TABLE No. 1.

STATES.	Population.	No. of Sheep.	Lbs. of Wool.	No. of Woolen Factories.	Value of Wool, Manufactured Goods.
Virginia.........	1,239,797	1,293,772	2,538,374	41	147,792
North Carolina ...	753,419	538,279	625,044	3	3,900
South Carolina ...	594,398	232,981	299,170	2	1,000
Georgia	691,392	267,107	371,303	1	3,000
Florida..........	54,477	7,198	7,285		
Alabama	590,756	163,243	220,353		
Mississippi.......	375,651	128,367	175,196		
Louisiana........	352,411	98,072	49,283		
Tennessee	829,210	741,593	1,060,332	26	14,000
Kentucky	779,825	1,008,240	1,786,847	40	151,246
Total..........	6,261,336	4,478,852	7,133,187	114	320,938
New-York.......	2,428,921	5,118,777	9,845,295	323	3,537,337

The above is only given to indicate approximate general results; for,
as I shall subsequently show, the returns of the product of wool are inaccurate to the last degree.

The question now arises, whence the immense disparity in the growth
and manufacturing of wool in the State of New-York, comprising 46,000
square miles of territory, and less than two and a half millions of popula-
lation, and the ten States above enumerated, covering an area of 495,000
square miles, and exceeding six millions of population? Is the growing of
wool, (for we will first consider this, as the main question, leaving the subject of manufacturing for subsequent examination,) to its present extent,
profitable or unprofitable in the State of New-York? I contend, and shall
attempt to prove, that taking a term of say ten or fifteen years, it has been
the most profitable branch of industry carried on in the State. If this is
true, why is it not equally profitable in the Southern States?[*] Is there
anything in their climate which renders them less favorable to the health
or wool-producing qualities of the sheep—or is there anything in their
topographical features, soils, herbage, or other circumstances, which unfits
them for a natural and easy adaptation to sheep husbandry? Or have
they other staples so much more profitable that it is not an object to grow
wool?

Having bestowed some attention on these points, and having been practically familiar with the most minute details of sheep husbandry from my
childhood, I have thought that the conclusions I have arrived at, and the
facts on which I have based them, might not be uninteresting to you. To
bring these facts connectedly before you, I shall necessarily be driven to
repeat some matter from my own and the writings of others, which you
have doubtless before seen in the publications of the day.

Let us now take up the first of the two preceding questions; and first I
will call your attention to the effect of Climate.

Sheep have been bred, time out of mind, on the Eastern Continent,
from the Equator to the 65th degree of north latitude, from the burning
plains of Africa and Asia, to the almost perpetual frosts of Iceland. The
Merino, (the different families of which, as will be shown, constitute the
only varieties suitable for wool growing on a scale of any considerable
extent,) has been bred in Europe, for ages, as far south as between the

* When I use the words "Southern States," without farther specification, you will understand me to
mean the ten enumerated in Table 1st.

(651) C

36th and 37th parallels of latitude, and has, within the last few years, been acclimated with perfect success as far north as various points in Sweden

If any difficulty exists in the climate of the United States, rendering it unsuitable for the rearing of sheep and wool, it must be its heat; and this must affect the wool-producing qualities of the animal alone, and not its health, as the following facts will show. There were upward of 660,000 sheep in the five most southern States, in 1839. In Florida, they have been acclimated as far south as the 29th degree. In Louisiana, Mississippi, Alabama, and Georgia, they not only flourish in the northern and more elevated sections, but on the low, fenny, tide-water region which skirts the Gulf of Mexico. In the above five States there were, in 1839, upward of 190,000 sheep below the 32d degree of latitude, viz.: in Georgia 32,986 Florida 7,198, Alabama 22,053, Mississippi 56,780, Louisiana 81,627 * They graze with equal impunity the vegetation on the margin of the Great Okefinokee Swamp (in Georgia and Florida) and on that which rankly flourishes among the ooze at the mouths of the Mississippi. It may interest some less acquainted than you are, sir, with this subject, to know that in 1839 the county in which New-Orleans stands (Orleans) contained 1,807 sheep; Jefferson, on the opposite side of the river, 6,871; St. Bernard, extending from Orleans to the Gulf, 1,154; Plaquemine, almost surrounded by the waters of the Gulf, and comprising the delta of the Mississippi, 1,832; Lafourche Interior, on the Gulf, 1,253; Terrebonne, another Gulf county, 1,013; St. Mary's, another, 8,211; and La Fayette, another 2,622.†

No portion of the United States is lower, hotter, or more unhealthy than much of the preceding, and none, according to commonly received notions, would be more unsuited to the healthy production of sheep. Yet that they are healthy in these situations is a matter of perfect notoriety to all conversant with the facts. So far as health is concerned, then, we are assuredly authorized to assume the position that no portion of the United States is too warm for sheep.

We come now to the effect of climate on the wool-producing qualities of the animal. Assuming the census returns of the United States in 1840 as reliable data, they would furnish strong proof that the warmth of the climate has a marked effect in diminishing the weight of wool per sheep; and they have been adduced as furnishing conclusive evidence to that effect, by persons more accustomed to broad assertion than patient investigation.

The following will give the weight of wool per head in the States enumerated in Table No. 1, estimated from the census returns of 1840:

TABLE No. 2.

	Lbs.	Oz.		Lbs.	Oz.
Virginia	1	7 845	Alabama	1	4 146
North Carolina	1	2 221	Mississippi	1	4 227
South Carolina	1	3 539	Louisiana	0	8 040
Georgia	1	4 487	Tennessee	1	4 809
Florida	1	0 410	Kentucky	1	6 271
New-York				1 lb. 7 680 oz.	

But an examination of the census will show that so far as several of these States are concerned, it is entitled to very little credit, in this particular, and that it is correct in relation to none of them.

In Louisiana, in fourteen counties from which 30,261 sheep, or nearly one-third in the whole State, are returned, not a pound of wool is returned.

In Florida, four counties, returning 228 sheep, return no wool. Let us

U. S. Census, 1840.   † lb

examine the comparative products, per head, as set forth in some of the counties from which returns of sheep and wool are both made, with a view of testing their accuracy. Escambia returns 485 sheep, and 837 lbs. of wool; Walton 386 sheep, and 575 lbs. of wool; Leon 1,798 sheep, and 3,360 lbs. of wool—or an approximation to 2 lbs. of wool per head. Nassau returns 436 sheep, and 1,200 lbs. of wool, or about 3 lbs. per head. On the other hand, Gadsden returns 1,875 sheep and 512 lbs. of wool; Jefferson 752 sheep, and 300 lbs. of wool; Madison 223 sheep, and 50 lbs. of wool; Jackson 960 sheep, and 376 lbs. of wool, or not quite a third of a pound per fleece! Now Leon is bounded on the west by Gadsden, and on the east by Jefferson, and all lie in the same latitude, and do not differ essentially in their soil, herbage, or temperature! Madison lies immediately east of Jefferson, (though its southern angle extends somewhat farther south,) and Jackson joins Walton. Nassau is in the same latitude. Hamilton, returning no sheep, returns the product as 20 lbs. of wool! In Mississippi, eight counties returning 15,227 sheep, return no wool; and there are repeated instances of the same glaringly obvious errors that have been exhibited in the statistics of Florida. For example, Smith county returns 741 sheep, and 1,067 lbs. of wool; Wayne 921 sheep, and 1,466 lbs. of wool. Jasper, bounding Smith on the east, returns 1,848 sheep, and 418 lbs. of Wool; and Clarke, bounding Wayne on the north, 1,199 sheep, and 188 lbs. of wool! By this, the sheep of Clarke shear less than 3 oz. per head, while those of the next county shear over a pound and a half per head. There are various other instances of under returns in the State.

In Alabama, two counties returning 2,138 sheep, return no wool; and in eight counties there are the same glaring instances of under returns with those given above.

In Georgia there is but one omission to return the wool, where the sheep are returned. In that county there were 3,360 sheep. There are eight or nine instances of obvious under returns, but these in the aggregate of the State are partly balanced by two gross cases of over returns. Cobb county returns 3,524 sheep, and 36,057 lbs. of wool and Richmond 758 sheep, and 3,032 lbs. of wool!

In South Carolina there are at least six instances of under returns.

In North Carolina there are no returns of wool in one county, containing 2,163 sheep, and in another 7,260 lbs. of wool are returned, and no sheep. There are ten cases of obvious under returns. In one of them 16,340 sheep are made to yield but 12,686 lbs. of wool.

In Virginia there are no omissions, and no obvious under returns. There are several over returns.

In Tennessee there are no omissions, but there are seven obvious under returns and two or three over returns.

In Kentucky and New-York there are not sufficient erroneous returns to materially vary the aggregate.

The foregoing facts show that the Marshals in many counties in the most southern States entirely neglected their duty in returning the product of wool; and where over or under returns have been made, it is probable that, by a misapprehension of duty, the amount of wool on hand was ascertained and noted down, instead of the annual clip.

And there is another and general error in these statistics, throughout all the States, by the census including in the number of sheep the lambs of the current season, which had not, of course, been sheared at the time of taking the census. A, at the time of taking the census, owned a flock of 200 sheep over one year old, and 100 lambs. He would give in his flock to the Marshal as 300, of course, for the census makes no distinction between lambs and grown sheep. He gave in 600 lbs. of wool, which would be 3 lbs. per head for those which had been sheared. But by the lambs being included in the census returns, it is made to appear that his sheep sheared but 2 lbs. of wool per head. In the next census the lambs and sheep should be separately returned, not only to obtain accuracy, (without which such statistics are valueless,) but the annual increase thus indicated would be, of itself, an interesting and valuable statistic.

In the preceding enumeration of erroneous returns, I have set down none as under returns where the product of wool has not been given as less than a pound per head; and where it has fallen under that amount, the returns from contiguous counties, possessing the same natural features, exhibiting a far superior product, as well as the general complexion of the returns throughout the State, have authorized me beyond a reasonable doubt so to consider it. I may add, that it is a fact of universal notoriety that there is no variety of sheep in any section of the United States, which shears but a pound of wool per head.* A careful inspection of the census, moreover, will not fail to satisfy any one that there are a multitude of under returns, (not specified by me, as the product is given over 1 lb. of wool per head,) in most of the States. This is shown by the same kind of comparisons which have already been alluded to. These are far more common in the extreme Southern States, where wool growing had not yet (in 1839) been reduced to any system, and where sheep had been little looked after or regarded. These errors grow less, as we approach the wool-growing regions of the north and north-west.

Taking those returns which we are authorized to consider correct,† it will appear that there is no great difference in the average product of wool, per head, in States separated by from ten to fifteen degrees of latitude, and no more than is clearly referable to incidental or extraneous causes, unless we come to the conclusion that the difference is in favor of the Southern States. In proof of this, the following table is offered, giving the products of some of those counties in each of the States enumerated in Tables No. 1 and No. 2, which exhibit the highest averages per head, (excluding those obviously over returned.)‡

TABLE No. 3.

STATE.	COUNTY.	Average Weight of Wool per Sheep.		Total average of the Counties given‖	
		Lbs.	Oz.	Lbs.	Oz.
Virginia	Fauquier......	2	2$\frac{3250}{4011}$		
	Harrison......	2	0$\frac{9056}{3115}$		
	Ohio	2	2$\frac{1050}{6689}$		
	Rockingham...	2	1$\frac{1849}{4791}$	2	1$\frac{1}{4}$
North Carolina	Currituck	2	7$\frac{443}{459}$		
	Person	2	5$\frac{913}{1077}$		
	Perquimans ...	1	12$\frac{2473}{2543}$		
	Tyrrel........	1	8$\frac{320}{1947}$	2	0

* I consider such to be under returns, independent of the mistake made by including lambs in the enumeration.
† With the exception of the error arising from the return of lambs—which perhaps would not greatly vary the proportionable result.
‡ It is proper to say that though I designed to take the highest averages, I did not go through a formal reckoning of the average in every county in the eleven States. I took those which appeared the highest, after a somewhat careful general inspection.
‖ Excluding the fractions of the ounces in preceding column.

TABLE No. 3—(Continued.)

STATE.	COUNTY.	Average Weight of Wool per Sheep.		Total average of the Counties given.	
		Lbs.	Oz.	Lbs.	Oz.
South Carolina.....	Beaufort	2	$12\frac{1094}{1375}$		
	Darlington	2	$4\frac{504}{3041}$		
	Georgetown ...	2	0		
	Richland	1	$12\frac{1180}{1511}$	2	3
Georgia	Bibb	2	$3\frac{306}{1178}$		
	Murray	1	$15\frac{14}{578}$		
	Twiggs	2	$10\frac{574}{2013}$		
	Ware	2	$1\frac{33}{49}$	2	$3\frac{1}{4}$
Florida	Escambia	1	$11\frac{97}{485}$		
	Leon	1	$13\frac{809}{800}$		
	Nassau	2	$13\frac{20}{169}$	2	$1\frac{2}{3}$
Alabama	Blount........	1	$15\frac{19}{51}$		
	Jackson.......	2	$6\frac{390}{651}$		
	Autauga	1	$15\frac{193}{1889}$		
	Greene	2	$0\frac{608}{4749}$	2	1
Mississippi	Claiborne	2	$7\frac{2711}{6527}$		
	Hinds	3			
	Perry	2			
	Warren.......	2	$7\frac{185}{953}$	2	$7\frac{1}{2}$
Louisiana	Concordia.....	2	$15\frac{751}{767}$		
	St. Helena	1	$7\frac{361}{513}$		
	St. Tammany ..	2	$1\frac{23}{55}$		
	Washington ...	1	$14\frac{10}{21}$	2	$1\frac{1}{4}$
Tennessee	De Kalb	2	$2\frac{82}{729}$		
	Franklin	2	$7\frac{179}{531}$		
	Smith	2	$0\frac{891}{1309}$		
	Wilson	2	$2\frac{5046}{19369}$	2	$2\frac{3}{4}$
Kentucky	Bourbon	2	$1\frac{3897}{23719}$		
	Clarke........	2	$11\frac{9133}{20053}$		
	Lawrence	2	$13\frac{955}{6521}$		
	Scott	2	$6\frac{983}{1119}$	2	$7\frac{3}{4}$
New-York	Westchester ..	2	$10\frac{3074}{20043}$		
	Ulster	2	$5\frac{895}{6355}$		
	Saratoga	2	$3\frac{5766}{6041}$		
	Orleans	2	$2\frac{2154}{69563}$	2	5

Taking these averages as a test, it would appear that the difference between the average products of the Southern States and New York is, in some instances, in favor of the former. Kentucky in the middle, and Mississippi in the extreme South, exceed the average of New York.

It is proper to say, however, that various local circumstances may have effected these results, and that taking the average of a considerable number of counties, in the several States, would essentially vary them. Notwithstanding this, the testimony which they offer is important, and becomes more so regarded in connection with another circumstance. The comparative statistics of the extreme Southern States themselves show that in a majority of cases their best products of wool come from their Southern and warmer counties.

Of the four counties in Louisiana, the product of which is given above, the most northern lies between the 31st and 32d degrees of latitude, and all of the other three south of latitude 31°. St. Tammany, which exhibits the second best product, borders on Lake Pontchartrain. The State extends north to latitude 33°.

In Florida, all the counties given lie in the same latitudes.

Mississippi extends north to latitude 35°. All the counties given are south of 32° 30', and one in latitude 31°.

In Alabama, extending north to the same latitude with the above, two of the counties given are in the northern or mountain region, and two of them south of 33°.

In Georgia, (same northern line,) one of the counties given is in the extreme north, two south of 33° and one in latitude 31°, being the county in which lies the major portion of the Great Okefinokee Swamp!

In South Carolina, two of the counties (both bordering on the ocean) are in the low, marshy, tide-water region; and the other two are in the central region.

In North Carolina, two of the counties given join the ocean; one is on Albemarle Sound, while one lies in the central and northern portion of the State.

In comparing the product of wool in the Southern States with that of the Northern—and more particularly with that of New-York—we must not lose sight of the fact that in the latter wool growing has become an important business, and is reduced to a system. The sheep are kept in pastures, and are sheared at regular intervals. In the Carolinas, Georgia, and the Gulf States, precisely the reverse of all this was generally true at least up to the year 1839. The sheep were little cared for; were suffered to breed promiscuously; and they roamed at large through forests, where, as the warm season advanced in the spring, their wool, beginning to detach itself, was left on shrubs and brambles, and in not rare instances considerable portions of it were thus lost prior to shearing.*

Giving their due weight to the preceding facts, the defects in the census, etc., it is, I think, undeniable that they account for all the deficiency in the average product of wool per sheep in our most southern States, compared with that of New-York, as set forth in Table No. 2. Indeed, sir, my own convictions are decided, and the facts reported appear to fully sustain them, that warmth of temperature, at least to a point equaling the highest mean temperature in the United States, is not injurious, but absolutely conducive to the production of wool. The causes of this are involved in no mystery. Warm climates afford green and succulent herbage during a greater portion of the year than cold ones. Sheep plentifully supplied with green herbage keep in higher condition than when confined to that which is dry. High condition promotes those secretions which form wool. Every one at all conversant with sheep well knows that if kept fleshy the year round, they produce far more wool than if kept poor. A half a pound's difference per head is readily made in this way. Within the maximum and minimum of the product of a sheep or a flock, the ratio of production always coincides with that of condition.

I have dwelt on this point at great and perhaps tedious length, sir, as the results set forth in the United States Census, unexplained, would clearly point to a different conclusion from that to which I have arrived. To invalidate testimony, ostensibly so certain and reliable, as well as to combat deep-rooted prejudices, I have deemed it necessary to scan thoroughly the accessible facts in the case.

----
* I make no account of difference in breeds, as affecting the product of wool between the South and North. The grade Merinos, not uncommon in New-York, would produce far more wool than the "natives," the principal sheep in the South in 1839. But the latter would equal or exceed the product of the numerous Saxon flocks of New-York.

LETTER II.

EFFECT OF CLIMATE, CONTINUED.

Effect of Climate on quality of Wool...Warmth of Climate renders Wool coarser—Reasons...Effect of Herbage...Opinions of Youatt—Doctor Parry—English Staplers—Writer...Can the tendency to grow coarser be resisted?...Opinions of Youatt—Lasteyrie—Mr. Lawrence...Experiment in Australia—Cape of Good Hope—South of Illinois—Kentucky—Tennessee—Mississippi—New-York...Warm Climates render Wools softer and longer, thus adding materially to their value...Proved to be the case in Australia...Testimony of English Wool-factors and Staplers...Same effect produced in the United States...Testimony of Mr. Cockrill.

Dear Sir : We come now to discuss the effect of Climate on the quality of Wool. There can be but little doubt, other things being equal, that the pelage of the Sheep and some other animals, becomes finer in cold climates and coarser in warm ones. This is usually attributed, by theoretical writers, to the effect of cold and heat in contracting or expanding the pores. This may have some effect, but to suppose that the delicate tissues of the skin can act, to any great extent, mechanically, in compressing the harder and highly elastic ones of the hair or wool, or compel their attenuation so as to permit their escape through diminished apertures, like the process of wire drawing, is, it seems to me, to assume that matter acts contrary to its ordinary laws. I am rather disposed to look for the causes of this phenomenon, in the amount and quality of the nutriment received by the animal. It was stated, in my preceding letter, that warm climates, by affording succulent herbage during a greater portion of the year, maintain in greater activity those secretions which form wool, and thus increase the quantity or weight of the fleece. The weight is increased by increasing the length and thickness of the separate fibres, just as plants put forth longer and thicker stems on rich soils than on poor ones.

Mr. Youatt, in his excellent and much quoted work on Sheep, after discussing and admitting, to a certain extent, the influence of warm temperatures in rendering wool coarser, says :

" Pasture has a far greater influence on the fineness of the fleece. The staple of the wool, like every other part of the sheep, must increase in length or in bulk when the animal has a superabundance of nutriment; and, on the other hand, the secretion which forms the wool must decrease like every other, when sufficient nourishment is not afforded. When little cold has been experienced in the winter, and vegetation has been scarcely checked, the sheep yield an abundant crop of wool, but the fleece is perceptibly coarser as well as heavier. When frost has been severe and the ground long covered with snow—if the flock has been fairly supplied with nutriment, although the fleece may have lost a little in weight, it will have acquired a superior degree of fineness and a proportionate increase of value. Should, however, the sheep have been neglected and starved during this prolongation of cold weather, the fleece as well as the carcass is thinner ; and although it may have preserved its smallness of filament, it has lost in weight and strength and usefulness. These are self-evident facts, and need not be enforced by any labored argument."[*]

Doct. Parry, a correct and able English writer, remarks :

" Sheep breeders have observed a sort of gross connection between the food and quality of the fleece. . . . The fineness of a sheep's fleece of a given breed is, within certain limits, inversely as its fatness, and perhaps also (although I am not certain on this point) as the quickness with which it grows fat. A sheep which is fat has usually comparatively coarse wool, and one which is lean, either from want of food or disease, has the finest wool ; and the very same sheep may at different times, according to these circumstances, have fleeces of all the intermediate qualities from extreme fineness to comparative coarseness."

[*] Youatt on Sheep, p 70.

In an examination before the English House of Lords, in 1828 various eminent staplers* testify, most decidedly, to the deterioration (in fineness) of the British wools and their increase in length of staple, "since the introduction of artificial food and the adoption of the forcing system." My own observations fully corroborate these positions. I have examined, in repeated instances, with a good microscope, the wool of individual sheep in my flock, taken in seasons when they have maintained a high condition, and in others, when, from some incidental cause they have been in ordinary or poor condition, and the difference in length and fineness is, uniformly, distinctly perceptible

If the sheep breeder in warm climates can take advantage of the tendency to produce greater quantities of wool, following that supply of succulent herbage throughout the year which Nature has placed at his disposal, and at the same time, by any unexpensive means which ho can employ, combat the correlative tendency to increased coarseness of fibre, he has most assuredly, other things being equal, an entire advantage over the breeder in colder regions.

We come now to the important inquiry, Can this latter tendency be successfully combated; or, in other words, can wool of any desirable fineness be produced in countries as warm, for example, as Louisiana, Mississippi, &c.?

Let us examine Mr. Youatt's testimony on this point also. He says:

" Temperature and pasture have influence on the fineness of the fibre, and one which the farmer should never disregard; but he may, in a great measure, counteract this influence by careful management and selection in breeding. . . . A better illustration of this cannot be found than in the fact that the Merino has been transplanted to every latitude on the temperate zone, and some beyond it—to Sweden in the North and Australia in the South— and has retained its tendency to produce wool exclusively, and wool of nearly equal fineness and value."†

Mr. Lasteyrie, equally good authority, uses the following language. When he speaks of the preservation of the breed in its " utmost purity," we are undoubtedly to understand him to refer as much to the fineness of the wool as any other point, this being the distinguishing mark or excellence of the breed.

' The preservation of the Merino race in its utmost purity at the Cape of Good Hope, in the marshes of Holland, and under the rigorous climate of Sweden, furnish an additional support of this, my unalterable principle: fine wool sheep may be kept wherever intelligent breeders exist."‡

Samuel Lawrence, Esq. the head of the great Lowell Manufacturing Company, in Massachusetts, who, by his vast purchases of fine wool in all parts of the United States for a long term of years, and his intimate practical acquaintance with the quality of the article, is entitled to have his opinion on this point regarded as of as great weight as that of any other individual, says:

" That the properties of wool are affected by herbage and soil, I have not a doubt, and were it not invidious, I would name some sections where wool growers are greatly favored by Nature. One thing is certain, whatever may be the character of the soil, where there are good shepherds there is sure to be found good wool. By judicious selections and crossing, I believe a breed may be reared which will give four pounds of exquisitely fine wool to the fleece."‖

This last sentence of this important extract, though not bearing so particularly on the point under examination, is recorded in its original connection for subsequent reference.

Australia and the Cape of Good Hope being cited by the distinguished

* Youatt: on Sheep, p. 71, where the names and testimony of these individuals are given; and more at length in Bischoff on Wools, &c., vol. II. pp. 118—200.   † Pp. 69—70.
‡ Lasteyrie on Merino Sheep, p. 101.   ‖ Letter of Mr. L., published in ' American Shepherd," p. 436.

English and French writers above quoted, as offering instances of the perfectly successful acclimation of Merinos, without deterioraticn of their wool, in warm climates, it may be well to inquire a little more particularly what the climate of those countries is; and what, if any, the other circumstances connected with them, having an influence on the quality of the wools grown in them.

Port Jackson, in Australia, in the vicinity of which the Merinos were first introduced, and are now extensively bred, lies in 33° 55' South latitude, corresponding as nearly with the latitude of Georgetown, South Carolina, as that of any othei important point in our country.* In describing this region (New South Wales) Malte Brun says :

" The coast itself is high but not mountainous ; and it is partly shaded by trees of gigantic size. Toward the south-east a great part is covered with coppice; much also is occupied with marshes. About Botany Bay† the soil is black, rich and exceedingly productive in plants. The north-east part seems lower. The coast is covered with mangroves. . . . The heat of December rises to 112° Fahrenheit. The forests and the grass have been known spontaneously to take fire.‡ The North-west wind, like the Khamseen of Egypt, scorches the soil and reduces it to a light dust. . . . Notwithstanding these disadvantages, the climate is very healthy, and very favorable to population. . . Those parts in which different trials have been made have rather too warm a climate for common barley and oats, though these grains have been found to succeed tolerably well on the poorer soils. . . All the vegetables grown in England are produced in the English colony. . . Peaches, apricots, nectarines, oranges, lemons, guavas, loquets, cherries, walnuts, almonds, grapes pears, pomegranates and melons attain the highest maturity in the open air."‖

The country, most of it, is remarkably deficient in water,§ though many portions are subject to destructive inundations.¶ Its drouths are unequaled for their duration and intensity in, perhaps, any inhabited portion of the globe.** Its vast plains, occasionally highly fertile, but more usually, only n detached spots, afford pasture throughout the year.

The physical features of this country, its system of sheep husbandry, etc. will be more particularly alluded to hereafter.

The English first introduced into this remote possession the coarse hairy sheep of Bengal. In the short space of three years these were so far changed by the effect of the climate and other circumstances, that their hair was entirely gone, and was succeeded by a fleece of wool.†† The South-Down and Leicesters were subsequently introduced, and their crosses with the Bengal sheep soon became as fine as the pure bloods of the former. At length some Merinos were imported by the colonists, and, says Mr. Youatt, " The experiment was satisfactory beyond their expectation. The third or fourth cross with the then prevalent sheep of the colony produced an animal with a fleece equal to that of the pure Merino in Europe ; and the wool of the pure blood seemed to improve as rapidly as the native breed had done."‡‡ In 1810, the export of wool from Australia and Van Dieman's Land was 167 lbs.; in 1833, it had reached 3,516,869 lbs.‖‖ In 1843, it amounted to 16,226,400 lbs.§§

The following, from a table in McCulloch's Dictionary of Commerce, will show the current prices (reduced to American currency) of some of the imported and domestic wools, in London, March, 1834 :

	$ cts.	$ cts.		$ cts.	$ cts.
SPANISH............per lb.	— 60 to	— 77	ENGLISH : North & South-Down. per lb.	— 44 to	— 48
PORTUGUESE	— 44 "	— 62	Leicester...................	— 33 "	— 44
GERMAN, SAXON, &c.........	— 48 "	1 15	Lincoln, Cotswold, Romney		
AUSTRALIAN	— 50 "	1 00	Marsh................	— 40 "	— 44

* Georgetown is perhaps half a degree nearer the Equator.
† This place is twelve miles south of Port Jackson.
‡ Malte Brun cites Collins (an author frequently quoted in relation to New South Wales) for this strong and, perhaps, exaggerated assertion. ‖ Malte Brun, vol. i. pp. 600—605.
§ Spooner, Youatt, etc. ¶ Malte Brun. ** See McCulloch's Commercial Dictionary.
¶ Youatt on Sheep, p 184. Spooner, Diseases of Sheep, p. 62.
†† Ib. p. 184. ‖‖ Ib. et Spooner. §§ Spooner.
(659)

D

It will be seen from this, that the best Australian wools already excelled the best Spanish, and were more than midway between them and the best Saxon. When we consider the almost infinite difference in the care, both in breeding and management, bestowed on the Saxon and Australian sheep it shows most conclusively the adaptation of the climate of Australia to the production of the finest wool—or, at least, that the adverse effects of its warm temperature, and the incidents to that temperature, are easily overcome. In Saxony, sheep are numbered, oftentimes their separate pedigrees registered, and each breeding ewe is stinted to a ram carefully selected with reference to her individual qualities. In Australia, where less capital and labor are employed, flocks of about three hundred breeding ewes*—where the country is destitute of timber, sometimes a thousand†— roam from one fertile and watered spot to another over the vast plains, in charge of the convict shepherd; and this system is followed throughout the year, including the tupping season. Three flocks are always penned together at night,‡ so that as many as nine hundred breeding ewes, of varying quality, must be promiscuously bred to, say, from thirty to thirty-five rams, running promiscuously among them.

The Cape of Good Hope is in south latitude 34° 23′ 40″.

Mr. Youatt, in describing the sheep husbandry of this region, overestimated, I think, the heat of the climate. Separated by lofty mountain ranges from the interior of Africa, the fertile regions adjoining the coast are not swept by its scorching winds, and the temperature is comparatively mild. "In a meteorological register kept at Cape Town, from Sept. 1818 to Sept. 1821, embracing a period of three years, the highest heat marked is 96°, the lowest 45°, Fahrenheit. The mean and annual temperature scarcely 68°—of winter 61°, of summer 89°."|| But sheep and their wool suffer from the fine sands which are lifted and driven by the prevailing winds. Says Malte Brun, "the wind blows often from the south-east with great violence. Nothing can be secured from the sands which it drives before it; they penetrate the closest apartments and the best-closed trunks. At this time it is not prudent to go out without glasses, lest the eyes should be injured."§

Though the climate can scarcely be designated a "torrid" one, as Mr. Youatt speaks of it, the mean temperature of its winter (61°) conclusively shows that cold can have nothing to do here with rendering the wool finer by a contraction of the pores. If, therefore, it can be shown that the wool of the fine breeds does not deteriorate in quality, it sufficiently proves that Australia is not an incidental exception in the testimony which it presents on the point under examination, but that it illustrates the uniform operation of the physical laws which pertain to the growth of wool.

After one or two unsuccessful attempts, the Merinos were acclimated at the Cape by the English colonists. In 1804, the colony numbered 536,-634 sheep. In 1811, there were 1,293,740. In 1810, the import of wool into Great Britain was 29,717 lbs.; in 1833, it was 93,325 lbs.¶

In Willmer & Smith's "Liverpool Annual Wool Report," for 1846, it is stated, "The shipments from this quarter (Cape of Good Hope) show great improvement, amply testified by the high rates the best flocks have commanded during the season. . . The best parcels now take rank with those from Australia."** The system of breeding and general management at the Cape closely correspond with those of Australia.

Let us now, sir, turn to the experience of our own country. I do not

---

* Cunningham's "Two Years in South Wales."   † Ib.   ‡ Ib.
|| Malte Brun, vol. ii. p. 112.   § Ib. vol. ii p. 112.
¶ Youatt on Sheep, p. 184.   ** Willmer & Smith's European Times of Jan. 4, 1846.

know that a sufficient number of experiments have been made as near the Equator as Cape Town and Port Jackson, to have their testimony regarded as entirely decisive on the point under consideration, but those have been made which throw much light on this question, if, indeed, any more is considered necessary. In the south of Illinois (Edwards county), in about latitude 38° 30', the finest varieties of sheep were introduced by Mr. George Flower, about twenty years since, from which he has bred up an extensive flock. That gentleman says :* "No deterioration in the wool has taken place; on the contrary, the wool fibre is somewhat finer." I have myself seen various specimens of Mr. Flower's wool, of the clip of 1844, and it bore an excellent character for fineness and evenness.

In a letter which I received from Hon. Henry Clay, in 1839, he says : " I have for some years had only the pure Saxony at my residence; but I am now satisfied that I should have derived more profit from sheep producing a wool less fine.† The climate of Kentucky is, however, well adapted to the Saxon sheep." Mr. Clay's residence is in about latitude 38°.

Mr. Mark R. Cockrill, of Nashville, Tennessee, in a letter published in the American Shepherd,‡ says :

" I have about a thousand head of fine sheep, and from 400 to 500 long-wooled or mutton sheep. My Saxon sheep were imported in 1824 or '26—I cannot say which—and I find as yet no falling off in quantity or quality of their fleeces; on the contrary, I believe a little improvement on both points, and a little more yolk, when well provided for, which, you know, does not abound much in the Saxon breed. In addition, the fleeces are a little more compact than formerly—hence more weight; and, from our mild climate, *the staple has become longer*. . . I assert it to be a fact that the cotton region I am now in [Mr. Cockrill dates from Madison county, Mississippi, where a part of his sheep are kept], in about latitude 32° north, is better than any country north of it to grow wool, as the sheep can be kept all the time grazing, by sowing small grain; for, if grazed off, it quickly grows again in a few days; and the wool of the fine Saxon sheep in this climate is softer and more cotton-like than any I have ever seen, although I have samples from all parts of the world. I have traveled from this very place to Boston, sampling all the sheep of note on the way, and I found nothing on my journey or at Boston as good as the wool I had grown, and so said all the wool staplers whom I met with, and they were not a few. I presumed, in reality, that the blood of my sheep was no better than many I saw, but the superiority of my wool I ascribed to our climate, and the provision for the sheep of succulent food the year round.— The weight of my fleeces is fair—say from 3 to 5½ lbs. each. . . Tennessee is not the true grass climate; about 28° north is the most congenial for grass: notwithstanding, our State is fair for pasture; blue and orchard grass, white and red clover, prosper pretty well. . . There is much country in Tennessee and other Southern States not fit for the plow, and would do admirably well for fine-wooled sheep, and can be profitably so employed. A small capital thus appropriated here in Mississippi would do better than cotton growing at present prices."

Nashville is in about latitude 37° 15'; and Madison county, Mississippi is about half a degree farther north than mentioned by Mr. Cockrill, viz extending from 32½° to 33°; its county seat (Canton) being more than a degree nearer the Equator than Port Jackson in Australia, and about two degrees nearer than the Cape of Good Hope !

Mr. Morrel, the compiler of the "American Shepherd," has obtained specimens of Mr. Cockrill's wool, and he says of them, " Judging from the samples, the conclusion is inevitable that little or no deterioration has been produced by the climate."||

This testimony of Mr. Cockrill is very important, both from the length and extent of the experiment. I have no doubt of the perfect correctness of his assertion that his wool has improved in those low latitudes; but the cause assigned by him cannot be received as the correct one, so far as the increased fineness of the fibre is concerned. The improvement in this particular, under a system of feeding which has "increased" both the "quan-

---

* In a letter published in the Prairie Farmer.   † Mr. Clay here alludes to the Merinos.
‡ P. 409. || American Shepherd, p. 41.

tity" and the "yolk" of the fleece, cannot be ascribed to the climate, nor to the feeding itself. It is undoubtedly owing to Mr. C.'s system of breeding and selection, a point which will be fully discussed hereafter.

If feed or condition exercise the principal influence on the fineness of wool (that is, within the range of variation to which it is subject on the same individuals), it follows that the effect may be produced in any climate, for high condition throughout the year is attainable in the most rigorous ones, by the supply of plentiful and rich food. The wool secretions are incident on condition—fatness, however superinduced.* It again follows that if wool of the highest degree of fineness (for the breed) can be produced in New-York, although the sheep is kept in a decidedly fat state throughout the year—quite as fat as it would become, grazing on green feed all the year round—that wool of equal fineness can be produced by the observance of the same influencing conditions (apart from feed) by the southern breeder.

Here again, undeterred by any considerations of what I deem a false modesty, I shall offer facts founded on my own personal experience as a sheep breeder; and I would remark, once for all, that throughout the whole of these letters I shall never so far prefer *beliefs* founded on the assertions of others, to *actual knowledge*, based on facts repeatedly and constantly brought under my personal inspection, for a number of years, as to suppress the latter, to rely solely on the former.

I have succeeded, in repeated instances, in producing an exquisite quality of wool, decidedly above the average of the breed (Merino) in the heavy fleeces of sheep kept fat the year round. I have made it a sort of a test latterly, in the selection of rams, to choose only those which not only carry heavy fleeces, in any condition, but which, in the highest, yield a wool equaling the choicest samples to be found on this variety. These facts will, by and by, be placed in a definite and tangible form, by the recorded testimony of the scales and the microscope.

But though the natural effects of warm climates and their incidents, to increase the bulk or coarseness of the fibre, is one which can easily be resisted, they work a change of another kind in the character of wool. They cause a longer fibre and a greater softness of staple. The effect of succulent nutriment during the year in increasing the amount of the wool will exhibit itself; but the skill of the breeder can so far regulate its action, that the increase is in the length, rather than in the diameter or bulk of the fibres. It is not difficult to conjecture why a staple of more rapid growth, supplied to excess with the secretions which enter into its composition, unexposed to great and rapid variations of temperature, should retain a greater degree of softness than one produced under opposite conditions. But, whatever the causes of these phenomena, their existence is placed beyond a doubt.

The increased *length* of staple, resulting from the nutriment of warm climates, has been sufficiently adverted to. The following statements made by some of the most eminent wool-factors, staplers, etc. in England, before a Committee of the House of Lords, in 1828, place the other point beyond controversy.†

Mr. Henry Hughes, wool-broker, London, says:

---

* No one has asserted, so far as I am informed, that dry feed will produce less wool than green feed, if the same degree of fatness is kept up. On the other hand, the rich cereal grains, oil-cake, &c. (without some of which a high degree of fatness cannot be maintained, on dry feed alone, during the four or five months' winter in latitudes north of 42°), might be supposed to be quite as conducive to the production of wool as grasses.

† For extended minutes of this very interesting investigation into the state of the wool-trade, &c. &c. in Great Britain, see Bischoff on Wool &c., vol. ii. p. 118 to 200.

"Latterly they (Australian and Van Dieman's Land wools) have been of varied qualities, but all possessing an extraordinary softness, which the manufacturers here so much admire that they are sought for more than any other description of wools, from that peculiar quality, which is supposed to arise from the climate alone. They are known to require less of the milling or fulling power than any other descriptions of wools. . . They are better adapted than the German wools to mix with British wools, because the superior softness which I have stated gives a character, when mixed with English wool, that the other does not, from the hardness of the fibre."[*]

Mr. Stewart Donaldson, merchant, London, says:

"I have no hesitation in pronouncing that the wools of New South Wales and Van Die-man's Land are decidedly preferred to the apparently similar descriptions of German wool . . . They have a softness and silkiness about them which, when worked up into cloth, shows itself more distinctly than in the raw material. I conceive that it is dependent on the climate alone. I am of opinion that wool of that quality could not be produced in any part of Europe."[†]

Mr. Thomas Legg, wool-stapler, Bermondsey, says:

"There are some of these wools of very beautiful quality, as good as any of the German wools."[‡]

Mr. Thomas Ebsworth, wool-broker, London, says:

"The peculiarity of the climate of New South Wales appears to have a very great effect on wool, so as to reduce it from a harshness to a very fine texture."[||]

This was the substance of all the testimony on this particular point; and when it is understood that the investigation was an issue between rival interests, where all the facts were thoroughly sifted, the fact that the above assertions were undisputed shows that they were considered of an undisputable character.

Allusion has already been made to the loose and careless system of sheep-breeding, etc. in Australia, compared with that in Germany. Taking this into consideration; taking also into consideration that the flock furnishing the best wool in Australia (Capt. McArthur's) is composed of grade sheep (Bengal and English, graded up with Merino and Saxon rams), the trifling effect of climate is made more strikingly to appear.

The statements of Mr. Cockrill in relation to the softness of the wools grown in Tennessee and Mississippi, sustain and are sustained by those above given; and they go to show that it is the result of a general law and not of any peculiar local influences peculiar to Australia

---

[*] Bischoff on Wool, &c. vol. ii. pp. 123-3.   [†] Ibid. 153-4.   [‡] Ibid. 184.   [||] Ibid. 184.

LETTER III.

ADAPTATION OF THE SOILS, HERBAGE, &c. OF THE SOUTHERN STATES TO SHEEP HUSBANDRY. 1. OF THE LOW OR TIDE-WATER REGION

Natural Features of the Southern States—Divided into three Zones...The Natural Features, Soils, &c. of each...The Tide-water Zone—Its destitution of Artificial Pastures and Meadows...Causes—Small amount of Domestic Stock kept—Unsuccessful Experiments in raising Clover and Grasses...Reasons why those Experiments were unsuccessful—Land too much Exhausted by Severe Tillage—System of Tillage compared with that of the Grazing Regions of New-York—Experiments unsuccessful, also, because improper varieties of Clover and Grass were tried...Much of the Land adapted to Grass—Shown by its Natural Pastures—Statements of Col. Allston—Opinions of Mr. Ruffin—of a Committee of the S. C. Agricultural Society...Land compared with that of Flanders—also with some parts of New-York...Climate perhaps unfavorable to certain Northern Grasses and to Red Clover—Opinion of Mr. Ruffin—Statements of Milton (S. C.) Agricultural Society...Clover not indispensable...Experiments suggested...Valuable indigenous and acclimated Grasses—Crab Grass—Millet—Bermuda Grass—its great value—Statements of Mr. Affleck...Peas—Their great value in the Southern States as a Green Crop Manure—Sprengel's Analysis of them—The Value of their Straw as a Manure compared with various substances—Table of the Value of Manures by Payen and Boussingault...Oats, Rye and Barley—Corn Blades—Sweet Potatoes...Conclusions from foregoing.

Dear Sir : Having discussed, in my previous letters, the effects of warm climates and some of their incidents, on the health of sheep, and on the quantity and quality of their wool, we come now to the second branch of my original inquiry—Is there anything in the natural features, soils, herbage, &c. of the Southern States, which unfits them for a natural and easy adaptation to sheep husbandry ?

The vast region south of the Ohio and Potomac, and west of the Mississippi—comprising an area considerably exceeding that of France, Spain and Portugal*—is distinguished, by its natural features, into three distinct zones, parallel to each other and to the Atlantic coast.

The lower or tide-water zone, which skirts the Atlantic, is a low, flat, sandy, and oftentimes marshy plain, from 50 to 100 miles wide, comparatively recent (tertiary) in its formation, and covered with pine forests over the greatest portion of its extent. The soils on the dry lands are generally light, and sometimes too sterile to admit of profitable cultivation ; that in the swamps and river bottoms, where the sand is replaced by a rich alluvion, is exceedingly fertile. The middle or hilly zone rises from the level of the preceding, first into gentle hills, and finally into high and oftentimes broken ground, as it approaches the mountains. The width of this does not greatly vary from that of the preceding. The formation is almost exclusively primary ;† and the soil varies, sometimes being poor, but more generally ranging, in its natural state, from medium to highly fertile. The forests consist of oak and other deciduous trees. The third or mountain region is formed by the different chains and groups of the great Apalachian range of mountains, and occupies not far from 70,000 square miles of the central portion of the territory under consideration.‡ It comprises the middle of Virginia, the west of North Carolina and South Carolina, the north of Georgia and Alabama, and the east of Tennessee and Kentucky. Its formation on the eastern declivities of the Blue Ridge (the most eastern chain) is primary, and thence to the Alleganies the rocks belong to

* Spain contains 170,000 square miles, Portugal 40,000, France 200,000—in all 410,000. Allowing 10,000 square miles of Louisiana to be east of the Mississippi, the area of the region referred to is 456,000 square miles.
† There are one or two interrupted belts of *new red sandstone*—vide McClure.
‡ Estimated not far, I think, from correctly, by myself. I can find no authority on this point.

the Transition order.* Its soil varies from thin and light to that of exu-
berant fertility. West of the mountains, the hilly zone rests on Transition
rocks and coal measures, and is succeeded west and south of Virginia by
the vast rolling or level plains which extend to the Ohio and Mississippi;
and which, instead of the silicious sands of the eastern coast, exhibit rich
and varying soils resting on limestone and other Transition and Cretaceous
rocks. In Virginia, the hilly region, which is one vast coal measure, extends
to the bottom lands of the Ohio; and its soils, taken as a whole, range from
ordinary to meager.†

We will now proceed to examine the capabilities and adaptation of each
zone, separately, for the purposes of sheep husbandry. It has already been
shown that sheep are healthy, and produce as heavy, and *may be made* to
produce as fine fleeces as elsewhere, in the tide-water zone. They are
easily kept—finding, in a climate so mild, considerable succulent food even
in the winter; and, south of North Carolina, large numbers would subsist
during the entire winter on the hardier wild herbage which continues green
in the forests and swamps. If this region was stocked with sheep, to the
extent alone to which they could find subsistence, summer and winter, on
wild herbage—or, in other words, get a living without costing their owners anything—the present number would be largely increased, and their
wool and mutton would add materially to the annual income of the owners of the soil. But a better system would undoubtedly be not to depend
upon wild herbage alone, but to have pastures or sheep-walks seeded with
the best grasses which will flourish on them, and provision made for a quantity of dry fodder, or some substitute for it, for winter use.

Can this summer and winter feed be produced, in the region under examination, to any considerable extent, at an expense which would render
its conversion into wool and mutton profitable? There are patches of
good natural pasture in many parts of the tide-water zone, apart from the
salt or fresh water marshes. But artificial pastures and meadows have
rarely been attempted. The planters in this portion of South Carolina, for
example, actually import hay! "Many of the cotton and rice planters .
. . in some cases buy hay from New-England. . . . Northern and
(in some cases) European hay is even carried up to supply Augusta and
Columbia, along rivers which flow through swamps covered with natural
grass, so rank and luxuriant as to be almost impenetrable."‡

This neglect of grass culture springs from several causes. Little farm-
stock, comparatively speaking, is reared or kept by the rice and cotton-
planters, from the fact that most of the labor on such plantations is performed by men; and the few animals kept are fed on wild herbage, or the
offal of crops which are raised for other purposes. The carriage and
draught horses and mules are fed in the winter on the leaves or "blades"
of corn; and the neat stock get their living in the swamps, and in the
corn fields, where the greatest portion of the stalks are usually left standing.

Nor is it to be denied that various unsuccessful experiments have been
made in the cultivation of the grasses and clover, which have discouraged
farther efforts, and led many to infer that the soil or climate, or both, are
decidedly uncongenial to them. That the soil or climate is as favorable to
the production of rich, thick swarded pastures or meadows, as in many

---

* So termed by Werner. Though little used now by geologists, I resort to it as the shortest descriptive
epithet which will include all these rocks, unless it be the Hemilisyan of Brongniart, the Submedial of Co-
nybeare, or the Graywacke of De la Beche—neither of which is so familiar, nor, it appears to me, any better. The Transition rocks are equivalent to both the Cambrian of Prof. Sedgwick, and the Silurian of Mr
Murchison—whose nomenclature is adopted by Lyell, Phillips, Mantell, &c.
† Dr. Morse, Mitchell, &c.   ‡ Ruffin's Agricultural Survey of South Carolina, 1843, p. 73

parts of the Northern States, I do not contend. Some of these soils are
doubtless, naturally too barren to be made to produce good yields of grass,
without an expenditure which would more than counterbalance the profits
accruing from them. Others have been sunk nearly to the same level by
wasting and improvident tillage; and it is on lands of the latter class,
mainly, that the experiments in introducing the grasses and clover have
been made. As long as they would produce cotton or corn, these crops
were annually taken from them, with perhaps an occasional year of rest
(i. e. lying without any crop being sown on or taken from them); and,
when reduced to such a degree of barrenness that the crop fell short of re-
paying the cost of producing it, clover or grass was resorted to in the vain
hope of suddenly repairing, through their instrumentality, the ravage and
desolation of years. The following is from the report of a Committee of
the Fishing Creek Agricultural Society, Chester District, South Carolina,
made to the President of the State Society in 1843; and, though this district is not in the tide-water zone, the system of cropping described is more
or less the prevailing one* throughout much of the cotton growing region:

" We generally plant cotton on fresh land four or five years in succession—then corn—
then wheat or oats—again corn and cotton; and, after it will produce little else, we sow it
in rye, and let it rest two or three years. There are no fixed principles observed in the rotation of crops. . . . We have no data whereby to fix the expense of cultivation accurately. We know this, however, that at the price of produce for the last two or three years,
we are sinking money."†

I ask what would be expected, in the way of grass or clover, from some
of the best grazing lands of New-York, after being cropped with grain
crops from ten to twelve years consecutively, with little or no manure?—
However carefully seeded with the best grasses, or with clover, they would
not form meadows worth mowing, nor pastures where an acre would summer a sheep—though, as now managed, an acre is poorly grassed that will
not summer five or six sheep. Take the map of New-York, Sir, and draw
a right line from Buffalo to a point a little south of Albany—say Coxsackie
—and all the region, speaking in general terms, south of this line and west
of the Catskill Mountains, is mainly devoted to grazing. It is the best
grazing region of the State, and much of it is equal to any in the Northern
States. The best farmers in no part of it take off to exceed three grain or
root crops before seeding down to grass; and, unless the soil is unusually
rich, it is customary to give barn-yard manure to one of these crops. This
is almost invariably the case where the land was in meadow when broken
up. Where no manure is given on meadow lands, or even on lightish pasture lands, two grain crops are considered sufficient by the most provident
farmers—it being an axiom among such, that all ordinary or thinnish soils
should be nearly or quite as rich when seeded down as when broken up.
In other words, they draw from the soil only what is equivalent to the
strength or fertilizing properties of the sod, and of the manure given.—
When seeded down to grass, these lands are usually depastured by cattle
or sheep several years before they are again broken up. If converted into
meadow, they are top-dressed from time to time with gypsum, and sometimes with stable manures.‡ The poorest soils, rocky hill-sides, declivities
much subject to washing and gullying, are rarely broken up after being
once properly seeded down. I repeat it, Sir—take all the grazing lands
of New-York, and crop them as severely as it is reported above to be done
in Chester District, South Carolina, and they would become so sterile that,

* Id est, so far as constant cropping without returning anything to the soil is concerned.
† See Ruffin's Agricultural Survey of South Carolina, 1843—Appendix, p. 6.
‡ It is not considered good economy, however, to top-dress any meadows with stable manures which
are dry and arable, and can thus be subjected to the regular rotations of the farm.

unless resuscitated by copious applications of manure, they would not yield grass enough to pay the expense of keeping them under fence, until they had lain waste for a quarter of a century.

Another cause of the failures which have attended some of the efforts to introduce the culture of clover and the grasses on the tide-water zone, in the Southern States, may, and probably has, existed in the improper selection of the varieties sown. As the first crop on a very meager soil—red clover, for example—is not appropriate in any region. In Flanders, the natural soils of much of which so closely resemble those of the zone under examination, it is not sown until the land is enriched and got in condition by several preparatory crops. The different grasses seem to be affected by various conditions in the soil or atmosphere, or both, which it is frequently difficult or impossible to detect. Timothy grass *(Phleum pratense)* is decidedly the favorite meadow grass of the grazing regions of New-York. White clover *(Trifolium repens)* invariably comes up spontaneously on those lands. Red clover *(T. pratense)* is sometimes sown with Timothy in meadows, and generally in pastures. Red Top* *(Agrostis (stricta) vulgaris)* is preferred on wet lands, where it comes up spontaneously. It is considered a prime pasture and meadow grass in such situations. June or Spear grass *(Poa pratensis)*, the Blue grass of the Southern and Western States, so prized there and also in England,† is considered an unprofitable intruder in our meadows, where it comes up spontaneously, and ultimately drives out the Timothy. The meadows are then said to be "run out," and are broken up. I have never known the seed of this grass sown in a single instance! The favorite Rye grasses of England *(Lolium perenne var. bienne)*, Lucern *(Medicago sativa)*, Sainfoin *(Hedysarum onibrichis)*, Orchard grass *(Dactylis glomerata)*, and various others equally celebrated in England and on the Continent, have been tried in New-York, and the experiments are generally regarded as decided failures. None of them, at all events, have obtained a footing among the grasses sown by our best farmers. On the other hand, the Red Top of New-York is but little regarded in England,‡ and Timothy was not in much better repute until the Woburn experiments demonstrated its great value for hay. Even now it is considered inferior, in general value, to many other grasses.‖ All this goes to show that even the hardiest grasses have their favorite situations; and that we are not authorized to pronounce against the practicability of forming pastures and meadows in a given region, because we have failed in a trial with two or three grasses, out of a list of as many hundreds.

. It has already been remarked that there are patches of good natural pasture on the dry as well as the wet portions of the tide-water zone These are frequent and extensive, and could be rendered infinitely more so by simply clearing the land. In your Memoir on the Cultivation of Rice, furnished to Mr. Ruffin, while making the Agricultural Survey of South Carolina, in 1843, you say :

"At first, rice was cultivated on the high land, and on little spots of low ground, as they were met with here and there. These low grounds being found to agree better with the plant, the inland swamps were cleared for the purpose of extending the culture. In the process of time, as the fields became too grassy and stubborn, they were abandoned for new clearings ; and so on, until at length was discovered the superior adaptation of the tide-lands, and the great facilities for irrigation afforded by their location. For these, the inland plantations were gradually and slowly abandoned, until now, that the great body of land, which

* Sometimes known as "Upright Bent grass," and in the Southern States as Herds-grass.
† Pronounced by Sole the best of all the grasses.
‡ *Agrostis vulgaris* is pronounced "a worthless or rather a mischievous plant," by Sir George Sinclair !
‖ "Our opinion," says Loudon, " is that neither Timothy nor (some other grasses named) is ever likely to be cultivated in Britain."

E

little more than a century ago furnished for exportation over 50,000 barrels of rice, now lies utterly waste, constituting, where trees have not overgrown it, the finest natural pasture which could be desired."*

Mr. Ruffin in his Report of the Survey, of the same year, asserts :

" Few countries possess greater natural facilities, or which are more improvable by industry, for producing in abundance, grass, hay and live-stock, and their products of meat milk and butter, all of which are now so deplorably deficient."†

The Committee appointed by the State Agricultural Society of South Carolina to take into consideration the scheme of reducing the quantity of cotton grown,‡ in their Report observe :

" Millions of acres in South Carolina, including the lower country, are admirably adapted to the raising of rich grasses. This might be added as another branch of industry, from which reasonable profits could be realized, and might very well be added to the cotton planter's income."

Corresponding statements, on equally indisputable authority, might be indefinitely multiplied, not only in relation to that portion of the tide-water zone lying within the limits of South Carolina, but in all the Southern States. South Carolina occupying a central geographical and latitudinal position, in reference to this zone, and its soils on it, about averaging, so far as I can learn, with that of the other States, it is not necessary to pursue the inquiry.

Where fine natural pastures spring up spontaneously on deserted lands, more or less impoverished—probably in most instances considerably so—how little difficulty would there be in forming, almost immediately, the best artificial pastures and meadows on millions of acres of just such land, (only that it is in its virgin state, and consequently far better,) now in unproductive forest ! And how small would be the amount of skill requisite to convert millions of acres more of cotton lands—which do not now yield *emunerating* crops—into pastures and meadows, which, as I shall show, would yield their owners a handsome remuneration !

And the culture of the grasses need not stop with these comparatively good and medium lands. They can be made to stretch their carpet of green over the poorest of your sands—over those now covered with stunted pines, or which, scorched and naked, reverberate back fiercely the burning heat of a southern sky.

There are few regions in the tide-water zone possessing poorer soils than some *cultivated* portions of New-York. In the vicinity of Albany, (between that city and Schenectady, for example,) the same loose, silicious sands, the same, though perhaps rather more stunted, growth of pines, would almost compel you to fancy yourself somewhere between Richmond, and Wilmington, on the route of the great Southern Railroad ! Denuded of their meager covering of dwarf pines, and the cohesion produced by their interlacing roots, these sands would be lifted and driven about by the winds. Yet on such a soil as this, you find the farm of the late celebrated Jesse Buel ! And fertile grass fields, dotted here and there with splendid mansions, are every year stretching out farther and farther among the arid sands. How are these rapid transformations in the fertility of the soil accomplished ? The stables, and mews, and cesspools of Albany can give the answer !

The following description of the natural soils of Flanders, now proverbial for its fine crops and rich pastures and meadows, is from the pen of that able English agricultural writer, Rev. W. L. Rham :

---

* Agricultural Survey of South Carolina, 1843. Appendix, p. 14.   † Ib. p 73.
‡ The Committee consisted of Whitemarsh B. Seabrook, Esq., John D. O'Neill, Esq., and W. J. Alston Esq.—and the Report was made, I believe, in January, in 1846.

" The greater part' of the land in Flanders is naturally poor; and in extensive districts, which now have the appearance of the greatest richness at harvest time, the original soil was once little better than the blowing sands which are met with in the neighborhood of the sea. Neither is it a genial climate which brings forward the fruits of the earth in abundance; for the climate is inferior to that of France or the southern parts of Germany. The soil may be divided into two classes. The first consists of the alluvial' clay loams near the coast; tne second, of various sands and light loams which are found in the interior. The most fertile is that of the low lands which have been reclaimed from the sea by embankments; it is chiefly composed of a muddy deposit mixed with fragments of marine shells and fine sea sand . In the interior of East and West Flanders the soil varies considerably; but the principal part is of a sandy nature. The sand, and a heavier loam which scarcely deserves the name of clay, are found much intermixed, which is owing to an alternation of layers of sand and loam, which are found by digging to a considerable depth. These layers are not of great thickness, and the accidental circumstance of the washing away of the sand in some places and the depositions from rivers in others easily account for this variety. Some of the elevations, which are nowhere considerable, consist of a very poor sand, and suggest the idea of their having once been the sands of the sea blown into hills, as is observable on the coast. These hills, if they may be so called, are naturally so barren that they were, not very long since, covered with heath, or at best planted with fir trees; but they have gradually been cultivated and improved, and only a few remain in their original state of heath and wood. The poorer sands have been brought into cultivation chiefly by the persevering industry of small proprietors and occupiers."

Have we not here a good general description of much of our southern Atlantic coast—the tide swamp and sandy plain—and even a graphically minute account of the "Sand Hill" region of South Carolina ?

Instances of the reclamation of such lands might-be indefinitely multiplied.

I do not offer the above facts to prove that it is either profitable or expedient to reclaim all the sterile lands of the southern sea-board by the same means that have been resorted to about Albany, or in Flanders. Except. in the vicinity of cities, where manures are plentiful and cheap, and uncommon market facilities are offered, it would not be profitable, unless it can be accomplished by less expensive means.

But it proves one and an important position : that it is the sterility of such soils—or perhaps their loose and "blowing" character in some places,. their sun-baked hardness in others—which prevents them from spontaneously producing esculent herbage ; and nothing in them, as has been frequently fancied, positively deleterious to vegetation. And it follows,. hence, that whenever it is *profitable* to convert them into grass lands, it is, *practicable* so to do by the proper application of manures. But do I hear: some of your South Carolina neighbors, of the anti-improvement school,. if you have any such,) say, " If our *soils* are, or can be made, generally,. suitable for the production of the grasses, our *climate* cannot ? " This position is obviously incorrect, as warmer climates, as, for example, Australia, the Cape of Good Hope, and various others, produce, where the soils are favorable, a luxuriant growth of grasses; and South Carolina herself, as has been already shown, produces them bountifully in situations where neither the latitude nor the elevation abates one jot of the heat of our fervid climate.

It is not impossible that the climate of the States farthest south—south, say, of North Carolina—may be unfavorable to certain grasses and clovers ; and perhaps so to the favorite ones of the Northern States. In relation to red clover, however, the acclimation of which is regarded by many as so important to those States, it seems Mr. Ruffin thought otherwise. He says :

"Perennial, or other permanent grasses, of which, doubtless, there may be found some peculiarly suited to the warm climate, (South Carolina,) would still more serve to give the great benefits of changed condition to the fields, independent of the much needed benefits of grass husbandry for feeding of live-stock and giving rest and manure to the land.. The grasses whose value has-been fully established by long experience in more northern coun tries, should be tried—not because they are from the North, (which in itself is a strong objection,) but merely because their good qualities are known, and possibly some such grasses may as well suit a more southern clime. And such, I trust, is red clover, the best of all green and manuring crops. For although this was long held to belong to the North only, I have fully experienced that its locality and the perfection of its growth *are fixed much more by peculiarity of soil than by latitude.* Not more than twenty years ago it was as general a belief in *Lower* Virginia, as now in South Carolina, that there the soil was too scanty and the sun too hot to raise red clover. But since marling and liming have made many of these soils calcareous, it is found that neither the sandy soil nor hot and dry climate forbid the raising excellent and profitable crops of clover. And so hereafter it will be found in South Carolina."*

In a Report by a Committee of the Milton Agricultural Society, (embracing adjacent parts of Laurens and Newberry Districts, S. C.) made to the State Society in 1843, they state :

" Our native grasses, except the crab grass, are of the poorest kind, principally sedge. Of the artificial grasses, some trials have been made with red clover and herds-grass.† On rich lots the first appears to succeed very well. For alternating with tillage crops we do not know of its having been tried ; but our impression is, that without manuring more highly than is customary here, it will not answer. We are not aware that it has ever been sowed with gypsum. The herds-grass, as far as it has been tried, appears to succeed very well on the bottoms that border our branches and creeks."‡

Lawrence and Newberry are not in the tide-water region, but so far as the effect of climate alone is concerned, their testimony has an equal bearing.

I have little doubt that red clover may be cultivated on good, rich soils even in the States south of North Carolina, and may possibly become, under some circumstances, a profitable crop in their rotations ; but, as has been already remarked, it will not do as a *first* crop on very meager soils, in any climate—and still less so, I apprehend, on such soils south of latitude 34°. It is not, therefore, the crop which you need, to *cheaply* ameliorate your poor and exhausted soils, to fit them either for grazing or for tillage. Grant that such soils can be fitted to produce it, as Mr. Ruffin suggests, by the application of lime or marl,‖ these manures will be found expensive, can be but slowly obtained in quantities sufficient to apply to large tracts, and, besides, when the soil is sufficiently ameliorated to carry clover, it will carry most if not all of your ordinary tillage crops. Though clover would aid materially in the rotation, in *sustaining* or even improving the fertility superinduced by lime or any other fertilizer, it is not, and cannot be made the *original* fertilizer on the sterile sands of warm climates. When we talk, therefore, of the initiatory steps by which such soils shall be brought from a state of barrenness to a state of production, clover does not come within the category of appropriate agents.

Though red clover ranks in the first class, if not the first in that class, on appropriate soils, as a grazing and manuring crop, I have never regarded it as indispensable—as what the *lawyers* would style a *sine qua non*—even in *sustaining* fertility anywhere except on rich calcareous wheat lands, where a severe and exhausting rotation is resorted to. Where wheat is taken from the soil *at least* every alternate year, for ten, fifteen, or twenty years, without any manure, excepting the intervening crop, and the droppings of animals depastured on it, clover will *better* sustain the land in the ultimately fatal struggle, than perhaps any other green ma-

* Ruffin's Agricultural Survey of S. C., 1843, p. 81.
† This should be the *Agrostis stricta* or *vulgaris*—the Red Top of the North. Some writers designate as the one species, some as the other.
‡ Ruffin's Agricultural Survey of S. C., 1843 ; Appendix, p. 9.
‖ Unless, however, the soil contains more *organic* matter than I suppose to be the case with many of your *sandy* soils, theory and practice both show that lime will not prove the proper manure. Though exceedingly valuable in its place, experience shows that it is no agricultural *panacea.* I shall allude to this subject more fully in a subsequent letter

nuring crop. But on the silicious grazing soils of Southern New-York, it is rarely used exclusively as a manuring crop, and is but little used, excepting slightly admixed with timothy, for pasture or meadows.* I think it should be used more ; but the fact stated shows that clover is not ro garded by practical men, who are perfectly familiar with it, as that indispensable crop, in all situations, which some of its more extravagant panegyrists would lead us to suppose. The conclusions which I would have you deduce from the above facts and statements are, simply, that if clover is found to flourish with you without extra trouble and expense, you will do well to make use of it in your rotation ; if not, it is chimerical, in my judgment, to engage in an expensive struggle with natural disadvantages to force its cultivation.

The herds-grass (red top) spoken of by the Milton Society, is a good grass on moist (but not boggy) soils, and having been found to succeed with you, is worthy of trial in such situations, but on dry soils, especially on arid sands, it would entirely fail. Nor have I much confidence in either timothy or spear (blue) grass, in such situations, in your latitude—none at all in the former.

It would be well, probably, to try limited experiments with all grasses, domestic and foreign, which have succeeded well on *soils similar to your own ;* as among these, some may be found which disregard climate, or are even better fitted to your climate than their indigenous one, as was tho .:ase with timothy at the North. The same remark is also true in relation to certain other esculents which are used as substitutes for the grasses, and for green manuring crops.

Notwithstanding the evident propriety of such experiments, I am strongly inclined to the opinion that it is to your own native grasses and esculents, or those of some kindred climate, you must look mainly for the basis of your grazing husbandry—and through this, the amelioration of your poor and exhausted soils.

I regret that I can find no list of those native grasses which sward over the deserted lands of the tide-water zone, and flourish with a tropical luxuriance in its swamps. You allude to them as 'native" grasses, so does Mr. Ruffin. Mr. Seabrook, in his Report on Cotton Culture,‡ speaks of "crop grass," by which I suppose he means Crab grass, *(Panicum sanguinale,)* coming up spontaneously after spring-sown peas; but farthei than this, neither of you specify varieties.|| Among these indigenous ones, particularly those which spontaneously make their appeaiance on dry lands, it would be exceedingly singular if there are not several very valua ble grasses *for your soils and climate*—grasses the seeds of which should form a part, if not suitable for the whole sowing, on the same kinds of soils on which they are found flourishing.

Crab grass grows in all parts of the southern States, and is a fair, though not a *very* superior pasture and meadow grass.

Golden millet *(Panicum milliaceum)* is a great pioducer and withstands

---

ı know of but very few farmers excepting myself, in this, (Cortland,) one of the best of the grazing counties, who sow unmixed clover seed. I confess myself decidedly partial to the crop. You may ride ten miles or more in many directions from my house, where half and frequently more than three-fourths of the fields are in pasture or meadow, without observing five acres of unmixed clover.

‡ For this elaborate and exceedingly able Report or Memoir, see Farmers' Library, 1845, October, November and December Nos.

|| Since writing the above, I have received from a South Carolina correspondent the following list of grasses and other esculents which flourish in the lower part of that State. Crab grass *(Digitaria sanguisalis),* earlier—the " Crowfoot" *(Eleusine Indica),* a little later, are, he says, the best grasses for hay, and thrive in cultivated grounds from the month of June till frost. The "Wild Okra" *(Viola palmata),* the "Partridge Berry" *(Mitchella repens),* the Wild Pea Vine, and several other esculents, obscure and unknown by name, flourish in most natural pastures from early spring till November.

a warm climate, but it requires a good soil. It has been cultivated with great success by Mr. Affleck in (Adams county) Mississippi.

Bermuda grass* *(Cynodon dactylon)* I have been led to consider, from the representations of Mr. Affleck, as the best grass, both for pasture and meadow, on the sterile sands of the tide-water zone. If half this enthusiastic admirer believes of it is true, it is of inestimable value to the South, and for *permanent* pastures and meadows, is by far the best grass in the United States. Mr. A. says :

" We are fully aware of all the objections made to the spreading of this grass, and have a practical knowledge of all the trouble it occasions ; and having also had several years' experience of its great, its incalculable value, we have no hesitation in stating that the latter is manifold greater than the former. The time is not far distant when all the rough feed consumed on plantations will be made from this grass ; and when the planter will consider his hay crop as of more importance than his sugar or cotton. . . . . . The excellence of this plant for pasturage is evinced by two circumstances. It is preferred by stock of every description to all other grass, and it grows luxuriantly in every kind of soil. It possesses an additional advantage, that of binding the loosest and most barren sandy tracts. But when it has once taken possession of close, rich soil, its extirpation is so difficult as almost to defy all the skill, industry and perseverance of farmers. It is used to bind the levees on the banks of the Mississippi, and of railroads. We saw it at Macon, Geo., Charleston, S. C., and so on, as far north as City Point, Virginia, where it partially covers the wharf. One hundred pounds of grass afford *upward of fifty of dry hay;* and *we do cut,* as a regular crop, five tons of hay per acre each season. Were we to state *how much more* has been cut, we might strain the belief of our readers. No other grass will yield such an amount of valuable hay ; surpass it in nutritive qualities ; support on an acre of pasture such a quantity of stock ; will improve the soil more quickly ; or so effectually stop and fill up a wash or gully. But, on the other hand, its *extirpation,* when once well established, is almost impossible ; though to check and weaken it, so far as to grow a grain or cotton crop, is easy enough. To do this, pursue the course of the best farmers of Kentucky in their management of a blue-grass sod— with a good breaking plow, having a wheel and coulter, and a stout team, turn over evenly and nicely a sod four inches thick and as wide as the plow and team are capable of, follow in the same furrow with another plow which casts the dirt well, and throw out as much of the fresh earth on top of the sod as possible or the depth of the soil will admit of. The crop that follows can easily be tended without disturbing the sod, and its gradual decay will greatly increase whatever crop may be planted on it—and that should be a shading one, corn and peas or pumpkins, or winter oats followed by peas. Good farmers will understand that heavy crops of hay cannot be removed, for many successive years, from any land, without some return in the shape of manure. To the careful, judicious farmer, who wishes to improve his land and his stock, and who does not expect to grow any crop without trouble, and who uses good plows, and keeps a stout team and that in prime order, we earnestly recommend to try an acre or two of this grass, in a situation where it cannot readily spread. To the careless farmer we say touch it not."†

The same gentleman writes me under date of Dec. 10th, 1846 :

" Bermuda grass well set, which affords the finest and most nutritious pasturage I have ever seen, will keep almost any number of sheep to the acre—three or four times as many as the best blue-grass ! "

Unless this is gross and willful exaggeration,‡ here you have a grass which is not only highly palatable and nutritive, but which will yield more than double both of pasturage and hay, than the best grass or clover of the Northern States ! || It has been tried as far south as New-Orleans, and the climate found no detriment to it. It will flourish on dry and almost barren sands.§ What can the farmer on the dry lands of the tide-water zone ask more ? Its inextirpable character I regard as decidedly in

* Cumberland Grass—Wire grass of Virginia—Creeping Panic grass.
† See Norman's Southern Agricultural Almanac, for 1847.
‡ Neither of which are we permitted to suspect, from the well-known character and intelligence of Mr Affleck.
|| People here in the North sometimes *talk* of getting three tons of timothy and four tons of clover (at two cuttings) per acre, but it is not done on one acre in ten thousand, on the best meadows ! Two tons is a good, and by far above a medium yield, of timothy, and three, of clover. The large amounts of Bermuda sometimes cut, which Mr. A. does not mention for fear of " straining the belief of his readers " he has stated to me personally, to be *eight tons !* :—equivalent to the yield of *three* first-rate acres of timothy on the best grazing lands of Southern New-York.
§ Mr. Affleck informs me he has repeatedly seen it growing well in such situations.

its favor on millions and millions on the thinner and poorer soils of that zone—as once admitted, it will put an end to the unprofitable tillage practiced on them, and remove all temptation to resort to it on others, as they are gradually rescued from barrenness. It will thus compel the adoption of that pastoral system which can alone make these lands profitable, or save them, if the forebodings of those who have been reared on them and are deeply attached to them, can be credited, from ultimate desertion.*

You have another fodder crop—and which may be made a green manuring one, in no respect inferior to clover. The pea is to the South what clover is to the North.† There is something in your soil or climate, or both, which seems to be specifically adapted to the development of this plant—for it flourishes with you under a much greater variety of soils and circumstances than at the North. A leguminous plant, like clover, it draws much of its aliment from the atmosphere; and it is perhaps as sensibly affected by the same cheap manure, plaster. Its haulm or straw, if cut and cured greenish, and well taken care of, makes a good, rich fodder relished by all kinds of stock. Peas are greedily eaten by neat stock, swine, and sheep, for which they form a healthy and highly nutritious food. The white field pea of the North is considered equivalent to our corn,‡ by measure, in fattening swine. For sheep, and particularly for breeding ewes, there is probably no feed in the world equal to nicely cured pea haulm,‖ with a portion of the seed left unthreshed.§ It gives them condition and vigor—and prepares them to yield a bountiful supply of rich milk to their young.

Though the pea is an annual, it becomes in effect a perennial, South, when it is desired, by suffering it to stand until some of the grain shells out.¶ It will mature in a southern climate, sown late in the summer, so that one, and even two preceding crops of it might first be plowed in as a manure. It will ripen among Indian corn, sown after that plant has ceased to grow, and there have been successful experiments of sowing it late with wheat, oats, &c., to have it obtain its growth (to be plowed under as manure) after those crops have been harvested.

Sprengel gives the following analysis of the pea. 1,000 parts in the common dry state yield

	Seed.	Straw.		Seed.	Straw.
Potash and soda	15·50	2·35	Sulphuric acid	0·52	3·35
Lime and magnesia	1·95	30·70	Chlorine	·38	0·00
Phosphoric acid	1·90	2·40	Silica, iron, &c.	4·40	10·85

* Statements of this kind have been repeatedly made in the pages of the Monthly Farmer by southern gentlemen.
† I had labored under the impression that the so-called pea—cultivated as a manuring crop in the South ern States, was in reality a variety of the bean; but Mr. Ruffin in his Agricultural Survey of South Carolina, (see Report of 1843, p 81,) and Hon. W. B. Seabrook in his Memoir on Cotton Culture, (see Monthly Journal of Agriculture, Dec., 1845, p. 287,) speaks of this crop—the former again and again—as peas, without the qualification which would be expected from gentlemen of so much learning, in case they were speaking of a plant by a vulgar misnomer, instead of its real name. The peculiar value of the crop at the South in the particulars described, I find asserted by Mr. Ruffin, Mr. Affleck, and various other writers and Agricultural Societies, in the strongest terms, and therefore it makes little difference, practically, whether the name is correct or not, but if not, the following analyses, &c., are misplaced. The bean resembles the pea in its qualities and value, but is rather inferior to it.
‡ The small, hard corn of the North contains more nutriment per bushel than the large southern corn.
‖ That is, cut and cured so that it will come out of the stock or mow bright, and with the leaves looking green—instead of having the ferruginous hue of over-ripe clover.
§ If cut greenish and well cured, the greener pods will not thresh out readily, and then they are in exactly the proper condition for breeding-ewes. If the crop is very light, cut it when all the pods are quite green, and feed it out without threshing.
¶ This is, however, poor economy in any case. If the object is peas, it is wasteful to the crop, and the quantity sown is uncertain; besides, the haulm is ruined for fodder. If the object is manure, the loss is still greater. Plants in drying lose the nitrogen contained in their sap, give up their saline matters, and are "resolved more or less completely into carbonic acid, which escapes into the air, and is so far lost."—See Liebig on this subject, and also the clear and able remarks of Johnston. (Johnston's Agricultural Chemistry, vol. ii. p. 176, et supra.)

The following table of the comparative value of manures, deduced from analyses made by Payen and Boussingault, will show the remarkable comparative value of the pea as a manuring crop, and it will be found otherwise useful for reference :

TABLE No. 4.

Kinds of Manure.	Water per 100.	Nitrogen in 100 of matter.		Quality according to state.		Equivalent according to state.		Remarks.
		Dry.	Wet.	Dry.	Wet.	Dry.	Wet.	
Farm-yard dung	79·3	1·95	0·11	100	100	100	100	Average of Bechelbrunn.
Dung water	99·6	1·54	0·06	72	2	127	68	Washed by the rain.
Wheat straw	19·3	0·30	0·24	15	60	650	167	Fresh of Alsace, 1838.
Rye straw	12·2	0·20	0·17	10	42·5	975	235	Of Alsace.
Oat straw	21·0	0·36	0·28	18	70	542	143	do.
Barley straw	11·0	0·26	0·23	13	57·5	750	174	do.
Wheat chaff	7·6	0·94	0·85	48	212·5	207	47	do.
Pea straw	8·5	1·95	1·79	100	447·5	100	22	do.
Millet straw	19·0	0·96	0·78	49	195	203	51	do.
Buckwheat straw	11·6	0·51	0·48	27	120	361	83	do.
Dried potato tops	12·9	0·43	0·37	22	92·5	453	108	
With'd l'ves of beet-root	88·9	4·50	0·50	230	125	43	80	Of mangel-wurzel.
Do. of potatoes	76·0	2·30	0·55	117	137·5	85	73	Withered top and leaves.
Do. of carrots	70·9	2·94	0·85	150	212·5	66	47	
Do. of heather	7·0	1·90	1·74	97	425	103	23	Dried in the air.
Do. of oak	25·0	1·57	1·18	80	293	125	34	Leaves fallen in autumn.
Do. of poplar	51·1	1·17	0·54	66	134	167	74	do.
Do. of beech	39·3	1·91	1·18	78	294	102	34	do.
Clover roots	9·7	1·77	1·61	90	402·5	100	25	Dried in the air.
Burned sea-weed	3·8	0·40	0·38	20	95	488	105	
Oyster shells	17·9	0·40	0·32	20	80	488	125	
Sea shells		0·05	0·05	3	13	3750	769	Dried sea-shells of Dunkirk
Sea-side marl	1·0	0·52	0·51	26	128	377	78	
Solid cow-dung	85·9	2·30	0·32	117	80	84	125	
Urine of cows	83·3	3·80	0·44	194	110	51	91	
Solid horse-dung	75·3	2·21	0·55	113	137·5	89	73	
Horse urine	79·1	12·50	2·61	641	652·5	15½	15½	The horse drank but little, the [urine was thick.
Pig dung	81·4	3·37	0·63	172	157·5	58	63	
Sheep dung	63·0	2·99	1·11	153	277·5	65	36	
Pigeon dung	9·6	9·02	8·30	462	2075	21½	5	Of Bechelbronn.
Guano	19·6	6·20	5·00	323	1247	31	80	Imp. into Eng. in its ord. state.
Do.	11·3	15·73	13·95	807	3487	12½	28½	Imp. into France, do.
Fresh bones	30·0		5·31		1326		7½	As sold by the melters.
Feathers	12·9	17·61	15·34	903	3835	11	2¼	
Woolen rags	11·3	20·26	17·98	1039	4495	9¼	2¼	
Horn shavings	9·0	15·78	14·36	809	3590	12½	3	
Coal soot	15·6	1·59	1·35	81	337·5	122	30	
Wood soot	5·6	1·31	1·15	67	297·5	149	35	
Picardy ashes	9·2	0·71	0·65	36	162·5	275	62	

It will be seen that pea straw is worth, as a manure, from 5 to 9 times as much as the straws of the small grains—is better than clover roots, and actually equals farm-yard dung!

Rye, oats and barley send up a good growth of straw, in many parts of this zone, even where the product of grain is small; and, sown in the fall, they afford sweet green pasturage, during the entire winter, in the more southern latitudes. This is a very important and a very favorable consideration in an economical system of sheep husbandry. All winter green feed (roots) in the Northern States must be cultivated, harvested, protected from the frosts of winter in cellars, and daily fed out—which necessarily renders it expensive. Where winter field crops can be depastured on the ground, it saves the greatest proportion of this expense ; and, though winter green feed is not indispensable to sheep, it promotes their health, early maturity, and is especially valuable to breeding-ewes. All the crops above named, too, can be profitably made use of as green manure.

Blades of corn, well cured, are relished by sheep, and they thrive on them.*

The sweet potato is also readily eaten by them, and it fattens them perhaps as rapidly as any other root crop. Although it might be regarded as too valuable for sheep feed, in regions where the whole force is given to he culture of cotton, there are others where, I cannot but believe, it might be occasionally if not regularly resorted to with profit, unless rye, oats, barley, &c. can be provided so much more cheaply that it is no object so to do. It is so cheaply planted by slips, and tilled with so little trouble, and it so admirably prepares land for subsequent crops,† that, on rich and otherwise favorable soils, my impression is strong it is, at all events, as cheap a winter feed for stock in the South as the Irish potato is in the North. Its average yield is about two-thirds that of the latter. The Irish potato is universally regarded as one of the cheapest feeds that can be given to all kinds of stock, to which it is adapted in the North. It is true that it is not fed so much as it would otherwise be, with us, in the winter by reason of the *cold*. It is difficult to protect this root from freezing, and at the same time leave it accessible for daily feeding, without putting it in dwelling-house cellars, which are usually at some distance from the feeding barns and yards; and besides, the conversion of this citadel of a northern matron's culinary stores, into a great, dirty root pit, would be a most grievous infringement on all the canons of good housewifery!

The foregoing facts show that the Southern States have already all that is necessary to feed stock and fertilize their fields. Their pea, take it all in all, is a full equivalent for the clover of the North.‡ By means of it—of Bermuda and some other grasses—aided by the droppings of sheep, and other cheap and convenient manures, a large proportion of the tide-water zone, now so unproductive, can be converted into grazing lands, which will yield as good a per centage on present capital and investment as the best cotton uplands, and produce wool *at a less expense per pound than any region of the United States north of the Potomac.*

---

* A friend of mine wintered a few Merino sheep on not only the blades, but the *stalks*, of our northern corn, chopping the whole up together, and adding a little bran or shorts. He found it cheap feed, and the sheep got fat enough to slaughter before spring.

† After the crop is harvested, swine are turned in, and they root the ground over so deeply and thoroughly that it is in a better state of tillage than could be produced by mere spring plowing.

‡ Mr. Ruffin, the great advocate for clover, admits that in the South it is not fitted to precede Indian corn, on account of the destructive *cut worms* it harbors, unless the land be plowed "early in winter," or other precautionary steps are taken. The pea is not liable to this objection. See Ruffin's Ag. Survey of S. C., 1843, p. 78.

LETTER IV.

**THE ADAPTATION OF THE SOILS, HERBAGE, &c. OF THE SOUTHERN
STATES TO SHEEP HUSBANDRY, CONTINUED. 2. OF THE MIDDLE OR
HILLY ZONE. 3. OF THE MOUNTAIN REGION.**

Climate, Soils, and Productions of the Middle or Hilly Zone—Its evident Adaptation to Sheep Husbandry
...The Mountain Region...Altitudes of different Ranges and Penks—Their general Shape—Freedom from
Rocks, Precipices, &c...Table Lands—Their Geological Formations—Products...Mr Clingman's Letter
describing the Roan and other Mountains in North Carolina...Mr. Buckley's Counter statements...Mr.
Earle's Description of the Mountains in Henderson and Rutherford Counties, North Carolina...Col. Colston's Statements in relation to the Mountains in Berkley County, Virginia...Hon. A. Stevenson's in relation to the Blue Ridge, and the Mountains in the South-west of Virginia—Hon. W. L. Goggin's in relation
to the same...Judge Beatty's Account of Sheep Husbandry on the Cumberland Mountains—Mr. Kramer's
...Mr. Buckley's Views in relation to the North Carolina Mountains examined and objected to...Climate
of the Roan and others compared with that of the Grazing Lands of New-York...Statistics showing the
Forwardness of the Seasons and the Temperature in New-York...Effect of Elevation on Temperature
On Vegetable Productions.

Dear Sir: The middle or hilly zone is high, dry, healthy, and has a
mild and, compared with the North, equable climate.* Its soils possess
the ingredients due to its formation—disintegrated granite—and are far
more fertile than those of the lower zone. Sometimes on the summits of
the hills they are poor and thin, and there are occasionally extensive ranges
of poor land, as in Virginia; but as a general thing, they vary from fair to
good; and on the bottom lands of some of the rivers and larger creeks,
they possess remarkable fertility. The valleys, however, are generally
narrow, and are everywhere the bed of streams, which abundantly water
this whole region, and furnish inexhaustible facilities for mills and manufactories. The slight cohesion of the soil, aided by the face of the country
and the system of tillage pursued in many parts of it,† render it peculiarly
subject to washing by heavy rains. The hill-sides are frequently cut into
deep gullies, rendering aration difficult,‡ and the surface soil is washed
into the valleys and into the beds of the creeks, not only impoverishing the
high lands, but, by impeding the courses of the streams, in some regions
converting those of the valleys into unhealthy marshes.||
Grasses suited to the climate flourish *when sown*, and on lands not utterly worn out, throughout all this region; and there is little doubt that
every variety which could be acclimated on the sands of the lower zone,
could be more readily acclimated here—and probably various others. The
pea succeeds in nearly every situation; oats also form a valuable manuring crop in some parts; while on many of the alluvial bottoms, such,
for example, as the Blackjack lands of South Carolina—rye grows luxuriantly, answering a valuable purpose either for grain, manure, or for winter

---

* The range of the thermometer is sometimes 60° to 75° in a single month (March or April) in New-
York !
† That is, a constant succession of clean tillage crops, such as cotton, corn, and tobacco.
‡ The Fishing Creek Agricultural Society, in their Report before quoted from, say : "The only really
waste land we have is our old fields, many of which are so washed and gullied as to be absolutely irre-
claimable." Mr. Ruffin says that "the destruction both of soil and of fertility has been enormous" from
this cause.
|| "The country was, at first, as its features indicated, nearly free from malaria and all its noxious effects
But as soon as the incessant and injudicious use of the plow caused the soil to be washed from the hilly
grounds into the bottoms, the before unobstructed clean bordered channels of all the small streams were
filled and clogged with earth, and vegetable rubbish, and finer matter, and the adjacent low lands were
thereby rendered swampy. The washing of the high land earth into the valleys so altered the original surface level as to kill the trees; and their decay, and, later, the obstructions by their fallen trunks, increased the
general evil. . . . . I infer " that these causes " have mainly served to nourish malaria and increase the
malignity of disease." [Ruffin's Ag Survey of S. C., 1843, p. 26.]

feed for stock. In this last particular, it would be, as I have before said, an important auxiliary in sheep husbandry.

The adaptation of most of this region to sheep husbandry is too obvious to require extended comment; and it becomes, therefore, simply a ques-tion of profit and loss, whether it is expedient to introduce it.* Let us turn therefore, to the adaptation of the mountain region to this branch of industry

The altitude of the southern mountains, with a few exceptions, is not very considerable. The loftiest, the Black and the Roan, in North Caro-lina, are respectively 6,476 and 6,038 feet in hight. The Peaks of Otter, the highest, and summits of the Blue Ridge in Virginia, are 4,250, and the highest Alleganies 2,500 feet high. Table Mountain in South Carolina is about 4,000, and the terminal masses of the Blue Ridge in Georgia are about 1,500.† The hight of the Cumberland Mountains, the most western chain, I nowhere find stated, but they are not reputed as high as some of the preceding. It will be seen, therefore, that none of the southern moun-tains rise above the range of the grasses. They are usually broad at the base, easy of ascent, and rounded or flattened on their summits, instead of rising from narrow bases into steep pyramidal forms with conical peaks; and from their geological formations and their shape (resulting probably from that formation,) they are uncommonly free from exposed rocks, preci-pices and abrupt acclivities. With the exception, perhaps, of the Cum-berland chain, large, exposed rocks abound far less, on most of these mountains than in many parts of New-England, or even the Old Red Sand-stone region of Pennsylvania, which are not only pastured, but plowed! Indeed, a side-hill plow, drawn by oxen, could be used on very many of the southern mountains, if cleared, to their very summits; and this is true, singular as it may appear, of some of the loftiest of them.‡ The Cumber-land Mountains are spoken of by Doct. Morse, as "stupendous piles of craggy rocks," and in these statements he has been followed by more re-cent geographers. But if this description applies to some portions of the chain, it certainly does not to others, as I shall have occasion to show.

On the sides, and sometimes on the summits of the mountains in differ-ent parts of this whole region, extensive plains or table lands, already pretty well covered with wild and domestic grasses and nutritious escu-lents, not unfrequently occur. Esculents suitable for sheep are to be found in greater or less quantities on nearly all of them.

West of the summit of the Blue Ridge, the geological formations, as has before been stated, belong to the Transition period—a rather unusual circumstance in mountain ranges, and undoubtedly more indicative of fertility in the superincumbent soils than the ordinary Primary formation.‖ Indeed, they are the same with those of the best grazing lands of South-ern New-York, and subtracting climatic and other§ effects of elevation, they should possess a general correspondence in their properties and pro-ducts, with the latter.¶

* This question will be fully discussed in a subsequent letter.
† For these altitudes, I am indebted to Professor Mitchell.
‡ For example, the Roan.
‖ It is true that soils formed from Primary rocks, when sufficiently fertile to sustain herbage of any kind are peculiarly adapted to the production of sweet grasses; but mountains of this formation are usually steeper, from the slower decomposition of granite, gneiss, and other Primary rocks, and their steepness expose them to increased abrasion, or washing. Hence their soils frequently but thinly cover the rocks and are of a meager and lixiviated character.
§ To wit, abrasion and denudation by rains. And, moreover, the "northern drift" of New-York has added a little lime to the soils formed from these rocks, and thus supplied, measurably, a want existing in all of them for most tillage crops.
¶ For example, the "Slate Hills," which rise on the west of Augusta, Rockingham, Shenandoah, Frod-erick and some other counties in Virginia, are composed of the same rocks (Hamilton group, including Genesee slate of the New-York system,) which underlie some of the best soils in New-York; and much of the land between these hills and the Alleganies rests on the same rocks, (Chemung,) which underlie the southern grazing region of New-York.

In ascertaining the particular products of these mountains, their climate, and general adaptation to sheep husbandry, I will first call your attention to the often quoted letter from Hon. T. L. Clingman, of North Carolina, to John S. Skinner, Esq., in 1844. Mr. Clingman says :

"You state that you have directed some attention to the Sheep Husbandry of the United States, in the course of which it has occurred to you that the people of the mountain regions of North Carolina, and some of the other Southern States, have not availed themselves sufficiently of their natural advantages for the production of sheep. Being myself well acquainted with the western section of North Carolina, I may perhaps be able to give you most of the information you desire. As you have directed several of your inquiries to the county of Yancey, (I presume from the fact, well known to you, that it contains the highest mountains in any of the United States,) I will, in the first place, turn my attention to that county. First, as to its elevation. Dr. Mitchell, of our University, ascertained that the bed of Tow River, the largest stream in the county, and at a ford near its center, was about 2,200 feet above the level of the ocean. Burnsville, the seat of the court-house, he found to be between 2,800 and 2,900 feet above it. The general level of the country is, of course, much above this elevation. In fact, a number of the mountain summits rise above the hight of 6,000 feet. The climate is delightfully cool during the summer; in fact there are very few places in the county where the thermometer rises above 80° on the hottest day. An intelligent gentleman who passed the summer in the northern part of the county (rather the more elevated portion of it) informed me that the thermometer did not rise on the hottest days above 76°.

' You ask, in the next place, if the surface of the ground is so much covered with rocks as to render it unfit for pasture? The reverse is the fact; no portion of the county that I have passed over is too rocky for cultivation; and in many sections of the county one may travel miles without seeing a single stone. It is only about the tops of the higher mountains that rocky precipices are to be found. A large portion of the surface of the county is a sort of elevated table-land, *undulating*, but seldom too broken for cultivation. Even as one ascends the higher mountains, he will find occasionally on their sides flats of level land containing several hundred acres in a body. The top of the Roan (the highest mountain in the county except the Black) is covered by a prairie for ten miles, which affords a rich pasture during the greater part of the year. The ascent to it is so gradual that persons ride to the top on horseback from almost any direction. The same may be said of many of the other mountains. The soil of the county generally is uncommonly fertile, producing with tolerable cultivation abundant crops. What seems extraordinary to a stranger is the fact that the soil becomes richer as he ascends the mountains. The sides of the Roan, the Black, the Bald, and others, at an elevation even of five or six thousand feet above the sea, are covered with a deep, rich vegetable mould, so soft that a horse in dry weather often sinks to the fetlock. The fact that the soil is frequently more fertile as one ascends is, I presume, attributable to the circumstance that the higher portions are more commonly covered with clouds; and the vegetable matter being thus kept in a cool, moist state while decaying, is incorporated to a greater degree with the surface of the earth, just as it is usually found that the north side of the hill is richer than the portion most exposed to the action of the sun's rays. The sides of the mountains, the timber being generally large, with little undergrowth and brushwood, are peculiarly fitted for pasture grounds, and the vegetation is in many places as luxuriant as it is in the rich savannah of the low country.

"The soil of every part of the county is not only favorable to the production of grain, but is peculiarly fitted for grasses. Timothy is supposed to make the largest yield, two tons of hay being easily produced on an acre, but herds-grass, or red-top, and clover succeed equally well; blue-grass has not been much tried, but is said to do remarkably well. A friend showed me several spears which he informed me were produced in the northern part of the county, and which by measurement were found to exceed 70 inches in length. Oats, rye, potatoes, turnips, &c., are produced in the greatest abundance.

"With respect to the prices of land, I can assure you that large bodies of uncleared, rich land, most of which might be cultivated, have been sold at prices varying from 25 cents to 50 cents per acre. Any quantity of land favorable for sheep-walks might be procured in any section of the county at prices varying from one to ten dollars per acre

"The few sheep that exist in the county thrive remarkably well, and are sometimes permitted to run at large during the winter without being fed and without suffering. As the number kept by any individual is not large enough to justify the employment of a shepherd to take care of them, they are not unfrequently destroyed by vicious dogs, and more rarely by wolves, which have not yet been entirely exterminated.

"I have been somewhat prolix in my observations on this county, because some of your inquiries were directed particularly to it, and because most of what I have said of Yancey is true of the other counties west of the Blue Ridge. Haywood has about the same elevation and climate as Yancey. The mountains are rather more steep, and the valleys somewhat broader; the soil generally not quite so deep, but very productive, especially in grasses. In some sections of the county, however, the soil is equal to the best I have seen.

"Buncombe and Henderson are rather less elevated; Ashville and Hendersonville, the county towns, being each about 2,200 feet above the sea. The climate is much the same, but a very little warmer. The more broken portions of these counties resemble much the mountainous parts of Yancey and Haywood, but they contain much more level land. Indeed the greater portion of Henderson is quite level. It contains much swamp land, which, when cleared, with very little if any drainage, produces very fine crops of herds-grass. Portions of Macon and Cherokee counties are quite as favorable, both as to climate and soil, as those above described. I would advert particularly to the valley of the Nantahalah, in Macon, and of Cheoh, in Cherokee. In either, for a comparatively trifling price, some ten or fifteen miles square could be procured, all of which would be rich, and the major part sufficiently level for cultivation, and especially fitted, as their natural meadows indicate, for the production of grass.

"In conclusion, I may say, that as far as my limited knowledge of such matters authorizes me to speak, I am satisfied that there is no region that is more favorable to the production of sheep than much of the country I have described. It is everywhere healthy and well watered. I may add, too, that there is water-power enough in the different counties composing my Congressional District to move more machinery than human labor can ever place there—enough, perhaps, to move all now existing in the Union."

A writer in the Albany Cultivator, Mr. S. B. Buckley, of Yates county, New-York, who has visited these mountains, thus objects to the views of Mr. Clingman:

"These mountains have a cold, damp climate, the summits of the highest being covered with clouds and mists a large portion of the summer season. Cold rains are of frequent occurrence, doubtless causing the deep vegetable mould alluded to by Mr. C. A large portion of the county of Yancey is an elevated table-land which is so damp and cold that the inhabitants do not raise corn sufficient for their own consumption. . . . . Mr. Husted informed me that in many seasons there was scarcely a month in the year without frost . . . . that he had been on the top of the Roan on the 25th of June, when a snow storm arose and completely covered the mountain, and that there were few days in the year but that it was foggy on the Roan. . . . . I have ascended most of the high mountains in that State, and rarely without encountering a storm, or finding their tops covered with mists, which disappeared in the cool of the evening, to be resumed by the warming rays of the morrow's sun. In encamping on the mountains, I generally found the thermometer to range from 45° to 60° and on the high mountains, during the day, it seldom rose above 65°. The inhabitants of the valleys pay great attention to the raising of cattle and horses, which, in the summer season, are turned upon the mountains in what is termed 'the range,' which consists of tall weeds, native grasses, and in many places white clover has become naturalized. * * * *

"These remarks will apply more or less to the mountainous region of Haywood and Macon counties, from which we conclude that they are not suitable to the raising of fine-wooled sheep, judging from their elevation, damp and cold climate, which, as before remarked by Mr. C., creates a deep vegetable mould, in which a horse will sink up to the fetlock. And would not sheep sink in also, and be liable to have the foot-rot? And in yeaning time would not many lambs be lost from the frequent cold rains so common there during the month of May?"*

In a previous communication in the Cultivator the same writer says:†

"On the 12th of May I arrived at Ashville, (the capital of Buncombe county,) intending to visit Mt. Pisgah, a high conical mountain in full view, about twelve miles distant, overtopping its neighbors. I was told that the season was not far enough advanced to bring vegetation forward on the high mountains. . . . The climate of this region is not much, if any, warmer than that of Western New-York. During the summer of 1842, the thermometer ranged generally from 70° to 85° in the valleys, while on the mountains it was frequently about 60°, and sometimes much lower. . . . . When I left the southern portion of Alabama, it was the middle of March; the woods were green, with their full expanded leaves; in about a week I had reached the elevated region south of Huntsville, in the northern part of the State, where the leaves had not yet attained half their usual size. From the 1st to the 10th of April, in Middle Tennessee, the leaves were nearly full grown and the inhabitants were busy in planting corn; but at the middle of April, for thirty miles on the table land of the Cumberland Mountains, the trees had just begun to put forth their leaves, and the ground was white in the morning with a severe frost.

"On descending into the plains of East Tennessee, the country was green with verdure, and the farmers were there also busy in planting corn, and now, the middle of May, among

* See Albany Cultivator, 1846, p. 242.   † Ib., 1846, p. 174

the mountains of North Carolina, I found myself where vegetation had scarcely clothed the plains and woods with green, while the leaves of the high mountain trees were about half grown. I should also remark that the spring of 1842 was from two weeks to a month earlier than usual."

I record a portion of the last extract for subsequent reference; and the object of these communications being to arrive at the *truth*, and not to ride a favorite hobby, or advance a preconceived theory, I have thought it proper to give the substance of all this gentleman's remarks, embodying as they do *all* the objections, real or supposed, which exist against the *highest mountains in the whole Southern States* for the purposes of sheep husbandry.

Per contra, we have the following statements of Henry M. Earle, Esq., of Pacolett, Rutherford Co., North Carolina : *

" On the question whether wool-growing will succeed in North Carolina or not, I would say that it depends entirely upon the exertions used, as I am thoroughly convinced that the country and climate are altogether favorable. The objections raised by Mr. Buckley, if they existed in all the mountain region, might be considered serious; but as they can only be offered against a few very high mountains, situated in the midst of many other mountains, and far from any level or plain country, such a hiding place as he speaks of would not be such a place as persons raised in civilized or refined society would wish to settle in. The Roan and Black Mountains were selected by Mr. Thos. Clingman, because they were the most elevated and noted mountains in Yancey Co., and not, I presume, because he thought they would afford the best pasturage for sheep; if so he was mistaken. On those mountains and in their vicinity are the finest grazing lands for cattle; and so there is in the low, marshy land of South Carolina; but neither location is favorable for sheep. I agree with Mr. Buckley, ' that a large portion of the county of Yancey is an elevated table land, which is so damp and cold that the inhabitants frequently do not raise corn sufficient for their own consumption.' This is partly owing to the climate; but mostly to the character of many of the inhabitants of those sparsely inhabited regions, where they too frequently depend upon the success of the chase for the largest portion of their subsistence.

" But if Mr. Buckley, or any other gentleman of observation, will come 60 miles farther south—on the line of the Blue Ridge, into Henderson and Rutherford counties, about the Tryon Mountain, which is the first that he will ascend in rising up from the level country east of the Blue Ridge, along the Howard-Gap Turnpike—high on the acclivity of the Tryon he will find a bench of land which possesses a very peculiar characteristic. At night, generally, there is a pleasant breeze, and for several miles along the mountain side there is never any dew to be found, and it is very rare that they have frost except in winter; and when the whole country above and below is covered with sleet, along this mountain side there is none. Here grow the finest native grapes that I ever saw, and the fruit crop never fails. And here are grown the heaviest wheat and rye in all the country. Here the inhabitants have the first dawn of the morning sun, and persons unaccustomed to the view fancy that they can almost see him coming up from the watery deep. On the eastern side of this mountain is the earliest pasturage in spring, and the latest in the fall that is found in the whole range of mountains.

" This location is about 46 miles E. S. E. from Ashville, and 20 miles S. S. W. from Rutherfordton. Here two of those ever persevering men from the North, called Yankees, have commenced to wall in a vineyard, and to cultivate the broom-corn for manufacturing brooms. They have the purest water that flows out of the earth, and around them are beautiful cascades more than a hundred feet high, and above them the toppling peak of the Tryon.

" Thousands of persons throng this mountain region during the summer, to enjoy the pure, bracing atmosphere, which on the eastern face of the mountain is dry and healthful; but farther back, in the mountains of the French Broad, there is much more dampness and heavy fogs.

" You may readily conclude that along the eastern slopes of these mountains, the climate and country are finely adapted to the growth of wool, as may also be seen by many of the fine flocks of native unimproved sheep, which wander here untended, regardless of wolves or dogs, their greatest enemies.

" For two hundred miles along the eastern slopes of these mountains, south, there are situations well suited for large flocks of sheep, and land is cheap. In many places it does not cost more than 20 cents per acre, and very fair land may be had for 40 cents per acre. '

In an Address,† remarkable for the force and pertinency of its sugges-

* See Albany Cultivator, 1846, pp. 335-336.
† Delivered in Martinsburg, Va., Oct. 30th. 1845, before the Berkley County Agricultural Society, published in the Valley Farmer, Dec. 1845, and Jan. 1846.

tions, Col. Edward Colston, of Berkley county, Virginia, makes the following statements:

"The western part of our county, containing perhaps 30,000 acres, is mountainous. I have ridden there for ten miles without seeing a human habitation, and although from its abundant herbage it might sustain for its owners 20,000 head of sheep, not a single one is to be found grazing on its surface. In this region may be found, also, much land fit for cultivation, with fine meadows and abundant water. Yet all this is worthless to our community, and a dead capital to the proprietors. There is territory and grass enough here to be divided into three or four sheep-walks, each sustaining from 3,000 to 4,000 sheep during the summer, with meadow and arable land enough, at a small expense, to provide amply for winter sustenance."

Hon. Andrew Stevenson, of Virginia, in a letter to Mr. Skinner,* says:

' Virginia has many advantages for breeding sheep, not surpassed in the United States The middle part of the State, and especially the whole range of the south-west Mountains and Blue Ridge, afford the greatest facilities for fine sheep-walks. Hills covered with fine herbage, extensive inclosures, abundance of running water, and well sheltered by trees against the heat and sun of summer."

The following extracts are from a communication in the Monthly Journal of Agriculture,† by Hon. W. L. Goggin, who recently represented the District he describes in Congress:

"Bedford, the county in which I reside, is bounded on the south side by the Staunton River, on the north by the James River, while its western extremity, the whole length, reaches the top of the Blue Ridge. . . . . The Peaks of Otter‡ are situated in this county, on the north-west corner—they are not only beautiful themselves, when seen as they are in the distance, but the whole range of the Blue Ridge presents, perhaps, here, the most interesting view of the kind in the State. These mountains afford an unlimited range for stock, and the advantages for sheep-walks (mild as is the climate, combined with the productiveness of the soil) that are nowhere equaled, as is believed, except by similar situations in the neighboring counties. . . . . Ranges for sheep may be had at a very reduced price on 'he mountains, and where, too, could be produced all the grasses in which they delight, such is the red and white clover, the meadow fox-tail, short blue meadow-grass, lucern, rye-grass, &c. These advantages, and then the beautiful, clear streams which abound in all the mountain regions, invite a pastoral life." Speaking of Amherst and Nelson counties, he says: ' The ranges for stock here, too, are extensive, and the beautiful, rich mountain sides interspersed with farm-houses, some of them even elegant mansions, betoken an independence among the inhabitants that is often found in such situations. Many of the mountains, to their very summits, are covered with the richest verdure." Of Madison and Greene counties he says: " Here, too, are abundant ranges, and the wonder is that sheep husbandry is not introduced."

The character of the loftier mountains of Virginia and North Carolina, for the production of grasses, would seem to leave no doubt, in this particular, in regard to the *lower* ones which form the prolongation of the same chains in South Carolina, Georgia, and Alabama. Let us now turn our glance to the great western chain—the Cumberland Mountains—in Kentucky and Tennessee.

The following extracts are from a communication published by Hon. A. Beatty in the American Agriculturist:

"But it is not upon our high-priced rich lands alone that we can carry on sheep husbandry to advantage. Kentucky has a belt of hill and mountain country, bordering on the Virginia line on the east, and on the rich lands of the State on the west, averaging about seventy-five miles in width, extending from the Ohio River and Big Sandy, latitude 38° 30', to the Tennessee line, 36° 30' north. The whole of this region is admirably adapted to sheep husbandry; the most northern part but a few minutes north of my residence, and extending about two degrees farther south. The lands are very cheap: the State price of those not yet appropriated only five cents per acre, and those purchased second-hand, more or less unproved, may be had from 25 to 50 cents per acre, and still less when unimproved. This country in a state of nature furnishes, during the spring, summer, and fall months, a fine range for sheep, and is susceptible of great improvement by clearing up and sowing the cultivated grasses for winter feeding. This whole country is finely adapted to the Spanish

* Monthly Journal of Agriculture, July, 1845, pp. 37–39.   It . October 1845, pp. 181–182
‡ The loftiest mountains, as before stated, of Virginia.

mode of sheep husbandry. Very large flocks might be driven to the mountain region, some thirty to sixty miles from the rich lands, immediately after shearing time, grazed till late in the fall, and then brought back to be sustained during the winter on the luxuriant blue grass pastures of the rich lands of the interior.

"A very intelligent friend, residing in the southern part of the above district of country, speaks of it in the following terms: 'One of the strongest proofs of this region of country being favorable to the growing of sheep stock is that we are situated in the same degree of north latitude with the sheep-raising parts of Spain—Leon, Estremadura, Old Castile, &c.—only that our mountains are more richly and abundantly clad with luxuriant wild grasses and fern, pea vine, and shrubbery, than the mountain regions of Spain, where they raise such abundant stocks of sheep. Wayne County, with a few adjoining counties, affords more fine water-power than any country of the same extent that I have ever known; and for health, and fine, pure drinking water, no country excels it on the face of the globe. Now is the time to commence the business of sheep husbandry, while land can be got almost for nothing. It is worthy of remark that our sheep, which are suffered to roam and graze in the mountains altogether, produce about *one-fourth more wool at a shearing than the sheep that are raised and grazed altogether on our farms, and of much better quality.'* In another part of his letter he says: 'The tops of the mountains of Spain are sterile, without verdure, producing no food for sheep, or other animals, to graze on. Our mountains are quite different. They are thickly clad from bottom to top, and all over the top, with fine rich wild grasses and shrubbery of every variety, for stock to graze on. In the midst of our mountains are to be found a great abundance of salt water and stone coal of the finest quality, together with a great variety of mineral waters and pure springs.'

"Another friend, residing in Knox County, writes to me: 'My sheep upon my farm, adjoining Barboursville, do not thrive, even with pasture and winter food, like the sheep in the extremities of the county, which have neither pastures nor winter food, except what they get in the woods. Without cultivated grasses of any description, sheep will live and do well all the winter, subsisting on the spontaneous growth of the country.'

"Another friend, residing in the northern portion of the above-described mountain region, writes that 'the counties of Carter and Lawrence, and the eastern portion of the State, are admirably adapted to sheep husbandry. There are several flocks of sheep in this neighborhood that thrive and increase wonderfully, *running at large,* at little cost or trouble to their owners. Many flocks have no other reliance, during the winter, but what they get in the woods. The great advantages of this country for sheep husbandry are, the cheapness of the land, it adaptation to grasses, grain, and roots—its healthfulness. Sheep delight in mountain or hilly land; the natural evergreens and shrubbery upon which sheep can feed and subsist on in winter; though it is not safe to rely altogether upon these.'"

Mr. C. F. Kramer of Woolverly Farm, Marion Co. Tennessee, in a communication in the Nashville Agriculturist,* says:

"After having spent part of the years '43 and '44 on different parts of the Cumberland Mountains—the part of Tennessee more particularly recommended by all writers in your journal, and others, for sheep-walks—I have, since last fall, settled on a portion of them near Jasper, Marion Co. and will, as briefly as possible, give you the result of my experience, which will, I believe, fully remove any erroneous impressions hitherto made.

"First, as to climate: The extreme salubrity of the mountains makes them the general refuge of the sick. Sheep here are remarkably healthy, and exempt from disease. The temperature is very even, varying during summer seldom more than from 75° to 80° of Fahrenheit, nor in winter more than from 45° to 30°. Snow during the two winters, little as there was of it, never remained forty-eight hours on the ground.

"The forest, so far from being dense, seldom contains more timber, after cutting out the smaller growth, as dogwood, &c. than is desirable for woodland pasture.

"The rocks, as far as my rambles have extended, are 'few and far between.' The better spots of soil (and there are enough to provide every farm with sufficient remunerating arable land, under a provident and enlightened system of tillage) are covered with nutritious weeds, as pea-vine, &c. &c. which are nearly all greedily devoured by sheep and cattle, and on which they fare well. The poorer soil is covered with sedge-grass, which my sheep have invariably eaten with avidity.

"When our herds and blue grass lands, which we are laying down, will be fit for pasturing, the cost of wintering will be greatly reduced, as the former yields good grazing in February—the latter during the whole winter. Our young cattle kept in good condition on the winter-range and two ears of corn per head per day.

"Although the wolves of our mountains are larger than those of the prairies, and may be more difficult to exterminate entirely, yet, thanks to our good hunters, their ranks have been already so thinned that they mostly prowl about alone, or at most in pairs, committing their depredations by night, on the sheep and hogs that are left to shift for themselves. In the

* June, 1846.

two years that I have been here, I know of but two instances of their having attacked young stray cattle by night. By day, sheep are perfectly safe ; and I should presume that every good sheep-master would have his flocks, for inspection, home at night, when any common fence will be an ample safeguard for them."

To recur, for a moment, to Mr. Buckley's statements in relation to the Roan and some of the contiguous mountains in North Carolina—if we concede all his positions to be correct—it but proves that they are exceptions to a general rule. But a review of his facts, it seems to me, scarcely justifies his conclusions.

The vegetation which seemed so backward to him, coming from the warmer climate of Alabama and Lower Tennessee, was in fact but little, if any, later than that of the elevated grazing lands of Southern New-York. The following table* will show the average forwardness of the seasons at the location of fifty-eight Academies, scattered over New-York, for a term of fifteen years. And these Academies, as would be supposed, are rarely found on the high bleak hills. In fact, the number in the southern grazing region is but small, and they are mostly on the low bottoms of the larger streams. The same remark will also apply to the high region between the St. Lawrence and Lake Champlain.

TABLE NO. 5.

	Mean Date.	No. of Localities.	No. of Observations
Shadbush in bloom........................	May 1	48	168
Peach do.	" 2*	57	175
Currants do.	" 4	58	269
Plum do.	" 6	52	264
Cherry do.	" 7	52	250
Apple do.	" 15	59	374
Strawberries ripe.......................	June 12	58	210
Hay harvest commenced.................	July 18	34	127
Wheat do. do.	" 25	45	186
First killing Frost......................	Sept. 23	57	471
First fall of Snow......................	Nov. 5	..	536

* As the Peach does not grow in the northern part of the State, this date must be considered the mean for the southern and middle parts only, and hence is too early as compared with other trees.

The blossoming of the apple tree in the grazing regions of New-York takes place when the leaves of the forest trees are considerably less than half grown, as Mr. B. found them on the "high mountain trees" of North Carolina on the 12th of May.

Snow storms sometimes occur in New-York as late as the one recorded by Mr. B. on the Roan; cold, damp fogs are not found destructive to sheep in some parts of England and Scotland, where they prevail probably quite as much as on these mountains ; and there are *many* parts of the grazing region of New-York, and *good grazing* lands, too, where the inhabitants "do not raise corn sufficient for their own consumption." As Mr. B. gives neither the dates nor the altitudes of his own thermometrical observations, no conclusions can be deduced from them. Speaking of the region about Asheville, the more definite statement is made by him, that during the summer of 1842, the thermometer ranged generally from 70 to 85 degrees, (which he pronounces not much, if any, warmer than Western New-York,) "while on the mountains it was frequently about 60 degrees, and sometimes much lower." If by *frequent,* he meant *ordinary* temperature, the summer climate of these lofty mountains much resembles that of New-York in June—usually considered the month of the pleasantest

* This table was prepared by James H. Coffin, a tutor in Williams College, from the Report of those facts annually required to be made by the Academies to the Regents of the University. This and some other tables and statements of Mr. C.'s, which I shall have occasion to quote, appear in a very able paper from him on the Climate and Temperature of New-York, in the forthcoming volume on Agriculture, in the Natural History of the State : some sheets of which have been politely sent me by Doct. Emmons, the State Geologist, who has that volume in charge.

G

temperature of the year—equally removed from the chilliness of spring and the sultry heats of the last two summer months. But as the altitudes of the latter observations are not given, they present us nothing definite or tangible. A smart walk of a few moments up or down a mountain side, would carry one through a variation of temperature amounting to a degree. By the rule of Professor Leslie,[*] commonly adopted, 300 feet of elevation diminishes the temperature 1°; but the experiments of Humboldt, Gay-Lussac, and various other observers, have shown that this cannot be relied upon. One degree is usually equivalent to a greater ascent. Mr. Coffin (in the paper before alluded to) deduces the conclusion that in the State of New-York, the ascent necessary to decrease the temperature 1° is 350 feet. Taking the mean of the range of temperature of Asheville, as stated by Mr. B. it gives 77½° as the average summer temperature of that place, which, as will appear in the table below, is about 10° higher and warmer than that of New-York for the same season and year, (excepting on the beds of two rivers—the Hudson and Mohawk.) Applying the New-York rule to the region of Asheville, it would require, then, an elevation of something like 3,500 feet on the mountain sides above that place, to equalize the temperature with that of the greater portion of New-York.

To show the entire accuracy of the subjoined table of temperatures, I would remark that it is founded on the Annual Reports of the Academies to the Regents of the University. The observations are therefore made by correct instruments,[†] on fixed conditions, and by scientific men. I have selected the points indicated in reference solely to a fair latitudinal and geographical distribution over the State;[‡] and to enable you to find them on the map, the name of the *place*, instead of the Academy, is given :

TABLE NO. 6.

	Lati- tude.	Eleva- tion.	Temperature 1842. June.	July.	Aug.	Remarks.
Flatbush..	40°73'	40	64 28	72 16	69 97	Near the extreme southern point of Long Island...
Po'keepsie	41 41	....	61 29	76 83	71 56	On the Hudson. Elevation not given............
Albany....	42 39	130	63 85	72 66	70 23	
Potsdam ..	44 40	394	59 62	67 36	67 12	In St. Lawrence County; north part of State.......
Lowville..	43 47	800	60 51	67 52	64 46	On the Black River.........................
Utica.....	43 06	173	63 58	70 15	69 15	
Syracuse..	42 59	....	59 75	65 77	64 86	} Both in same county, but given on account of dif-
Pompey ..	42 56	1300	37 70	64 20	63 50	} ference in elevation.................
Homer....	42 38	1006	38 88	64 14	65 67	In the southern or grazing region.............
Ithaca.....	42 27	417	63 80	69 65	67 74	Do.　　　do.　　...........
Prattsburg.	.. ..	1494	56 83	63 24	68 71	Do.　　　do.　　...........
Rochester.	43 08	506	60 66	66 94	67 35	In the heart of the Wheat growing region.......
Wyoming.	42 49	800	59 97	71 50	56 99	Do.　　　do.　　...........
Fredonia .	42 26	345	63 42	69 60	68 71	In the grazing region; on the shore of Lake Erie..
Lewiston..	43 09	280	62 05	68 91	68 50	On Niagara River......................

The five last named places are in "Western New-York."

But there is one fact stated by Mr. Buckley, in relation to the lofty mountains of North Carolina, which, irrespective of all thermometrical observations, demonstrates conclusively, to my mind, their adaptation to sheep husbandry. This fact is, that *white clover* grows (of course, spontaneously,) on them. Or perhaps I should rather say, that the mountains themselves become thermometers, their vegetation registering, by a well settled natural law, their temperate climate. Says Malte Brun :

" Under the burning climate of the torrid zone, we have only to ascend the mountains to enjoy the fruits and flowers of the temperate regions. Tournefort found at the base of Mount

---

[*] Prof. L.'s rule, however, was only made applicable by him to *tropical* regions.
[†] Half, probably, of the thermometers in common use are inaccurate !
[‡] For the records of temperatures given, see Report of the Regents, 1843, p. 240. For latitudes and elevations of the Academies, see Report of 1838, pp. 212 to 215, and map.

Ararat the common vegetables of Armenia; half way up, those of Italy and France; and upon the summit those of Scandinavia. Forster saw several Alpine plants upon the mountains of Terra del Fuego."

Mr. Mudie also remarks :

" If we take each mountain as the index of its own meridian, we shall find that each one expresses, by its vegetation, all the varieties of climate between it and the pole."[*]

Humboldt, and our own Doct. Forry, notice an equally striking develop ment of this law, on the Western Continent.[†]

This would go to show what I have little doubt is the fact, (my impressions, too, being strengthened by a comparison of latitude, elevation, and recorded thermometrical observations,) that on the sides of the Roan and other lofty mountains of North Carolina, and pretty well *up* on their sides, too, the climate is not greatly dissimilar from that on the high grazing lands of New-York and New-England. On the sweetest and best of the latter, white clover always comes up spontaneously, and will immediately re-sward any field thrown out of tillage. It sometimes flourishes on soils of ordinary fertility, but never on very sour or boggy ones, or on those the poachy character of which would render them liable to communicate hoof-rot or other diseases. It indicates, most decidedly, both a soil and climate fitted for sheep.

You will not understand, Sir, of course, that in the remarks made and facts stated, at so great length, in relation to three or four mountains, my object has been simply to refute the views of Mr. Buckley in relation to them. In a region of 70,000 square miles, the unadaptation of half a dozen mountains, or a much greater number, to this or any other branch of husbandry, would be of but little comparative importance. Anticipating, however, the croakings of the timid—the exaggerated counter statements of those rash and sanguine men who are ever ready to rush into whatever is *new*, without judgment to guide or perseverance to sustain them : who abandon their undertakings at the first obstacle, and apologize for their ficklety by magnifying the difficulties encountered by them : I deemed it expedient to lay before you some useful data for comparisons, (and conclusions,) which will be equally applicable in the case of all our southern mountains.

The hilly and level regions *west* of the mountains, and lying between them and the Mississippi and Ohio rivers, scarcely require a separate notice—particularly after the statements of Mr. Cockrill, given in my second letter. As a whole, they are undoubtedly more fertile, and better adapted to the production of the grasses, than those of corresponding latitude, in even the hilly zone, east of the mountains.

---

* Mudie's World.

† Since making the extract above from Malte Brun, I observe the following better, or, at least, more definite expression of the same fact by Doct. Forry : " In ascending a lofty mountain of the torrid zone, the greatest variety in vegetation is displayed. At its foot and under the burning sun, ananas and plantains flourish ; the regions of limes and oranges succeeds ; then follow fields of maize and luxuriant wheat ; and still higher, the series of plants known in the temperate zone. The mountains of temperate regions exhibit, perhaps, less variety, but the change is equally striking." See Forry's *Climate of the Unit f States.*

LETTER V.

PROFITS OF SHEEP HUSBANDRY IN THE SOUTHERN STATES.—I. DIRECT PROFIT ON CAPITAL INVESTED.

Different points of view in which the question of the profitableness of Sheep Husbandry in the Southern States is to be regarded...Direct profit on Capital invested first considered...Average prices of Wool in New-York...Average weight of fleece—Price of Sheep—Increase in Lambs—Amount of Manure...Price of Land...Number of Sheep supported per acre...Estimate of the Expenses and Profits of 100 Sheep, taking average prices of Wool for the last fourteen years...Present low prices of Sheep—Causes—Estimate of Profits of 100 Sheep, at present prices of sheep and Wool...Profits far below what they might be by breeding better Sheep...Writer's Flock—Annual yield of Wool—Prices sold at for six years—Statistics of Premium Flock...Show that Wool can be produced at a large profit in New-York at present prices... Healthfulness and economy of substituting Mutton for a portion of the Bacon consumed in the Southern States...Economical advantages which Sheep possess over other animals—No risk by Death—Manure more valuable—Best clearers of Briery Lands—Improvers of Vegetation...The cost of producing Wool in the South, compared with the cost in New-York...Number of Sheep which can be supported per acre South—Greater number than on land of the same quality North, by reason of the winter growth of grains and grasses in the former...Col. Allston's statement—R. L. Allen's—Col. Hampton's—Hon. R. F. Simpson's in relation to the Atlantic States south of Virginia...Price of Lands in those States...Winter Vegetation in Tennessee, Kentucky and Virginia...Mr. Coles's statement—John S. Skinner's...Recapitulation... Estimate of Profits on 100 Sheep South—Compared with New-York...Profits on the Southern Mountains ...Doct. Brockenboro's statements—Mr. Murdock's...Economy of Migratory Sheep Husbandry...Advantages for it in the South compared with those of Spain...Drawbacks on Profits of Sheep Husbandry— Dogs and Wolves...Their depredations compared with those in Australia and the Cape of Good Hope... Remedy.

Dear Sir: In ascertaining the Profits of Sheep Husbandry in the Southern States, several considerations present themselves, apart from the mere question of direct annual profit or loss on a given investment in Sheep and in land for their subsistence. The more immediate and obvious profit is doubtless the first question; but in regarding the general advantages or disadvantages of this branch of husbandry—particularly in a region circumstanced in all particulars as the Southern States are—we are farther to consider the practicability and comparative economy of making it the basis of an effectual amelioration in soils naturally sterile, or those which have been rendered so by excessive and injudicious cultivation; and its comparative efficacy in giving to Southern Agriculture a mixed and convertible character, and thereby sustaining (or improving) all the present good tillage lands, in the place of continuing the "new and old field" system—(tilling land until it is worn out, then abandoning it and opening new lands,)—once so general, and even now by far too prevalent. And there is another point of no mean importance : whether, independent of preceding considerations, and even if the staples furnished by sheep husbandry proved no more profitable, in direct returns on capital invested, than some of the present staples, it would not be better economy, on the whole, for the South to produce the raw material and manufacture domestic woolens, particularly for the apparel and bedding of slaves, than to be dependent for them on England or Massachusetts.

To ascertain the direct and immediate profit on investment in sheep husbandry, let us appeal to well settled facts and statistics, instead of contenting ourselves with vague and general propositions. For the following Table of the average prices of good wool* in the State of New-York, which was published in my replies to Mr. Walker's "Treasury Circular" in

* Such wools as are used for the manufacture of broad and other cloths of good quality—ranging, say, from ⅜th blood Merino to pure Saxon—excluding native, grade (below ⅜th Merino) and all English wools

1845,* I was indebted to a most respectable and extensive purchaser of wool, and its accuracy is beyond question.

TABLE No. 7.

Year	Average price per pound.	Year.	Average price per pound.
1832................40 cents.		1839................50 cents.	
1833................50 do.		1840................33 do.	
1834................45 do.		1841................35 do.	
1835................48 do.		1842................30 do.	
1836................54 do.		1843................31 do.	
1837................30. do.		1844................40 do.	
1838................36 do.		1845................32 do.	

It will thus be seen that for a period of fourteen years preceding 1845, the average price of good wools was 39$\frac{4}{4}$ cents per pound.†

The average weight of fleece in sheep yielding this wool has been about 3 lbs.; the pure-blood Saxons less; but those bearing the coarsest wool included, in the average, more.

The average price of sheep of the quality under consideration, has been not less than $2 per head in the fall, and lambs half that price.‡ The annual increase in lambs would be about 80 per cent., or if less by reason of the number of wethers in the flock, the *growth* of the latter would give a corresponding increase in profit. One hundred sheep, properly littered, will make at least forty loads of manure during the one hundred and fifty days during which they are confined to dry feed, in our Northern winters.

The grazing lands of New-York, cut up as they are into small farms,|| and each being provided with dwelling and farm buildings, are worth from $15 to $30 per acre. Prime sheep lands will average about $20.§

In relation to the amount of land necessary to support a given number of sheep, the experience of a good many years has satisfied me that the rule commonly laid down on the grazing lands of New-York and New-England, that, on the average, one acre of land will give subsistence to three fine-wooled sheep throughout the year, is an accurate one.¶ On grain farms, it is considered good economy to keep one sheep for every acre of cleared land which the farm contains; on those where mixed husbandry is practiced, two; and, on those exclusively devoted to sheep, three.

In the following, and all similar estimates, I shall reckon the profits on the *land* and *expenditures*, instead of the *land* and the *commonly quoted prices* of grass, hay, &c., consumed. These prices, in the interior, are

---

* See Report of the Secretary of the Treasury, 1845, p. 461. I thought, and so stated to Mr. Walker, that the Table placed wools about 1½ cents per pound too high. But subsequent information has convinced me that I was in error. In my statement of the *average* profits of sheep husbandry, in those replies, I estimated the average price of wool by the prices paid by a local and much smaller purchaser, and for a comparatively limited term of years. I was not then aware of the utter defectiveness of the U. S. Census returns (pointed out in Letter II.) In relation to the annual product of wool, and therefore was misled in the average weight of fleeces; and, speaking from impression rather than experiment, I placed the value of the manure altogether too low. Those questions and replies have led me into experiments and inquiries, which have resulted in more accurate information. I allude to this subject, because I think it every man's duty to correct any errors or explain any discrepancies subsequently discovered by him, in his statements which have been thrown before the public, and thus are placed in a position to mislead.

† During 1846 it was from 30 to 32 cents per pound, but as this estimate is not based on extensive purchases, like the preceding, I have not placed it in the table.

‡ Including *grade* sheep, which form the greatest proportion of the whole number. There have been very few pure-blood Merinos in the State, and many of the Saxon flocks have been so miserably deteriorated in carcass and weight of fleece, that they have sold for low prices. But *good* Saxons sold much above this until within three or four years; since then, the Merinos have been rapidly driving out the Saxons, and those of good quality and undoubted pedigree have sold for from five to twenty-five times as much. The higher the price, the greater the profits, by reason of the value of the increase.

|| It would be my impression that the farms in the grazing regions do not, on the average, exceed 130 acres each.

§ Id est, in the grazing region.

¶ I say " fine-wooled sheep," because the larger and coarser Downs, Leicesters, Cotsw ,lds, &c. consume much more, as will hereafter be shown.

merely nominal, as they cannot be obtained for beyond a small portion of the annual crop. They do not, therefore, form a proper basis for correct general estimates.

The expenses and losses in keeping sheep, not already alluded to, are all set down below, as high as they will average on well managed farms.

Dr.	$ cts.	Cr.	$ cts.
100 Sheep to interest on purchase money....14 00		By 300 lbs. of Wool, at 39 4-7 cts. per lb.11½ 7½ 3-7	
To int. on 33½ acres of land at $20 per acre....46 66		" 80 lambs at $1 per head.............. 80 00	
" curing and storing hay on 11 acres of above.13 75		" 40 2-horse loads of winter manure at	
" expense of shearing................... 4 00		50 cents per load................... 20 00	
" salt, tar and summer care............... 4 00		" summer manure, calling it only equal	
" labor of foddering, &c., during winter, say. 5 00		to shearing and summer care*.... 8 00	
" loss by death 2 per cent. above the value of		Total...................$226 71 3-7	
pulled wool................... 4 00			
Total........................$91 41		Balance...................$135 30 3-7	

Making the net profit of $1 05, or 20¼ per cent. per acre on lands worth $20.

Since the passage of the Tariff of 1846, there has evidently been a panic among the wool-growers of New-York, and the rise in bread-stuffs, beef, pork, and dairy products, occasioned by the change in the British Tariff, and the famine which has prevailed in Europe by reason of the short crops of 1846, has tended farther to depreciate sheep, by offering inducements supposed to be very strong, to embark in branches of husbandry furnishing the former staples.† Sheep are consequently cheaper than they ever were before. Prime *grade* sheep, bearing wool of as good quality as the average of that embraced in Table 7, have in some instances sold for ten shillings per head, and coarse common sheep for one dollar—lambs half a dollar—making, in the ordinary proportion between lambs and grown sheep, about 75 cents per head, taking a flock through !

Wool of the quality embraced in Table 7 has fallen to an average of say 31 cents. Under the impression that sheep and wool have reached their minimum prices,‡ it becomes an interesting subject of inquiry whether they can yet be produced, at a profit, in New-York. The following figures I think, will fairly show :

Dr.	$ cts.	Cr.	$ cts.
100 Sheep, to interest on purchase money, at		By 300 lbs of Wool, at 31 cents per pound...93 00	
$1 25 per head.................. 8 75		" 80 lambs, at 62½ cents per head..........50 00	
To int. on 33½ acres of land at $20 per acre..46 66		" 40 2-horse loads of winter manure, at 50	
" cutting, curing and storing hay on 11 acres		cents per load.......................20 00	
of above....................13 75		" summer manure, calling it only equal to	
" expense of shearing...................... 4 00		shearing and summer care............ 8 00	
" tar, salt and summer care............... 4 00		Total......................$171 00	
" labor of foddering, &c. during winter, say, 5 00			
" loss by death 2 per ct. above the value of			
pulled wool.... 2 50			
Total.................$84 66		Balance....................$86 34	

Making $2 59, or nearly 13 per cent. *net* profit per acre on lands worth $20.

In the preceding estimates I have only regarded the profit of sheep husbandry, as it has averaged for a series of years, among those possessing *good ordinary* flocks.

---

* I place the summer manure, undoubtedly, considerably below its actual value. No experienced farmer will say that good solid sheep manure is worth less than 50 cents per load, and as the summer manure is at least equal in quantity, and is deposited immediately on the land, I see no reason why it is not equally valuable.

† That the diminution of English duties on these staples will give them a better and steadier market, there can be little doubt; but not the *very high* one of the past season, occasioned by the severe famine which has prevailed in many parts of Great Britain. Many, therefore, who have sacrificed their sheep, reckoning on such prices, will probably find that they have "reckoned without their host."

‡ I say this under the decided impression that our wools, at this price, *if properly washed and put up,* would triumphantly compete in the foreign markets with those of the wool-growing nations of Europe; and even with those of Australia, the Cape of Good Hope, and other Austro-oriental regions. For a more full examination of this point, see Appendix D

It falls far short of that realized by breeders and flock-masters, who started their flocks with the best pure-blood sheep then to be found in the country ; and who have subsequently continued to improve them by great care in breeding, and by a rigorous course of selection.

I have bred Merino sheep for a number of years, and latterly in considerable numbers : and in no case have my grown sheep averaged less than 5 lbs. of well washed wool per annum. The quality of the wool may be inferred from a comparison of the prices at which it has sold, with those in Table 7. In 1846, I sold for 35 cents per pound ; in 1845, for 33¼ cents ; in 1844, for 48 cents ; in 1843, for 33⅓ cents ; in 1842, for 35 cents, and so on.

To give more precise data, I select the following statement of the products of a flock, on which I drew the first premium offered by the New-York State Agricultural Society for " the best managed flock of sheep," in 1844 :

[From the Transactions of the N. Y. State Agricultural Society, 1844, p. 254.]

" In the winter of 1843–4, I wintered in a separate flock fifty-one ewes over one year old, two ewe lambs, two rams, one of them one and one of them two years old. Of the ewes over one year old, twenty-eight were full-blood Merinos ; twenty-three were half-blood Merinos and half-blood South-Downs ; the two ewe lambs were three-fourth-blood Merino and one-fourth-blood South-Down : and the two rams were full-blood Merinos. The flock were kept as follows through the winter: They were fed hay morning and night, and were, as a general rule, required to eat it up clean. At noon the flock were daily fed three bundles of oats and barley (which had grown mixed, say three parts oats and one part barley,) until the 25th of December—after which they received four bundles of oats. The grain was light and shrunken. They received no hay at noon during the winter, and usually consumed all the straw of the grain fed them. They had a good shelter, and access to pure water at all times. From this flock I raised fifty-three lambs. The full-blood Merinos, including two rams, and the two three-fourth-blood lambs, (in all thirty-two,) sheared one hundred and eighty-six pounds and four ounces of washed wool, which I sold at forty-eight cents per pound. Four of the full-bloods had two years' fleeces on. The half-blood Merinos and half-blood South-Downs (twenty-three) sheared eighty and one-half pounds of washed wool, seventy-one pounds of which I sold at thirty-eight cents per pound. During the summer of 1844, the flock were kept in good ordinary pasture, and salted once a week."

Thus, the Merino fleeces averaged 5 lbs. 13¼ oz. and sold for $2 79¾ each ; and the grades between Merino and South-Down averaged 3 lbs. 8 oz. to the fleece, and sold for $1 33 each.

It will be observed that four of the full-bloods (they were ewes) had two years' fleeces on. A two years' fleece will not weigh as much as two single years' fleeces from the same sheep. On the average, it will weigh about three-quarters as much.* On the other hand, the lot included two three-quarter-blood lamb fleeces, which would fall below the average weight of the others, and a portion of the flock were yearlings and two-year olds. The Merino never attains its maximum weight of fleece before three years old, and ordinarily not until four, and therefore the aggregate weight of wool of the 32 sheep, given above, does not, to say the least of it, give too favorable a view of the product of sheep of this quality. This is proved by the fact that my entire flock of full-bloods sheared about three-twentieths of an ounce over six pounds each, the succeeding year.

It would give me great pleasure to subjoin similar statistics of other carefully bred flocks, were authorized statements of them in my possession, or published within my knowledge.

It is sufficiently apparent from the above facts and estimates, that wool has not yet reached the lowest point at which it can be produced at an ample profit, on lands of the value indicated, *if the sheep are of the proper*

* That is to say f the single years' fleeces would equal 6 lbs. each. a two years' fleece, instead of weighing twice as much, or 12 lbs., will not exceed three-quarters of such aggregate weight, or 9 lbs. The wool wastes when it becomes so long, and perhaps does not grow so rapidly.

quality; and these facts farther suggest the expediency of relying on our own efforts to "protect" this interest, rather than the fickle support of National legislation.

For the production of a cheap, wholesome, and highly nutritious food, no animal excels the sheep. Theoretical considerations, as well as experiment, show the superiority of mutton to pork in the formation of vigorous muscle;[*] and its tendency is less, particularly in hot climates, to engender inflammatory and putrid diseases. The consumption of considerable quantities of fat is indispensable, in cold climates, to supply the necessary amount of carbon to support "combustion," as Liebig terms it, in the lungs, or, in other words, to maintain the animal heat. Hence the Laplander and the Esquimaux find a grateful diet in train oil, or the adipose parts of Arctic fish and mammalia. That the pork should be the favorite meat, in the Northern States, is not perhaps so singular, but that it (under the name of bacon) should constitute the principal one consumed in our warm Southern latitudes, and especially that it should constitute so large a proportion of all the food consumed,[†] is indeed a most anomalous fact, and is utterly unparalleled among the practices of other nations occupying the same latitudes. The tendency of this practice to produce disease, physical inertia, indisposition and incapacity to sustain continued activity, will not, I think, be questioned by the pathologist or the close observer.

Mutton and lamb are a favorite, if not *the* favorite food of the English of all classes. Notwithstanding all that has been said and written of the "roast beef" of "Old England," mutton is more eaten there by people of every rank.[‡] On the other hand, it is evidently *not* a favorite meat in the United States, though its proportionable consumption is evidently increasing. Whence the difference? Circumstances have led to habit, and habit, in a great measure, regulates appetite. It needs no other proof than is to be found in the experience of every individual, to show that the appetite is readily trained to relish what was even positively disgusting, and to become indifferent to what was once the most grateful.

That the preceding facts are well worthy of attention among those who are favorable to the introduction of sheep husbandry, among planters who supply not less than 3 lbs. per week of good bacon, or a full equivalent, to each slave, on plantations where the number ranges from ten to one hundred, and sometimes many more, there can be little doubt. Twenty-five slaves would thus consume 3,900 lbs. of bacon per annum; and the more common allowance of the opulent planter is about 200 lbs. per head, or 5,000 lbs. for twenty-five. If an equivalent for at least half of this was

---

[*] The theoretical considerations will be found sufficiently discussed in Liebig's "Animal Chemistry." For experimental evidence, I know of none that can be more depended on—which approaches any nearer actual demonstration—than that which is furnished by the English prize-fighters. To attain the proper condition to sustain the protracted and tremendous exertions of their brutal trade, their flesh must attain the hardness and toughness of whipcord, and they must, at the same time, maintain that physical elasticity (technically, "corkiness,") which adds agility to iron strength. These men, while training, are suffered to eat little or no adipose matter, and not even the *lean* of *pork*. Their animal food is exclusively beef or mutton, or both. Some trainers prefer the former, some the latter. I have seen this matter very fully alluded to, but do not now remember any more explicit authority than that contained in the following note to Carpenter's Principles of Human Physiology, (p. 357.)

"The method of training employed by Jackson, (a celebrated trainer of prize-fighters in modern times) as deduced from his answers to questions put to him by John Bell, was to begin on a clear foundation by an emetic and two or three purges. Beef and mutton, the lean of fat meat being preferred, constituted the principal food; veal, lamb and pork were said to be less digestible ('the last purges some men'). Fish was said to be a 'watery kind of diet;' and is employed by jockeys who wish to reduce weight by sweating."

[†] I mean this portion of the remark to apply more particularly to the non-laboring classes. The proportion consumed by the slave, though ample, is not excessive, when his laboring habits are taken into consideration

[‡] I state this on the authority of various individuals who have been much in England, and who have been placed in positions to form a pretty accurate opinion. Mr. Colman speaks of the "extraordinary" consumption of mutton in England, without, however, giving any comparative data.

made in mutton, it would be far cheaper, and, if I have not erred in previous statements, better for the slave.

There are two or three other highly favorable considerations to be taken
into account among the direct profits of rearing sheep.

The risk by death, by ordinary causes, is nothing. Two per cent. is allowed in the preceding estimates, as the full product of wool and increase
is carried out. But, in reality, the sheep never dies " insolvent." If the
colt or the bullock dies on our hands, after two or three years of trouble
and expense with it, the loss is nearly a total one. If the fine-wooled
sheep dies at any age, the wool then on it, or what it has already produced,
more than covers all the cost which it has ever made us.*

Not only is the winter manure of the sheep superior to that of any other
domestic animal, the hog and fowl excepted, but it practically becomes
still more so in proportion, in summer, when scattered over the pastures,
by reason of the conditions in which it is deposited. The soft porous excrements of the cow† or horse, exposed to the exsiccating action of sun and
wind, evolve most of their fertilizing properties into the atmosphere, and
this effect would increase in proportion to the warmth of the climate. The
excrements of the sheep, on the other hand, are deposited in small, hard,
rounded pellets, which fall down between the leaves of the grass, and are
thus in a great measure protected from the sun and wind, until they are
trodden into and incorporated with the soil.‡ Then, again, they need no
spreading,|| like the dung of the horse and cow. And finally, instinct, in
leading the sheep almost invariably to seek the summits of the elevations,
in warm weather, for its night quarters, leads it to deposit much more manure in proportion, where it is most needed, on the drier and more barren
hill-tops; and where, being more remote from water-courses, less of its
juices are liable to be washed away by rains, into the streams, or on to the
lands of others.

Sheep are also far more efficient than any other animal (if we except
the worthless goat) in clearing up new lands, or neglected old ones, of
those briers and shrubs which it is often difficult to eradicate without plowing; and they often abound on lands which cannot be plowed with profit.
And, when plowed, the shrubs in the fence corners must be left (to the
utter shame of all good husbandry), or the fence must be removed—sometimes at a great inconvenience. The sheep delights to browse on the buds,
and to strip the bark of most shrubs,§ and they thus soon destroy them. It
would be good economy for the farmer to keep his neighbors' sheep, without charge, on all very briery or coppiced unarable lands, if he could not
so stock them himself.

Finally, it is generally believed by experienced flock-masters—and observation has led me to fully coincide in the opinion—that sheep not only
improve the lands they depasture more than any other animal, but that
they exert an almost specific influence in improving *the character of the
vegetation*. All wild, poor grasses gradually disappear from their pastures

* I speak, of course, of the cost of rearing and feeding.
† Gazzeri found that 100 parts of recent cow-dung contain 25 per cent. of dry, solid matter, and that 5 per
cent. of this is lost in 40 days by exposure to the air. I do not think *this* indicates the full loss which would
be sustained in a southern latitude.
‡ These rounded pellets are covered, too, in the animal in good condition, with a coating of mucus, which
farther protects them from evaporation.
|| Their urine, also, is voided in quantities which render it highly beneficial; while that of the horse and
cow is voided in such large quantities in one place that it is not only in a great measure wasted, but in a
dry time (so that it is not diluted by the moisture in the soil), its rich salts, so far from benefiting, actually
kill the verdure.
§ This is particularly true of the blackberry or bramble *(Rubrus villosus)*, and the raspberry *(Rubus
idoeus)*, often great pests on new or neglected lands at the North. Sheep can even be made to attack the
elder *(Sambucus canadensis var. pubescens)*, and various other troublesome intruders, by turning them upon
them in thawing 'spells," in the winter, after they have been for some time confined to dry feed.

H

and are succeeded by the best ones; and the sward becomes remarkably dense and even. This is probably due to the richness and better distribution of their dung and urine.

If upward of twenty per cent. profits, over and above all expenditures, have been and still can be made, on lands worth $20 per acre, by wool-growing—on lands, too, where the reign of an iron winter confines sheep to dry feed at least five months of the year—how are we to estimate those profits on lands costing but a small part of this sum, which, though inferior to the former, will, by reason of the shortness and mildness of the winter support about an equal number of sheep per acre, and also save the expense of preparing dry feed, of foddering, and a large proportion of that laid out in barns, shelters, &c.?

It will be seen that, by assuming the data of the last of the two preceding estimates (with the exception of the loss by death), the gross cost of producing 300 lbs. of wool, on the grazing lands of New-York, is $82 16, or 27$\frac{28}{75}$ cts. per pound. This is undoubtedly as *low* as it can be produced where the fleeces do not exceed the average weight of 3 lbs. Let us now proceed to inquire what would be the gross expense per pound in the Southern States.

You inform me that " one or two—not more—" sheep find subsistence during the summer on the *natural* pastures of the *tide-water* zone in South Carolina.* The broad-tailed, and other large breeds, now mainly fed there, consume nearly double the amount of feed required by the fine-wooled sheep. But, to make our estimate perfectly a safe one, we will assume that two fine-wooled sheep only will consume the summer herbage of an acre. Fields of rye sown in September or October, you farther inform me, will support " two sheep and their lambs" per acre, " from the 20th of December to the 10th of March." Numerically, then, here you have the same stocking that is borne by the lands of New-York, viz. three sheep per acre. And, making the allowance already alluded to for the different consumption of breeds, an acre would sustain three full-grown Merino sheep. As the rye subsequently yields its crop, the wool is not chargeable with the expense of its tillage.

Rye will continue to grow in the winter on all lands not too sterile, or too elevated, south of latitude 36°, and, in favorable situations, at least two degrees farther north. Grass, and some other hardy esculents, also maintain a winter vegetation in many portions of the whole of this region.†

R. L. Allen, Esq., after a recent visit to the plantation of Col. Wade Hampton, near Columbia, S. C., thus speaks of the winter verdure in that region :

" Though everything like grass or weeds is rigidly excluded in the early stages of the crops, yet, as these approach maturity, the thick netting of crab and various other grasses and plants, which are ever struggling for existence in this warm clime, are allowed to come forward and mature; and their growth furnishes forage for cattle and sheep during the winter, and an important addition to the vegetable manures for turning under and adding to the fertility of the soil. . . . The sheep, together with the cattle, mules and horses, which are not at work, are turned into the natural pastures in summer, and, in addition to these, they have the run of the corn-fields in winter, and without seeing any other shelter against the severest storms than a thicket or hill-side, they thrive and fatten throughout the year.— This condition is secured by the mildness of the climate, and the consequent growth of vegetation during the entire winter."

* [These statements, and all others credited to Col. Allston, are, when not otherwise specified, contained
  in letters from that gentleman to the writer.]
† Among these, " a plant called 'Wild Rye,' affording excellent herbage during the winter months, springs
  up spontaneously on the rice-field banks, and between the cotton beds, on some plantations on the River
  Congaree, S ?"

John S. Skinner, Esq. thus writes me :*

"Col. Hampton's flock numbers 800, I believe. He kills the finest sort of mutton through out the winter and spring—very fat and excellent in all respects. He told me last summer, at Saratoga, that they never get a mouthful except what they can find in the woods and fields."

Hon. R. F. Simpson, Member of Congress, of Pendleton, South Caro lina, thus describes the region in which he resides, and some of the contig uous ones :†

HENRY S. RANDALL, Esq.   WASHINGTON, Jan. 22, 1847.

Dear Sir : I take much pleasure in answering your inquiries, and only regret that I have not more time to do full justice to the subject. If my answers fail to inform you with suffi cient clearness on any point, I shall be most happy to add to them, at your suggestion.

The Alleghany Mountains, as you are aware, run from N. E. to S. W. That part of them north of the S. C. line lies spread out in different chains or ridges to a distance of nearly 50 miles ; and the whole region is commonly called " on the mountains." The climate is healthy and the grass fine. Many of the valleys in this region are very rich, particularly on the wa ter-courses. The ground is covered with snow as much as four weeks annually. The range is good, but there may be too much humidity for sheep.‡ The land is cheap, say $1 per acre—but much can be bought at 50 cents. I have learned from good authority that sheep can be farmed out during the winter at *ten cents a head*, in any ordinary quantity. The farmers who take them, too, will be liable for loss by death, in many instances.

There is a strip of country lying east of the Blue Ridge, and parallel to it, from 20 to 30 miles wide, extending through North and South Carolina and Georgia, which I think espe cially adapted to sheep husbandry. The land is poor for the production of our southern sta ples, and is sparsely settled, but the pasturage is good. There is a perennial grass, known as " woods grass," which springs up in the woods after they are burned each winter, which makes excellent pasture for all kinds of stock. It starts vigorously in the spring, and sheep fatten on it by the middle of July. It lasts all the summer, and provides sufficient food for sheep during the entire winter, except when snow is on the ground, which is not more than two or three days at a time, and usually not more than ten days during a winter.

The few days during which the grass is covered up with snow are the only ones, during the entire year, when it is necessary to feed sheep. This is usually done with oats in the sheaf. . . . Supposing ten sheep equal to one cow, I think one acre would afford sub sistence to three sheep.

But few people mow here. In a few instances, herds-grass has been sown and mowed, but the product not weighed, to my knowledge. Both herds-grass and the natural ones, on our bottom lands, look much richer, and to all appearance would turn off a heavier crop of hay than any meadows to be seen on the line of travel through Virginia.

As I have before remarked, the land is poor, except the small bottoms on creeks and branches. The latter are rich, and will produce 30 bushels of corn and from 10 to 15 bush els of wheat per acre. They also produce oats and rye, but I do not know how much by measurement. I suppose from 10 to 20 bushels each. The land is valued low—from 50 cts. to $1 50 per acre—and it is only necessary to buy $500 or $1,000 worth of it, to embrace sufficient bottom to raise provisions, and oats to feed sheep when snow is on the ground.— The range‖ is very large, and everybody's stock has liberty to roam over it, without hin drance or compensation.

Our common method of managing sheep is as follows: The flock are kept in the planta tion during the winter by some; others turn out in the woods. In May they are sheared, the lambs marked, &c., and they are turned into the out pastures. When they come up. they are salted, and no other attention is paid to them until fall, when most persons *shear again.* They are rarely brought up unless to get a lamb for the table. This treatment ren ders them wild, and prone to jump into the owners' or neighbors' wheat fields, from which they are driven out with rocks and sticks, and sometimes with dogs. They are, in all re-

* Jan. 15, 1847.
† This letter would have been more appropriately included in my IVth Letter, but was not received in time, and it is by far too valuable and interesting to be omitted.
‡ The effect of *humidity* on sheep is, I think, often misunderstood and greatly exaggerated. Wet, cold soils are uncongenial to sheep, but they suffer no more from those ordinary fogs and vapors which prevail in insular positions, or which are attracted by mountain ranges, than other domestic animals. As has been before remarked, sheep thrive in the peculiarly foggy atmosphere of England—also in Holland. Their healthiness on mountains is proverbial, yet these elevations are usually subject to fogs, and clouds rest on the sides or summits of the loftier ones. As the southern mountains are cleared of their trees, their atmo sphere will be less humid, or d that soft vegetable mould (which excited the fears of Mr. Buckley) will ac quire the consistency which it always does on a *dry* foundation, when exposed to the sun and air; and it will be the means of supplying the sheep with rich vegetable nutriment, instead of poisoning them with "hoof-all."
‖ The provincial signification of this word, South, is the uninclosed pasturage in the forest and " out fields,"—*i. e.*, worn-out lands thrown out to commons.

spects, treated more like outlaws than domestic animals. When out, all the flocks in the neighborhood mingle together. From their disposition to ramble, and the incursions of dogs, they get scattered, and scarcely any farmer can get up to the fall shearing more than one-half of his count.

The region above described includes Pickens, Grenville and Spartansburg, so far as this State is concerned. Going east of this strip, you at once get into good land, where the settlements are frequent. Here snow is rare, and wheat, rye and barley are used for winter pastures for sheep, and they continue growing during the winter. Wood grass does not abound in this region, as the woods are not kept burnt."

Very respectfully, yours, &c.
R. F. SIMPSON.

The preceding statements give a sufficient idea of the expense of feeding sheep in the Carolinas, Georgia, and the Gulf States. In all of these, there is a striking similarity in soils and natural products, and also in climate—with, perhaps, the exception of North Carolina, which is a trifle colder. In all of them, as well as in all the other Southern States, land can be bought at the same low prices.†

The cost of the winter forage of sheep in Tennessee may be inferred from the statements of Mr. Kramer, (in Letter IV.) On even the lofty Cumberland Mountains, in that State, grass grows during the entire winter, and snow rarely covers the ground to exceed forty-eight hours! Judge Beatty's statements in relation to Kentucky (in the same letter) show that the luxuriant blue-grass pastures of that State will sustain sheep during the entire winter; and that they frequently obtain their whole subsistence on the grasses, even on the mountains. Let us now turn to Virginia, the most northern of the Southern States. In a recent letter to me, John S. Skinner, Esq. says :

"Hon. Mr. Coles, a Member of Congress from Virginia‡—a sedate, attentive and practical farmer—once informed me that his flock of 200 sheep, kept in good condition summer and winter, did not cost him $10 a year. . . . . You must know that they, in the general way, as I believe, never feed their sheep, winter or summer, except where the ground is covered with snow—which is rarely the case, and then the snow does not lie more than a day, or at most two days. . . . No doubt winter pasture might be provided by sowing rye in the proper season (the usual system is to sow it the last thing, and as long as the farmer can "catch a chance") and putting the ground in good condition; and in that way adequate provision might be made for any deficiency of natural pasture. . . . . When the snow does cover the ground in Virginia, they give the sheep corn-blades—an excellent fodder. I think the rule was when I was a boy (in the rare exigency alluded to) to give them a bundle of blades each. A bundle of blades compacted would be about as large as the upper part of your arm."

North-Western Virginia seems to be considerably colder than the corresponding portion of the State east of the mountains; and the winter foddering season is not greatly shorter—though the amount of fodder consumed *must* be far less—than in Western Pennsylvania, or in many portions of New-York.|| Yet, singularly enough, more sheep are bred here in proportion. probably, than in any other portion of the Southern States!

---

* Some other paragraphs from this letter are omitted for quotation under the heads of which they specifically treat.

† Hon. S. Strong, a Member of Congress from this (N. Y.) State, writes me, after consultation with various Southern Members, that " good lands may be purchased for $1 50 per acre, and in great abundance, in most of the Southern States."

Mr. Garret Andrews, of Wilkes Co., Georgia, in a communication in the American Agriculturist (April, 1844), says : "Several hundred acres (in the middle or hilly zone) are often sold for a dollar or less per acre. The usual rule is to sell the wood-land for what it may be thought to be worth, and give the purchaser the old lands and the houses for nothing. . . . . For $1,000 or $1,500, a comfortable house and out-houses, garden, &c. may be had, with several hundred acres of land, . . wanting nothing but a fair chance to become as fertile as may be desired. . . . There is no end of the materials for manure."

I recently saw it stated by a gentleman in a communication which was published in the N. Y. Farmer and Mechanic, that he was authorized to *give away* good land in the Cumberland Mountains to sober and industrious settlers.

The prices in the N. C. Mountains will be seen from Mr. Clingman's letter. (Letter IV.)

‡ Mr. Coles resided in Pittsylvania, a county adjoining North Carolina, in the middle or hilly zone.

|| Jesse Edgington, of Holliday's Cove, Brooke Co, Va., writes me: "Our average time of foddering is at least 4 months, and we generally provide provender equal to 5 tons of hay for each hundred grown sheep, for the winter."

This region being essentially Northern in its characteristics no allusion will be had to it in subsequent remarks.

It will be seen from the preceding statements that in many, if not most situations, throughout the whole Southern States, sheep will obtain sufficient food throughout the year from the pastures,* or from autumn-sown grains, excepting on the higher or more northern mountains. As has been before remarked, as the grain subsequently yields its crop, its tillage is not properly chargeable among the expenses of producing wool. The preparation of hay, and labor of foddering, are also dispensed with. By the rule of estimation followed in relation to New-York, the items on the debit side of the account would then be—interest on purchase money; interest on land; expense of shearing; salt, tar, and general supervision; and loss by death. The items on the credit side would .be the same with those of New-York.

Your own statements, Sir, as well as those of Mr. Simpson, show that, in many situations, both in the tide-water and hilly zone, three sheep can be supported on the herbage of an acre, without other fodder. His statements show that such lands can be bought at "from 50 cents to $1 50 per acre." The annual account then would stand thus:

Dr.	$ cts.	Cr.	$ cts.
100 sheep—to interest on purchase money, at $1 25 per head......................$8 75		By 300 lbs. of wool at 31 cents per pound...$93 00	
To interest on 33⅓ acres of land at $1 50 ..... 3 50		" 80 lambs, at 62½ cents per head......... 50 00	
" expense of shearing...................... 4 00		" Manure† ....;.................. 28 00	
" salt, tar, and general supervision 8 00			
" loss by death 2 per cent. over and above value of pulled wool 2 50		Total..........................$171 00	
Total........................$26 75		Balance $144 25	

Making $4 32, or *two hundred and eighty-eight* per cent. clear profit per acre, on lands worth $1 50!

By the respective estimates it will be seen that the gross cost of producing a pound of wool (allowing 3 lbs. to the fleece) is, in the Southern States, 8$\frac{1}{2}$ cents; in New-York 27$\frac{22}{43}$ cents‡—or *nearly three and a half times greater* in the latter! I have put down the expense of shearing the same in both cases, and the supervision, South, twice as high as the *summer* care, in the North. Shearing always costs $1 a day, per hand, in the North, and the summer care devolves upon the paid laborer whose every hour counts. The shearing would not be worth to exceed $2 a hundred on a plantation where slaves are kept, and the supervision or care could scarcely be considered an expense, when it could be borne mainly, if not entirely, by superannuated or decrepit slaves, or even by children. The real expense of growing wool on land of this quality and price would be about 5$\frac{1}{2}$ cents per pound;|| and calling the fleece 4 lbs. (which weight it always *ought* to he made to attain) it would but little exceed 3½ cents.§ This is above Mr. Coles's estimate of expense in southern central Virginia, and Mr. John S. Skinner has repeatedly expressed the opinion that it could be grown in various parts of the Southern States at 3 cents per

* This supply could be rendered far more certain and available, where desirable, by leaving a portion of the fields undepastured in the latter part of summer and autumn. This "fog" or after-grass would not only afford much food, of itself, but it also greatly favors the sprouting of the young grass underneath it, by the protection it offers from frosts and cold winds.

† I have put this down the same as at the North, because I suppose it is just as valuable at the South, and quite as much needed. Few are disposed to appreciate the value of manure when it is not presented to their view in bulk, as in the barn-yard; but it is worth quite as much, dropped in the first instance over the fields. I feel confident that I have not over-estimated its value either for the South or the North.

‡ To obtain these results, I divided the whole annual expense, as set down in the respective estimates, with the exception of the charge of 2 per cent. for loss by death, by the amount of wool produced. For reasons already given, I do not consider the wool chargeable with such loss by death, except in an estimate where the full product of wool and lambs is carried out.

|| In this estimate I call shearing $2 per hundred, salt and tar $1, and supervision nothing.

§ Estimated as in the preceding note.

pound.* My own impression, however, is that the land, properly inclosed, that will support 3 sheep per annum, will cost, except in occasional localities, not less than $4 or $5, let the amount be more or less; and this would bring the cost of production (with 3-lb. fleeces) to between 7 and 8 cents per pound. I shall hereafter assume it to be 8 cents.

On many of the more northern mountains of the Southern States, and on the high peaks farther south, neither the grasses nor grain grow sufficiently to support sheep, unless the range is very large in proportion to the number, during the winter.† Here, as in the Northern States, dry feed must be prepared for the winter subsistence of sheep. This can be readily done, as the best meadow grasses of the North and the clovers flourish on the sides of the mountains.‡ There is little doubt that sheep can be wintered on dry feed on many of the mountains, and yet, on account of the extreme cheapness of the lands, the cost of producing wool not exceed eight cents per pound.

In the circumstances of many of the lowland plantations, it would be a most economical arrangement to summer the sheep on the mountains, and then drive them to these plantations to be wintered on pasture, fog, or grain fields, according to convenience. After the lambs have reached a sufficient age in the spring, and the sheep are shorn, marked, &c., a flock might be sent thirty, fifty, or even a hundred miles to its summer range on the mountains, at a trifling expense; and large numbers could be kept there under the surveillance of a single shepherd and a brace or two of dogs. By this system the lowland plantation would be saved from maintaining pasture on more expensive lands; many of its less marketable products could be converted into wool, meat, and manure; and it would be enriched by the wintering of the sheep.

Such, you are aware, is the system of sheep husbandry in Spain. The sheep are wintered on the plains of Estremadura, sometimes reaching the north of Andalusia. Both of these provinces, though in a latitude corresponding with that of a portion of the United States, extending from Albemarle Sound to a little north of Philadelphia, are parched, during the summer, to a state of arid sterility, by the burning winds of Africa.‖ In

---

* See Monthly Journal of Agriculture.

† With sufficient range, however, they not only obtain subsistence, but get fat. John S. Skinner, Esq., writes me; " In the mountains of Virginia, viz., at the Warm Springs, Dr. Brockenboro told me that a flock of sheep which he had bought for use during the watering season, strayed, and got off beyond reach during the summer; that the winter after they were rarely seen: and that as chance offered they were *shot*; and that finer and fatter mutton he never desired to see." The Warm Springs are in Bath county, among the Western or Allegany Mountains, a few minutes north of latitude 38°.

‡ See Mr. Goggin's statements in Letter IV. Since the above was written, I have received the following statements from Mr. W. Murdock, of Asheville, Buncombe county, North Carolina:
" Excellent swards of grass are grown in this district from Orchard grass or Cock's-foot. • Timothy and Italian Rye grass I have found to thrive remarkably well. I never saw them do better in any country. I received my seeds from England, and they succeeded admirably, and in ground by no means favorable to a fair trial. Turnips succeed remarkably well here, and even 130 miles farther south, as I am informed by Mr. Edward Calhoun—the kinds I don't know—but here the Globe, Aberdeen, Norfolk, &c., do well. . . . . If grounds were reserved as you suggest, for the winter feeding of sheep, the full growth being underpastured, and if some of the stubbles were plowed up and sown broadcast with turnips mixed with rape or colza, very little fodder will be required, in fact only when snow is on the ground, which seldom exceeds fifteen or twenty days during the year." [This fully confirms the positions assumed by me near the close of Letter IV.]

" I think that Curled Kale would be excellent for the winter keep of sheep, or cattle of any kind. I got some seed from England and sowed it like any cabbage seed. I put out the plants two feet asunder in tolerable ground. It grew three feet high and two feet in diameter. That I planted in the open field the sheep got at in October, and ate it, stock, branches and all, to the ground. That planted in the garden has, like the rape, stood the severe frosts uninjured. It is a delightful vegetable all the spring, and stands a warm or a cold climate. . . . . This and rape are, I think, all the green food necessary to keep sheep through the winter, with the addition of a little hay. Rape may be sown broadcast in inois. weather in May or June, and mown off for the sheep, when required, about six inches above ground. If the shoots are not required for pasture, let them go to seed, and the feed will pay better than any other crop, for making oil and rape cake."

‖ Here is a notable instance of the want of correspondence between isothermal and latitudinal lines between the west of Europe and the eastern portion of our own Continent. The two Spanish provinces the latitude of which is above given, have a climate more resembling the scorched *llanos* of Caraccas than any portion, even the most southerly, of the United States.

the winter, however, they are covered with verdure. About the first of May the sheep start for the mountains.* Formerly many of them rested on the lofty *parameras* and mountain sides of Old and New Castile—the latter bleak, sterile and craggy, compared with the sides of our own Southern mountains. But a friend recently from Spain informs me that those once magnificent flocks (now, alas! thinned by confiscation,† the wholesale plunder of invaders,‡ and for the subsistence of adverse armies,||) do not at present stop in any considerable numbers on the Castilian mountains, but pass north to the Cantabrian, and that portion of the Iberian range north of Soria—or crossing the latter, spread over the Eastern Pyrenees, and the mountains of Saragossa north of the Ebro.

Anything like an elaborate comparison between the facilities for sheep husbandry furnished by the mountains of Spain and the Apalachians of the United States, south of the Potomac, would, perhaps, be out of place in this connection. But a glance at them may throw useful light on the question of comparative profit. If the Spaniard can grow wool at a profit, where the natural and physical features of the country gives him no advantage over us, we can certainly do so; for in every other respect we have the advantage.

The Eastern Pyrenees rise to a hight of 10,000 feet,§ more than double that of the Peaks of Otter, or that of any other portion of the Apalachian range, with the exception of a few summits in North Carolina. Mount Perdu, one of the Pyrenees, is 11,283 feet in hight,¶ or 4,807 feet higher than the Black, the highest mountain of the United States east of the Mississippi. Maladetta, Vignemale and others rise considerably above 10,000 feet.** Glaciers exist on different parts of the whole chain. "The acclivity of the Pyrenees on the side of Spain, is often extremely steep,†† present ing a succession of rugged chasms, abrupt precipices, and huge masses of naked rock."‡‡ Miñano, a Spanish writer of authority, in defending his countrymen from the charge of indolence, speaks particularly of the ef forts of the hardy peasantry on the "almost inaccessible mountains of the Asturias, Galicia and Catalonia." The vegetation on these mountains is extremely variable, in some places being as luxuriant as the best on our Southern Apalachians, but more frequently dwarfish and meager. On large portions of them it is entirely wanting. The northern acclivities are frequently swept by cold and piercing gales from the Bay of Biscay. On the whole, it will be seen that they do not compare with our southern mountains in the advantages which they offer for sheep husbandry.|| ||

---

* For singular and interesting particulars in relation to their march, &c., and the municipal regulations pertaining thereto, see Livingston on Sheep, p. 36 *et supra.*
† Some of the choicest flocks in Spain were confiscated by the Government during the great anti-Gallic struggle. In the winter of 1809, the Spanish Junto confiscated the great flocks of the infamously celebrated Godoy and several other nobles, and they were bought by foreigners for exportation.
‡ The French Marshals, not finding anything in Spain to benefit the *fine arts* of *la belle France,* as in Italy, condescended, it is said, to benefit her *Agriculture,* by driving home some of the best flocks of Spain. The Allied Armies compelled the restitution of the *marble* and *canvas,* but those *priceless flocks* either could not be re-collected, or they were not regarded as of sufficient importance to be returned.
|| The Commissariat of the English, French and Spanish armies,

"The foe, the victim, and the fond ally,"

found the great Spanish flocks a very convenient resort, and availed themselves of it fully. The Guerillas, contrabandists, and fugitive inhabitants, of course, did the same.
§ Malte Brun.   ¶ lb.   ** Encyclopædia Americana; art. *Pyrenees.*
†† Montserrat (in Catalonia), so famous for its monastic establishments, will occur to you in this connection—where the steepness is so great that the monks ascend from hermitage to hermitage by ladders or stairs cut in the rocks!   ‡‡ Encyclopædia Americana; art. *Pyrenees.*
|| || How much the associations of early life—early *reading*—dispose us to exaggerate even the physical extent of the region covered by these mountains, connected as they are with so many romantic and interesting remembrances! The whole chain, extending from Cape Finisterre to Port Vendres, does not exceed 250 miles in length; and the space covered by it is not, in Western parlance, a "circumstance" to that occupied by our Southern Apalachians! Yet, in the western *half* of this chain, Pelayo and his successors maintained their Visi-Gothic kingdom, overthrew the descendants of the Abassides and Ommiades, and finally wrested Spain from the Moorish yoke. Who remembers, without the imp under his eye, that Ban

The route pursued by the Spanish flocks from, say, the middle of Estremadura to the Cantabrian mountains (the western portion of the Pyrenees), cannot fall short of 300 miles. It equals 270 miles in a direct
line. In addition to the length of the journey, they are compelled to cross
the Castilian mountains, and if they come from the south of Estremadura,
also the mountains of Toledo. Their route to the eastern Pyrenees would
be farther and still more difficult. Every circumstance, then, excepting
municipal regulations,* gives our Southern States, on both sides of the
Apalachians, a manifest advantage over Spain, for the purposes of migratory sheep husbandry.

Before closing the investigation of the question of the direct profits of
wool-growing in the Southern States, it is proper to inquire if there are
any special local militating causes or disadvantages not yet adverted to
which should be taken into the account. Diligent investigation has satisfied me that there are no such causes—on the other hand, that there is a
remarkable exemption from them—with one exception. That exception
is the destruction caused by wolves and dogs.

Wolves are found in nearly all new, and particularly in mountainous
countries ; but they invariably rapidly give way before the extension of
population.† They have even now ceased to be very destructive in the
most sparsely settled regions of the South. Mr. Simpson, in the letter before quoted from, says :

"There are but few wolves in South Carolina, excepting on the mountains. Otherwise,
our sheep which roam at large untended by shepherd, and uncared-for by any one, would
soon be exterminated. The wolves are not numerous even on the mountains. They are
not so destructive as dogs, which every now and then attack and destroy the sheep. A
trusty shepherd, with a dog or two and a rifle, would prevent this."

These remarks would apply equally well to nearly all the Southern
States. Wolves do but little damage, and would soon cease to do any ;
but the miserable, prowling curs are, in many places, a serious detriment.
There is something singular in the fact that while so much complaint is
made of them in the Southern and Western States, in New-York, where
there is certainly a great surplus of them, we hear little, comparatively
speaking, of their depredations. I am inclined to attribute it to the fact
that dogs are here constantly familiarized with the sight of sheep. The
first even playful movement of the adventurous puppy toward them is
severely chastised, and he is thus educated to recognize them as within
the category of " protected " animals. The dog which slays or even pursues a sheep, finds a long *pedigree* or a *silver collar* utterly unavailing to
save him from immediate death.‡

But even in the South or West, the loss occasioned by the depredations

nockburn was fought and Flodden lost to defend a *Kingdom* of half the dimensions of a good-sized Ameri
can *State !* In comparing the agricultural capabilities—and especially in estimating the ultimate result of
agricultural competition between our own country and the European ones, we rarely take sufficiently into
view the great disparity in territorial dimensions.
* For the monopoly of privileges conferred on the flock-masters of Spain to the oppression and prostra
tion of every other branch of husbandry, see Lasteric, and also Livingston on Sheep.
† A bounty of $10 is paid for the destruction of every full-grown wolf, and $5 for a wolf's whelp in the
State of New-York.
‡ In New-York it is provided by law that every bitch over three months old shall be taxed $2 ; every ad
ditional one owned by the same man $5 ; two dogs over 6 months old $1 ; every additional one $3. The
avails of these taxes constitute a fund, out of which Supervisors of Counties are to pay for any sheep slain
by dogs whose owners are unknown. This is not often enforced.
Any person may kill any dog " which he shall see chasing, worrying, or wounding any sheep," unless by
direction of owner.
The owner or possessor of any dog on being notified " of any injury done by his dog to any sheep, or
his dog having chased or worried any sheep," must within 48 hours kill his dog, or forfeit $2 50, and the
farther sum of $1 25 for every 48 hours thereafter, unless " it shall satisfactorily appear to the Court that it
was not in the power of such owner or possessor to kill such dog." Revised Statutes of New-York, vol. I
chap. xx., title xvii.

of other animals, or the expense of guarding against them, would be light compared with that in some of the wool-growing regions of the Old World.

In Australia, the sheep are exposed to the attack of wolves, dogs, and *convicts*, and are constantly attended by a shepherd, and nightly folded, and guarded by a watchman with dogs and a fire.*

At the.Cape of Good Hope, the shepherd and folding system is also fol lowed. In addition to wolves, and wild dogs which hunt in packs, and from their superior sagacity are much more formidable than wolves,† the Cape sheep are preyed upon by a variety of animals, and when they pass the mountains to glean the herbage which springs on the banks of the streams on the vast and lonely Karoos, they are exposed to the attack of the lion, the panther, the leopard, and the whole Feline family, so abund-ant and so particularly formidable in Southern Africa.‡ And they have had, and probably yet have, an enemy more destructive than all of these, in the Bushmen, more wild, irreclaimable, and predatory than their con-geners, the Bedouins of the Arabian desert.||

I have seen it proposed§ to teach young cattle to protect sheep from dogs, in the following manner : Turn a few steers into the pasture with the sheep, and with them a cow or two, having young calves at their sides. Send a dog into the field, and immediately the cows, followed by the steers, will commence a furious onset on the dog, and gore him or drive him from the field. After this is repeated a few times, it is said the steers will suffer no dog to enter the inclosure.

This might do very well under some circumstances, but I should prefer to *rely* on the remedy proposed by Mr. Simpson : the *dog* and the *rifle*. There are no " shepherd dogs " large and powerful enough to encounter and *kill* wolves and vagrant dogs, excepting the great sheep-dog of Spain ; and he is so irreclaimably ferocious to all excepting his charge, that he might frequently bring his owner into difficulty, and even endanger human life. My impression is that a shepherd dog or two, to be on the alert, and a brace of mastiffs to capture and, if need be, slay wolf or cur, would be adequate protection for the sheep on a considerable range, and the expense of maintaining them would be trifling.

* Cunningham's "Two Years in New South Wales," vol. i. p. 251.
† Missionary Labors and Scenes in Southern Africa, by Rev. Robert Moffat, pp. 23-4.
‡ The following stanza from the spirited lines of Freiligrath—" The Lion's Ride "—will occur to you .

> "And the vulture scenting a coming carouse,
> Sails, hoarsely screaming, down the sky ;
> The bloody hyena, be sure, is nigh,
> Fierce pillager he of the charnel-house !
> The panther, too, who strangles the Cape-Town sheep
> As they lie asleep,
> Athirst for his share in the slaughter, follows ;
> While the gore of their victim spreads like a pool in the sandy hollows ! "

|| To these may be added the savage Kaffirs, who, in their recent struggle with the Colonial Government, destroyed and drove off immense numbers of cattle and sheep. In 1834, " the natives," says Youatt, " drove off or destroyed 80,000 cattle and sheep almost innumerable."
§ By a writer in the American Agriculturist.

LETTER VI.

PROFITS OF SHEEP HUSBANDRY IN THE SOUTHERN STATES—2. AS THE BASIS OF AMELIORATION IN NATURALLY STERILE AND WORN-OUT SOILS.

Feasibility of rendering the naturally sterile and worn-out Soils of the South productive...Means must be ample and cheap...Ordinary Animal Manures from Stables, &c., not attainable in sufficient quantity—too expensive if transported far by land carriage...Animal Manures of Commerce still more out of the question...Gypsum—not sufficient of itself...Wood Ashes—Leached Ashes—their great value, but limited quantity...Lime (marl)...Swamp Mud—inexhaustible quantity of each...Valuable Effects of Lime on Soils...Otherwise when there is a deficiency of Organic Matter...Opinion of Johnston, Brown, Lord Kaimes, Anderson, Morton, Thäer, Petzholdt, Chaptal...Southern Tertiary and Granitic Soils destitute of Organic Matter...Expensiveness of Marl—not very permanent in its effects...The best Swamp Mud worth more per load...This, too, an expensive manure...Both too costly for extensive ameliorations...Is there, then, any resort?—There is—it is to be found in a Mixed System of Green and Animal Manuring, the latter made attainable by Sheep Husbandry...Experience and Testimony of various English Farmers under analogous circumstances...Reasons why Sheep are preferred to Horned Cattle for this purpose...Considered more profitable in England, and by some in the United States, independent of Fleece...Singular Hallucination of Col Taylor on this subject...Sheep preferred as Improvers of Poor Lands in the Northern and Eastern States, but the end sought by different means from those employed in England...The English System—Reasons why it is inapplicable in the United States...System in the Northern and Eastern American States...Proper System in the Southern States, on Lands now partly Grassed, and on Naked Soils...Green Manuring—how accomplished—Proper Plants for the purpose—Practical Rules—Expensiveness...Should the Pasture Lands of the South be exclusively devoted to Sheep Grazing?—Should not...Home Demand should be supplied by Home Production, in the Staples furnished by all the Domestic Animals—Reasons therefor...As a surplus or exporting Animal Staple, Sheep furnish the one in which the South can best compete with other Producers.

Dear Sir : Let us now pass to the second point in reference to which we are to consider the profits of sheep husbandry in the Southern States, viz.: the practicability and comparative economy of making it the basis of an effectual amelioration in soils naturally sterile, or those which have been rendered so by excessive and injudicious cultivation.

The first of these classes of soils is confined, mainly, to the tide-water zone. The second is found both in this and the hilly zone, and, I need not say, in immense quantities.

How can these soils be profitably ameliorated ? It is certain that this can only be done by the introduction into them of substances fitted to become the food of plants—or which, by chemical combinations or changes, prepare other substances to become such food. On soils naturally too sterile to sustain useful vegetation, the quantity of fertilizing matter introduced must be comparatively large. Hence it must be cheap, or its cost will more than overbalance its advantages. There are various manures which separately, or in conjunction, would convert the worst acre of barren sand between Richmond and Raleigh, or, if you please, on the Desert of Sahara, into a fertile garden, provided it could have timely rains and be protected from the burying sands. But it is utterly useless to argue the *feasibility* of this means or that, without at the same time examining its *economy.*

The direct and profuse application of animal manures, for example, would probably effectually ameliorate any of these soils. But where are these manures to be obtained, in a region where the first necessary condition for their production, *i. e.* the vegetation necessary to support domestic animals, is wanting ? The quantity accumulating in the cities and villages of a comparatively sparsely populated region—in a climate where the preservation of putrefying substances would be incompatible with health, would be inconsiderable. And whether more or less, it would not pay the cost of transportation to any considerable distance by land carriage. Guano, poudrette, bone-dust, and all the expensive manures of commerce, are still more out of the question. Gypsum, on account of the smallness of the quantity requisite, is a cheap manure, and, reasoning from analogy, should be a valuable one, under proper circumstances, at least on the granitic soils of the South.* Mr. Ruffin states that it produces little effect in the Tertiary sands.† It is considered by practical men to be, at the best, rather an *aider* of organic manure than a *substitute* for it, and when repeatedly applied without any other fertilizing substance, it ceases to produce any visible effect. On an exhausted soil, the chemical constitution of gypsum shows that it could not replace *all* the substances abstracted by the plants ; and on one naturally sterile, there is small probability that it would happen to supply the only deficiency necessary to the production of vegetation. Wood ashes constitute a most valuable manure on probably every class of soils, and, unlike lime, gypsum, soda, etc., which afford only a limited number of those substances which constitute the necessary food of plants, they afford in a greater proportion than any other manure the inorganic substances which are taken up and assimilated by plants.‡ They are used with the most beneficial effect on the granitic soils of New-England, the calcareous and aluminous ones of Middle New York, the silicious ones of the southern or grazing region, and on the Tertiary sands of Long Island. On the latter, of the same geological formation with your tide-water zone—in fact but a continuation of it—even the leached or washed ashes bring a shilling per bushel (the same that is paid for the unwashed ashes by the soaper and manufacturer of pearl or pot ashes) for agricultural purposes.|| But the supply cannot be made sufficiently large for extensive agricultural ameliorations, without a destruction of the forests, which would inflict a grievous and utterly inexcusable wrong on posterity.

The Southern Atlantic and Gulf States possess two natural and inexhaustible deposits of fertilizing matter, which, it is supposed by many, would be fully adequate to the general "reclamation"§ of their barren and exhausted evils. The first of these is the marl, which underlies large portions of the low country of Virginia and South Carolina, and probably the-

* I refer here to the successful example of its use on the granitic soils of New-England. I have particularly specified this class of soils because your *barren* ones are limited to them and to the Tertiary. Gypsum is used at the North on nearly every class of soils with advantage—calcareous, aluminous, silicious and all; intermediate varieties. It will be found very valuable, I have no doubt, on your mountain lands, particularly in localities where the clovers flourish.

† Ruffin's Agricultural Survey of South Carolina, 1843.

‡ To show the value of ashes as the food of plants, and at the same time the difference between those made from different woods, I append the following analyses of those of two well-known southern trees. That of oak ashes is by Sprengel, that of pitch-pine ashes by Berthier :

Constituents.	Oak.	Pitch-Pine.	Constituents.	Oak.	Pitch-Pine.
Silica............	29·95	7·50	Potash	16·20	14·10
Alumina	} 8·14	11·10	Soda................	6·73	20·75
Oxide of Iron			Sulphuric Acid	3·36	3·45
Oxide of Manganese		2·75	Phosphoric Acid	1·02	0·90
Lime	17·38	13·60	Chlorine............	2·41	
Magnesia..........	1·44	4·35	Carbonic Acid......	15·47	17·50

|| This fact I consider an important hint to the planters of the tide-water zone, and it is to be hoped that it is one which will not be thrown away. Leached ashes are valuable also on every other class of lands. The southern portion of my farm (lying on Chemung rocks) is silicious. The northern part is covered with "northern drift," and is therefore calcareous. I use from 3,500 to 4,000 bushels of leached ashes per annum, without any discrimination as regards the soil, and on almost every variety of crops, and invariably with marked advantage. Doct. Emmons, our State Geologist, having in charge the volume on Agriculture, stated to me that he considered these leached ashes far more valuable by bulk than a rich marl (accessible to me) containing 90 per cent of carbonate of lime.

§ This word ("reclaim") has a provincial signification throughout the North, when applied to land. It means "to render productive." Unlike the words "fertilize," "enrich," etc., it implies *degree*, as well as *manner*. To "reclaim" land, therefore, is to fertilize or enrich it to such a degree that it will yield fair crops. I shall use the word both as a verb and a noun, to avoid the circumlocution otherwise necessary to express this idea.

whole Tertiary formation, or at least that portion of it extending through the Atlantic States. The second is the swamp mud, which, rich with the alluvial deposition of ages, fills nearly every depression of the surface capable of retaining water, in the whole tide-water zone.

Mr. Ruffin recommends the former as the best and most attainable fertilizer on both of the classes of soils under examination. He seems to think it adequate, of itself, to their full and permanent amelioration. I do not desire a word which I shall say to bear, or even seem to bear, a controversial tone toward the views of this ardent and enlightened friend of Southern Agriculture. In expressing my dissent from them, my limits and the occasion only permit me to allude to a few well-settled principles and facts on which I have based my opinions. Lime acts mechanically and chemically on soils. It stiffens loose and opens clayey ones. It is to a certain extent, one of the necessary constituents of plants; it neutralizes acid substances in the soil; it forms compounds, and promotes the dissolution of existing ones, to prepare suitable food for plants; and sometimes produces certain other minor beneficial effects. But its great, its chief object, is to produce the food of plants by its chemical action on the *organic matter in the soil.* Hence, says Johnston :

" Lime has little or no effect upon soils in which organic matter is deficient;" and he farther says: " Under the influence of lime the organic matter disappears more rapidly than it otherwise would do, and that after it has thus disappeared, fresh additions of lime produce no farther good effect; . . . it causes the organic matter itself ultimately to disappear."

" It is scarcely practicable," says Brown, ' to restore fertility to land even of the best natural quality, which has been thus abused ; and thin moorish soils, after being exhausted by lime, are not to be restored."

" An overdose of shell marl," says Lord Kaimes, " laid perhaps an inch thick, produces for a time large crops, but at last renders the soil capable of bearing neither corn (grain) nor grass, of which there are many examples in Scotland." " The same," continues Johnston, " is true of lime in any form. The increased fertility continues as long as there remains an adequate supply of organic (animal and vegetable) matter in the soil : but as that disappears, the crops every year diminish both in quantity and in quality."

" On poor arable lands, which are not naturally so, but which are worn out or exhausted by repeated liming and cropping, lime produces no good whatever." (Anderson, Brown, Morton.)*

Let us now turn to the opinions of some of the most eminent European Continental writers. The celebrated Thaër in his " Principles of Agriculture " (Section IV. Part I.) says :

" On no soils are the effects of lime so beneficial as on those which contain a great quantity of sour humus prejudicial to vegetation, or on those which have been supplied more or less abundantly with animal manure for a considerable period, without receiving an application of lime, or some other substance of a similar nature. In the latter case it is frequently much more efficacious than an amelioration of stable manure would be; but it soon impoverishes the soil so much that in a few years it becomes indispensably necessary to manure it abundantly with rich animal or vegetable matters. As some portion of the humus, although in all probability of an insoluble nature, always remains in arable land even when it appears to be much exhausted, it of course follows that an application of lime will always be productive of very marked effects even on the poorest soils, because it will call into action all the nutritive particles which they contain. A second amendment of a similar nature bestowed shortly after the first, will be productive of some, although in general of much less benefit ; and the effect of each subsequent amelioration of this nature will be progressively diminished unless the soil receives an additional supply of humus. . . . The effect produced by lime on land of this nature (reclaimed bogs and marshes) is much more beneficial and durable than that of any other manure. On the other hand, repeated ameliorations of lime will soon totally exhaust and impoverish poor and sandy soils, and reduce them to absolute sterility, even though each separate application seems to be productive of some good effect. . . . Many persons who have not rightly comprehended the cause of the effects produced by lime, prefer it to manure, and have believed in the possibility of doing entirely without the latter ; but the total exhaustion of the soil which such a course of proceeding must sooner or later produce, caused them to fly to the opposite extreme. . .   An co-

* See Johnston's Agricultural Chemistry, vol. ii. p. 139-142.

lightened and scientific agriculturist will soon perceive that the use of lime can never super sede that of dung, but that it renders this kind of manure more energetic in its action. . . .

In many places where its ameliorating effects were known and appreciated, many agriculturists have calculated that marl would prove a cheaper manure than stable dung; and have, consequently, determined to do without the latter altogether; and, therefore, have diminished their stock of cattle, and sold their hay and straw. It may easily be imagined that as soon as the chemical effects of the marl ceased to operate, as must be the case when the land no longer contained undecomposed or insoluble substances, the soil became sterile, and a second marling was incapable of producing any beneficial effects, there being no humus for it to act upon."

Petzholdt, in his "Lectures to Farmers on Agricultural Chemistry," (Lecture XVII.) says:

"Quick-lime greatly accelerates the decomposition of humus, whether of animal or vegetable origin, inducing a more speedy liberation of its salts than would otherwise take place. This is the reason quick-lime has proved so advantageous in the cultivation of bogs; the lime not only accelerates the decomposition of the humus, but it may be said altogether to be the cause of the decay of humus, which, as it exists in peat, is scarcely by itself undergoing the process at all. . . . Where there is neither humus in the soil, nor undecomposed silicates, the application of lime as manure will be useless. . . . So much, however, is deducible from all experience, that the mere application of marl to an exhausted soil is of no use whatever, unless it is carried on the field in such quantities as to constitute a new soil, covering the whole surface to the depth of a foot. . . . In a chemical point of view, marl is not of any value except where the soil requires a supply of lime. . . . The other mineral constituents of marl are far too inconsiderable in amount to be reckoned upon."

Chaptal, in his "Chemistry applied to Agriculture," (Chap. iii., Art. 2,) thus expresses himself:

"It is acknowledged that lime is principally useful upon fallow lands which are broken up; upon grass lands, whether natural or artificial, which are prepared for cultivation: and upon muddy lands, which are to be put into a fit state for culture. It is well known that in all these cases there exists in the land a greater or less quantity of roots, which, by the application of lime, may be made to serve more immediately for manure, by the solubility it will give to the new products formed by them. . . . Independently of this effect, which, in my opinion, is the most important, lime exercises other powers, which make it a very valuable agent in Agriculture."

These authorities might be multiplied *ad infinitum.*

On the alternately too loose or too hard soils of the dry and barren lands of the tide-water zone, lime would doubtless have two salutary effects— the mechanical one already noticed, and it would furnish *one* necessary food of plants. But of its power to render these soils, or the exhausted ones of the middle zone, anything more than transiently fertile, there is no probability, if they are, as I suppose them to be, generally rather, and sometimes very, destitute of organic matter. This destitution I infer from ocular examination;* also from the fact that they are covered with little vegetation, with the exception of the long-leaf pine, to produce by its annual decay a store of organic matter; and, finally, if this organic matter existed in these soils in any considerable quantity, *they would not be sterile.* They probably possess the ordinary inorganic constituents of dry Tertiary and granitic soils, and no properties directly deleterious to vegetation. Organic matter, then, in my judgment, is what they principally stand in need of to render them fertile. Now, by applying lime to them, it would undoubtedly do good in two ways, as before admitted; but the considerable temporary apparent amelioration, as evinced in some instances by the increased growth of vegetation, is factitious, for the lime is only acting with and exhausting the little organic matter in the soil, to leave it to greater eventual sterility. Hence the saying that "lime enriches the father but impoverishes the son," is a true one when the lime is applied to soils possessing but a small proportion of organic matter. On such, lime soon

* I have seen no analyses of these soils, and mean therefore as I say, simply, examination by the eye

leads to exhaustion, unless *organic matter is added to the soil in proportion
to the waste*. The theoretical and practical considerations which should
govern in the application of this fertilizer to soils are discussed more fully
and, in my judgment, more ably by Johnston, in his Agricultural Chemistry, than by any other writer. To him I take the liberty to refer you.

Marl raised from pits, as it must necessarily be (except when denuded,
or cut through, on the beds of streams, &c.) where it occurs only as an
under-stratum in a flat country—where the pits, too, often require machinery, or much manual labor, to keep them free from water while working—must be an expensive manure. From its tendency to sink in the soil
it is not so permanent a one as would naturally be expected. On reclaimed
swamp lands—as, for example, on the rice lands—abounding in vegetable
matter, it will be found a most efficacious manure, and, *when needed*, will
repay the necessary outlay; but I fear it will be found otherwise ultimately, if not immediately, on the barren sands and exhausted granite soils of
the South. Applied *with* swamp mud, it would constitute a fertilizer
scarcely, perhaps, admitting of a superior, even on the latter soils. In
their *single effects*, however, I cannot but believe that the best swamp mud
—that which is black and fetid by the long continued accumulation of organic substances (and especially if charged with shells, and the shields of
Infusoria)—would be worth more per load than the richest marl. The
mud, too, should be considerably cheaper than the marl, no deep excavations being required to obtain it.* Digging and draught, and, in the case
of the mud, draught alone, would render both decidedly expensive manures, relatively to the value of the land after being ameliorated by them,
even assuming that amelioration to be complete and permanent. On lands
immediately contiguous to conveniently reached depositions of mud or
marl, on a scale so limited that it could be carried on at spare intervals
without encroaching on the regular routine of plantation labor, it might
be good economy to haul out mud and marl, and thus gradually reclaim
small pieces of land.† It certainly would be better economy than to waste
those intervals in idleness. But in anything like an extended and speedy
system of reclamation—the fertilization of thirty, forty or fifty acres per
annum, instead of one, two or three—the means above adverted to are, in
my humble judgment, utterly out of the question. The labor would absorb all the labor of man and beast on the plantation; and it is exceedingly
questionable, in my mind, whether the land, when fertilized, would sell for
the cost of the manure.

Hard would it be for many a South Carolinian or Virginian to turn his
back on the *Lares* and *Penates* of his race—forgetting many a proud local
and ancestral association—but as a question of dollars and cents, sometimes a necessary one, and, at all events, usually the paramount one, I
think it past a reasonable doubt that it would be better economy to desert the worn-out or naturally barren soils of our South-eastern coast, and
purchase the virgin and fertile lands of the South-west (even including
the extra expense of building and fencing), than to attempt to reclaim the
former by means so expensive as those above indicated.

What, then, is the resort? Are there *any* means by which those lands
can be *profitably* reclaimed? I answer, Yes; and the resort is a mixed
system of green and animal manuring—the latter made attainable by sheep
husbandry. Experience is the best test of all theories. And we have had

* I am inclined to think, however, that this mud, if spread directly on the surface, would contaminate
the atmosphere with unhealthy miasma, generating agues and bilious diseases. If so, it would require incorporation with the soil, by plowing.
† It seems to me, however, that these expensive manures would be more profitably applied in *keeping
up* the fertility of the *best* lands, or as *assistants* to other and cheaper means of reclaiming the poor ones

so little experience in the premises, in our own country, let us turn to that **of the first** agricultural nation of the Old World. There is no foreign nation where so high a degree of intelligence is brought to bear on farming operations—where cause and effect are so carefully studied and accurately noted—as in England. This care and accuracy are indispensably necessary in a country where high rent and heavy taxation render good farming or bankruptcy unavoidable counter-alternations to the agriculturist. Prevailing conclusions among such a class of agriculturists—*undisputed* conclusions, too—are assuredly entitled to great respect, and may almost be regarded as settled facts. Now the farmers of England are perfectly familiar with every kind of manure accessible to our Southern farmers, unless it be swamp mud and cotton seed. Lime, for example, is plentiful and cheap, and is much used in Agriculture all over the kingdom. If either this, or any of the manures of commerce, were considered, of themselves, economical fertilizers of the poor, sandy or light upland soils of England, there is no country in the world where they are more plentiful, and, when the use of the soil and the price of products are taken into consideration, more cheap.

What the settled conclusions of the English farmers are, in relation to the profitable amelioration of those soils, will be seen from the following undisputed testimony of some of the most eminent and respectable of them, taken before the Committee of the House of Lords, charged with the inquiry into the state of the wool trade, &c. in Great Britain, in 1828, from which I have so freely quoted in preceding Letters.

Mr. WILLIAM PINKNEY, *Salisbury Plain:* Land such as I occupy could not be maintained without the aid of sheep. . . . The sheep are our principal dependence for supporting our crops; indeed, I could not occupy my farm without my flock.

Mr. JOHN ELLMAN, Jr., *Sussex:* I do not consider it possible for the light lands upon the Downs to be kept in cultivation without flocks. I could not keep the farm I now hold without sheep. . . On the South Downs the wool must be grown, let the price be what it will.

Mr. FRANCIS HALE, *Alringham, Suffolk:* The description of land I occupy could not be kept in cultivation without the aid of sheep.

Mr. HENRY KING, *Chilmark, Wiltshire :* The size of my farm is about 4,000 acres. I clip annually about 6,500 South-Down sheep. . . . Such lands as I occupy cannot be kept in cultivation without the aid of sheep.

Mr. JOHN WOOLLEDGE, *near Bury St. Edmunds, Suffolk:* An estate near the above place contains 8,890 acres, let to tenants, and consists principally of poor sandy and gravelly land, the produce of which in grain is very precarious, amounting in dry summers to little or nothing. The occupiers, therefore, depend almost entirely on their flocks of sheep for the payment of their rents and the employment and support of the population. . . . I am of opinion that two-thirds of the counties of Suffolk and Norfolk may be comprehended in the sheep districts, and that they produce two pounds and a half of wool, and three-fourths of a lamb, to the acre, upon an average. . . . The produce of the land depends materially upon the folding system; there is not sufficient straw for manure without the assistance of sheep.

Mr. WILLIAM ILOTT, *Abbey Milton, Dorsetshire:* I calculate the annual growth of wool in Dorsetshire at 10,000 packs of 246 lbs. each. It is estimated . . . that 800,000 sheep, or one sheep and one-seventh per acre, . . . are kept in this county. A considerable part of the county of Dorset is composed of light lands, and can only be kept in tillage by the aid of sheep.

C. C. WESTERN, Esq.: It is utterly impossible that the Down Districts can be cultivated to advantage without sheep. We never fold our Merino or other sheep; the land is too wet.

LORD NAPIER: If we had not sheep upon our lands (the highlands of Scotland), it would become the habitation of foxes and snipes, and return to waste; it would produce nothing but grouse and wild game of different sorts.

Is it asked, Why are sheep preferred to horned cattle? Many of the reasons are given in my preceding Letter. Then, again, the scanty and short pasturage of light lands, on which sheep will thrive, will not afford sufficient "bite" (as it is provincially termed in the Northern States) to

profitably carry large stock. And, finally, there seems to be a settled conviction among the English farmers that sheep give a better return for the
food consumed, and therefore better repay the extraordinary expenditure
necessary to bring poor lands in a fit state for the plow, than any other
animal. In an able essay in the London "Plough" (June, 1846), the following remarks occur, which may be probably regarded as an expression
of the prevailing opinion in England:

> It is justly admitted that, of all the domestic animals reared and fed for profit in Great
> Britain, sheep are of the greatest consequence, both individually and in a national point of
> view, and afford a better return than can be obtained either from the rearing or feeding of
> cattle; the very fleece shorn annually from their backs is worthy of consideration. . .
> Sheep husbandry deserves to be esteemed in all its different branches, and claims the prior-
> ity of consideration among agriculturists."*

The manner in which the "very fleece" is here spoken of, shows that
wool occupies but a mere, subsidiary place in making up the profits of
English sheep husbandry. I know many intelligent and experienced farmers in *this country* who think sheep feeding more profitable, or equally
profitable, with cattle feeding, leaving the wool out of the account.†

The experience of the English farmers accords with that of those of our
own Northern and Eastern States, in relation to the superior advantages
of sheep husbandry on poor and light soils. Observation has shown both
that such soils do not profitably carry bullocks or other large animals, and
that such animals are poorer manurers than sheep. But their methods of
availing themselves of the advantages of this husbandry are entirely dissimilar. The English farmer finds mutton and grain the marketable pro
ducts which pay best. The first returns a profit on the crop (turnips)
which produces it, and at the same time fits the land for the latter. The
high price and ready sale of mutton allows the English farmer to force the
growth of turnips on poor soils, by the application of highly condensed
manures.‡ In the fall the sheep are turned upon small patches of them,
surrounded by an inclosure of hurdles. The turnips are drawn, sliced, and
laid in troughs for the sheep. When one patch is consumed, the hurdles
are removed, and thus the field is gradually passed over—the sheep converting the whole crop into animal products and manure. The land is
then plowed for grain, and a succession of crops are taken from it. By
this means the land is soon reduced to its former level, and the same system is again entered upon.

---

* After reading this and the preceding testimony, one cannot look back without a smile on the unac-
countable monomania of that excellent man and public benefactor, Col. John Taylor, in relation to sheep.
In one of the essays of "Arator," he says:
"My conclusions are that they require and consume far more food, in proportion to their size, than any
other stock; that they are more liable to disease and death; and that they cannot be made a profitable object throughout the whole extent of the warm, dry climate and sandy soil of the United States, but by ban-
ishing tillage from vast tracts of country." . . . "It is probable that the hot constitution of sheep produces a rapid digestion, and that insatiable appetite, by which the fact is accounted for of their flourishing
only, to any extent, in fine meadows or extensive wildernesses. If this voraciousness is not gratified, the
animal perishes or dwindles; if it is, he depopulates the country he inhabits. The sheep of Spain have
kept out of existence, or sent out of it, more people than the wild beasts of the earth have destroyed from
the creation; and those of England may have caused a greater depopulation than all her extravagant wars.
It may be owing to this animal, the independence of one country is almost overthrown, and of the other
tottering." (!!!) He further expresses the opinion that England, "by the help of her moisture and verdure,
can raise wool cheaper than the United States." (!)
It would appear that Col. Taylor formed all his conclusions on a small flock kept by himself. They may
have been a bad and unthrifty flock. But it is strongly probable that he was influenced by deep-rooted
prejudices, imbibed before his judgment was ripened, or his experience formed; and that these, unknown
to himself, warped all his views. I can account in no other way for the evident and palpable hallucination
under which he made nearly every statement in his Chapter on Sheep.
† A gentleman who has been one of the most successful feeders of cattle and sheep in this State (P. N.
Rust, Esq. of Syracuse) recently remarked to me in conversation that he had invariably found that sheep
paid better for feeding than cattle.
‡ Bone-dust, and frequently guano or some other manure with it, is *drilled in* with the turnip seed, so
that much cost is obviated by making a little go a great ways; and there is a remarkable congeniality in
the climate and atmosphere of England to the growth of this root.

In the United States, much of this system would be inapplicable and unprofitable. Here wool, instead of mutton, is the principal object. Even in the Southern States, where the climate would render the English system practicable, the expense of producing either of these articles, by winter turnip feeding and folding, would add so much to its cost that it could not profitably compete with that grown in the ordinary manner. The same remark applies to the relative expense of the two systems of manuring. A constant repletion of rich succulent food, like turnips, would sensibly increase the amount of manure, and, by folding, it would be more evenly distributed. But neither of these considerations would begin to offset against the increased expense, in a country where good lands are so cheap and bread-stuffs so low. Besides, no good, but, on the other hand, positive injury, would result from thus annually *fattening* " store "* sheep, kept for the production of wool and for breeding.†

The system of improving poor lands in the Northern and Eastern States by sheep husbandry, is mainly by summer pasturage. The droppings of the sheep gradually enrich them,‡ and consequently increase their herbage. Thus, in a few years, poor and scanty pastures are converted into rich, productive ones. This might be far more rapidly done by giving these pastures also the winter manure of the sheep, made in the feeding yards. But it is generally thought more profitable to give the winter manure to the richer tillage lands, which are made to supply the grain and hay of the farm. The light pasture lands are thus kept permanently in pasture or are only plowed, by the provident, at very long intervals. This system is rendered necessary, or, at all events, convenient, by the topographical features of our farms. Here the poorer and lighter are generally the higher and more broken lands, which are less convenient of aration, and for the hauling on of manure, or the hauling off of crops.

In the Southern States, on lands which now yield even a smallish supply of esculent grasses, the northern system is all that is *necessarily* required. Those grasses will every year increase, and the land will be gradually fertilized, by the droppings of the sheep, without a cent's expenditure on it of any kind ; and every particle of herbage will be turned to its most profitable account, by being converted into wool, mutton and manure.

But where there is not sufficient existing verdure to form the germ, so to speak, of a future good pasture—or, in other words, to support a sufficient number of sheep to convert it, within a reasonable time, into good pasturage—some other course must be adopted. Proper plowing and seeding, simply, will, I have not a doubt, be found adequate in a great many instances where it would hardly be suspected. It is very natural to take it for granted that a soil, not spontaneously producing the grasses, is not fertile enough to produce them, even if properly sown upon it. But experience has amply demonstrated the contrary in several of the North-western States. There are various causes, besides a want of fertility, which may produce such nudity ; but this is not the place to enter upon speculations on this topic. Two very common and obvious causes are too great looseness or compactness of the surface, which prevents seeds from taking root, especially in a dry, hot climate. Plowing would always loosen

---

* This convenient word is provincially applied, in the Northern and Eastern States, to sheep and swine which are to be kept over the year, to breed from (and the former to produce wool), as contradistinguished from those which are fattening for slaughter.

† This point will again be adverted to. It is sufficient now to say that breeding-ewes, if brought to a high state of fatness, raise fewer lambs. The lambs are born weak, and are very apt to perish. There are also other objections.

‡ Aided by an occasional top-dressing with gypsum.

K

hard, and frequently stiffen loose surfaces.* The grass seed harrowed into a properly prepared soil, at the suitable season of the year, might so root itself as to withstand the subsequent heats, while those dropped on a hard or a loose surface by birds, or borne there by winds, would be exposed directly to the rays of the sun, which, if it did not entirely prevent germination, would dry up and kill the tender roots before they could strike deep enough to resist its influence. Much will depend, in this experiment, upon a proper selection of the variety of grass sown. That variety should be sown which is found to flourish best on similar soils, in the same climate, even though relatively it may be an inferior grass.†

By means as cheap and attainable as these, I have not a doubt that no inconsiderable portions of the nearly naked soils of the tide-water zone might be brought into pasture sufficiently good to make their ultimate conversion into prime pastures, by means of sheep husbandry, certain.

On the worn-out granite soils of the middle region, the once fertile red clay lands—now occupied only by dwarf pines, worthless broom-grass, etc.—*deep* plowing and thorough sowing (with the aid of steeps and the cheap top-dressing, before adverted to) would *generally*, I believe, bring these lands into pretty good pasture. These soils, having been subjected only to the one-horse plow, and hand tillage, *are worn out only on the surface.* This is proved, in innumerable instances, in Fairfax, and other northern counties of Virginia. Lands considered entirely worn out, and sold for a mere trifle, are subjected to the northern two-horse plow, and from one to three inches of earth, *never before disturbed*, is brought to the surface, which readily supports grass, and even grain crops—the latter temporarily. Thus, most fortunately, the means are still left, with the aid of pasturage, to make many of these lands profitably productive, and to restore them to much of their former fertility.

We come now to another class of lands which may, in many cases, be worth reclaiming, but which will not, by merely being plowed and sown produce sufficient grass to make their fertilization by sheep husbandry attainable—or attainable within a moderate period of time. These are the inferior (but not the worst) sands of the tide-water zone. Here green manuring must be resorted to, by means of plants which will better withstand the climatic and other difficulties in the way of their getting well-rooted, and which will flourish in poorer soils than the grasses. Both of these conditions are answered by various plants. Spurry *(Spergula arvensis)* and white Lupins *(Lupinus albus)* will flourish on dry, barren, and even shifting sands, and are extensively used as green manuring crops on such soils, on the Continent of Europe. From their rapid growth and extraordinary productiveness, they are admirably adapted to this end. The introduction of these plants would probably supply an important desideratum in Southern Agriculture, unless, as I have already expressed the opinion,‡ the pea leaves little to wish for, as a green manuring crop on every class of southern soils. Soaked in a solution of nitre—rolled in lime—top-dressed, after sprouting, with a slight sprinkling of ashes and gypsum‖—

* The sands of the tide-water zone are everywhere, at greater or less depths, underlaid by clay. These might in some cases be reached by the plow, and portions of them incorporated with the superincumbent soil.
† See Letter III.   ‡ In Letter III.
‖ Sprengel's analysis, in Letter III., shows the large amount of potash required for the seed, and of lime for the straw of the pen. The favorable effect of plaster on this, as on most other leguminose, is well known. Ashes, plaster and lime can be purchased here at an average of less than ten cents a bushel. A bushel of gypsum, mixed with say two bushels of ashes, makes a top-dressing which will pay for itself a number of times over, on any land to which I have ever seen it applied. In addition to rolling the seed in lime, a few bushels of it, or of marl, would make a good, and, where accessible and cheap, an economical top-dressing. When I speak of the price of lime here, I do not refer to marl. The latter, in its natural state, could be purchased at the beds for probably a shilling a load.

the pea would probably take root and flourish in any soil which the price of land in this country would justify an effort to render productive, now, or for a long term of years to come. Indeed, the capacity to produce this plant *may* afford the best practical test of the economy and expediency of attempting it in any given case. If a good green manuring crop can be made to grow on the soil without any more expensive aids than those above suggested, the *lever* of improvement—cheap, but effectual—is placed in the hands of the planter, and, if he possesses the least degree of energy, he has no occasion to seek a new soil and home by emigration.

Mr. Ruffin states, if I remember aright, that a few *quarts* per acre, of peas, are sown by the Southern planters. In the North, three *bushels*, at least, are sown; and this quantity would be little enough to produce the largest amount of green manure.

Theory would indicate that the crop should be turned under before it comes into full flower,* but experience and convenience both deserve consulting in the premises.

An active span of horses, with a Northern two-horse plow, and an expert plowman, would readily plow two acres per diem, on sandy soils, and plow it well.† The expense of getting in a crop of peas can then be readily estimated.

If one crop can be made to take root and grow, and is plowed under *when green*, the great point is attained, and there will be neither difficulty nor uncertainty subsequently. The organic matter thus deposited in the soil is the *basis* on which future improvements can be effected *ad libitum*. As far south as South Carolina, at least two, and probably three crops could be plowed in during a single season. This might be done in time for winter grain, and a crop of the latter sown as a covering crop with grass seeds. The grain would refund much of the previous expense.

Plowing in two or three crops in succession may, at first view, seem an expensive process; but, with the exception of the extra seeding, it is no more labor than is bestowed on *every* wheat crop by a large proportion of the farmers of Western New-York! When the ground is summer fallowed, the ordinary practice on our wheat lands is to have it three times thoroughly plowed and harrowed, and the first time a crop of clover is plowed in. All this is a light outlay compared with thorough marling, or manuring with swamp mud. And, after either of the latter processes, the land has yet to be plowed and seeded.‡

It would not be *necessary* to plow in as many as *three* crops of peas, to lay the foundation of ordinary pasture. Two, and possibly one, would suffice. The comparative utility of forcing forward the fertilization of land, rapidly or gradually, depends much upon the amount of capital which the landholder has to devote to this object. The amount of labor subtracted from the ordinary operations of the plantation would be very small, in any case, in proportion to the object to be attained. A single expert plowman, with a good team, could give even the three plowings to a large field.‖

* "Because flower-leaves," says Johnston, give off nitrogen into the air; and, as this element is supposed especially to promote the growth of plants, it is desirable to retain as much of it in the plant and soil as possible."—Ag. Chem., vol. ii, p. 185.
† Perhaps more. That amount is frequently exceeded here, on stubble lands.
‡ So that the expense to be offsetted against one of those processes (in estimating their comparative economy as a means of reclamation with green manuring) is plowing, harrowing, and seeding *twice*.
‖ I have attempted to fix no definite data on this point, because you, who are acquainted with plowing Southern lands, are better competent to do so. I would remark, in this connection, that my convictions are very strong that the introduction of the two-horse plow of the North would lead to a decided improvement in your Agriculture, from the superior manner in which it does its work, and by lending to deeper plowing. The wheel will cause it to run as shallow as a one-horse plow, however where the character of the soil renders it desirable.

Even in the case of either of the two classes of soils before treated of
(those now producing thinnish pasturage, or which can be converted into
pastures simply by plowing and seeding), one or more green manuring
crops would form a most excellent and accelerating initiatory step, and,
where sufficient capital is possessed, I have no doubt, a most economical
one, toward their fortilization.

In view of all my preceding statements, do you ask me if I advocate
sheep husbandry exclusively, on all the lands at the South which already
are, or should be devoted to grazing? Most assuredly not. I have already laid it down as a maxim that "agricultural production should be
controlled by the demand or want, and the adaptation of the country to
such production." By this rule, the South should, at least, never import a
horse,* a mule, a pound of beef, pork, butter, cheese or wool. She *wants*
them all, and she can *produce* them all more economically than she can
import them. That declared impossibility in politics, an *imperium in im-
perio*, should be in Agriculture, so far as it may be consistently with the
above maxim, the attitude of every farm and plantation. Each should be
independent to the greatest economical extent, so far as the production of
the necessaries, comforts and luxuries of life are concerned, of every other
farm or plantation in the *world !* This mixed and multifarious farming is
objected to by theorists, inasmuch as it trenches on the division of labor
principle. But it favors rotation, and thereby prevents the exhaustion of
soils—leads to a more bountiful use of the every-day comforts of life†—
and, finally, it is less hazardous. The one-crop farmer, if crop and market
are both in their most favorable state, realizes great profits. But if the
market is poor, or the crop small, the loss is proportionately large. The
farmer pursuing mixed husbandry will not generally fall greatly behind
the *best* profits of the other, and his losses are rarely considerable. It is
better to play for a *hit* than a *gammon*, where, as in the case of the small
capitalist, affluence or penury "stand the hazard of the die !"

If the above positions are true, the South is called upon to increase the
breeding of other domestic animals as well as sheep. To an extent sufficient to *supply her own wants*, I consider her imperiously called upon so
to do. I advocate the breeding of sheep specially—on a vastly more extended scale—because, as has been already shown, they are the best (if
not the only) reclaimers of your unproductive lands; and because in that
surplus of the products of grazing, which these extensive reclamations will
bring about, *they furnish you the exporting‡ article (wool) for which you
can find the largest extra-limital market, and in growing which you can best
compete with other producers.*

Let us suppose, for the sake of the argument, that these newly reclaimed
pasture lands would carry heavy beasts as well as sheep, and with equal
benefit to the land. After supplying the home demand, what would be
done with the surplus horses, mules and beeves ? To what markets in the
world could you export horses and mules, with the exception of some of
the West India islands—the markets of which a few thousand head of
these animals would annually glut ? Do you ask me what would prevent
your sending your surplus beef to England? Nothing. But neither the
South, nor the North, nor the East, can compete with the great North-west

* Unless for the improvement of breeds.
† I mean by this that the planter who raises all the necessaries of life will be more liberal of them than the one who *purchases* them.
‡ I do not use the word here in its technical sense. I mean carried beyond mere local limits for sale—whether that sale be effected in the same state, in some other part of the U. S., or abroad.

in producing beef (or pork*) for exportation to foreign countries. Its immense natural pastures—the profusion and cheapness with which Indian
corn can be produced on its virgin soils—give it an advantage which increased transportation by no means counterbalances. The question then
arises—Why, for the same reasons, cannot the vast North-western plains
produce wool more cheaply than the South, and undersell her in our own
and the foreign markets? In the first place, the western pastures—that
is to say, the wild or natural ones—which produce beef so cheaply, are.
by reason of the coarseness and rankness of their verdure, *not* adapted to
the growing of sheep. Secondly, the shortness and mildness of the southern winter give a decided advantage in wool growing, by affording green
winter feed—an advantage not profitably available probably, on an extended scale, with large grass-feeding animals. Again, in the North-west,
though there is less snow, the winter is about as long, for all the practical
purposes of husbandry, as in New-York.† Killing frosts come as early in
autumn ; the naked ground is frozen as solidly, and far more deeply; and
verdure puts forth but little if any earlier in the spring. The South then
possesses the same great advantage with the North-west in the production
of wool—*cheap lands ;* and, superadded to this, she has the short, mild
winters, which give her a decided advantage over both the North and
North-west. She has a marked advantage over the Northern and Eastern
States in *both particulars,* and, instead of importing manufactured wools
from them, she ought to supply them, by export, with at least the raw material. And she will do this at no distant day, unless her sons are content,
in the great struggle and battle of industrial interests, to sacrifice their
own by apathy or irresolution.

---

* I have not alluded to the rearing of swine any more fully, as they are but partially a grazing animal.
—But if the position assumed in the text be correct, it is another argument in favor of devoting your lands
to the production of *surplus* wool, instead of *surplus* corn.
† The winter feeding of sheep in New York has already been stated to average about one hundred and
fifty days.

LETTER VII.

**PROFITS OF SHEEP HUSBANDRY IN THE SOUTHERN STATES.—3. BY GIV
ING TO SOUTHERN AGRICULTURE A MIXED AND CONVERTIBLE CHAR-
ACTER.—4. BY FURNISHING THE RAW MATERIAL FOR THE MANUFAC-
TURE OF DOMESTIC WOOLENS.**

Expediency of Rotation in Crops...Consequences of omitting it on Wheat Lands of New-York...Mr. Gay-
lord's views...Consequences in the Southern States...Mr. Roper's Report in the Legislature of South
Carolina—[Cotton Statistics of that State—Comparison with other States—General Agricultural Resources
—necessity of new staples]...Judge Seabrook's Report to the State Agricultural Society of South Carolina—[Agricultural Statistics—Remedies proposed for present "distress"]...Singular omission of Wool as
one of the proposed new Staples...Southern prejudice on this subject—Causes...Impropriety of the one-
crop system—Diminishes crops—Deteriorates land—Multiplies insects...Fertility sustained by Rotation—
Causes..." Resting"—Its inexpediency...Some of the Crops of every Rotation must be converted mainly
into Manure—Superior economy of converting them into Animal Manure—Sheep the most profitable animals for this purpose...Leading principles of a profitable Southern Rotation—Six-shift Course proposed—
Five-shift Course—Six-shift Course for poor soils...Col. Taylor's Four-shift Course—Objections...Com-
parative profit of growing Wool, Cotton and Rice, incidentally alluded to...Economy of producing the raw
material for the Manufacture of Domestic Woolens...Cost of Slave Cloths per head per year...Prices now
paid for these Cloths—Cost of manufacturing them—Data for estimating such cost...Great profits of Manufacturers in the Northern States—Their Dividends—Their method of exchanging Cloth for Wool—Working Wool at the halves...Cost of Cloths obtained by these methods...The South may obtain the same advantages—Natural Facilities—Cost of Machinery—On what terms worked—Operations...Cloths spun and
wove by hand cheaper than the imported ones—Cost of the several processes of manufacturing them—
Estimate of Cost per yard at the North...Cost of establishing Carding and Cloth-Dressing Machinery...
Home-made Fabrics diminishing at the North—Causes...Same Causes will not operate to so great an ex
tent at the South—Reasons...Probable Cost of Home-made Cloths, South.

———

Dear Sir: The third great benefit claimed by me among the profits of
sheep husbandry in the Southern States was, "its comparative efficacy in
giving to Southern Agriculture a mixed and convertible character, and
thereby sustaining (or improving) all the present good tillage lands, in the
place of continuing the "new and old field" system (tilling land until it is
worn out, then abandoning it and opening new lands), once so general,
and even now by far too prevalent."

The first object of mixed husbandry has been already stated—the home
supply of the various necessaries of life. Its second, and still more important one, is the preservation of existing fertility in all soils fit for tillage.—
It certainly requires no proof or argument to demonstrate the superior expediency of maintaining the fertility of soils, if it can be done, by a rotation of crops, even though each of these crops is not, separately considered,
the one which would yield the greatest immediate profit. In the language
of the hackneyed aphorism, it is never expedient to "kill the goose which
lays golden eggs."

This constant cropping with one plant was once extensively practiced
on the wheat lands of New-York, as many of their present owners can
bitterly attest. Even now there can be no doubt that, on nearly all of
them, wheat returns too often in the rotation. These lands were once
rapidly, and are still, I fear, slowly declining in value; while the grazing
lands of Southern New-York, where men have been *compelled* to be more
discreet, have been constantly improving and approximating to the former
in market value.*

———
* This calls to mind a letter which I received from an old and valued correspondent, the late Willis Gay
lord, but a short time prior to his death. I had spoken of the advantages of his own, the wheat region,
over the grazing region in which I reside. Mr. G. combated this idea. He thought capital invested here

The same system has prevailed on the rice, tobacco, and cotton lands of the South, and has, for a variety of reasons not necessary here to. be discussed, been, in the case of the two latter at least, more fatally persisted in. I have already alluded to the exhaustion of your soils consequent on this course of culture, but to show the wide extent of the evil—its pecuniary consequences individually, and on whole States—the now admitted necessity of a rotation of crops—the equally conceded necessity of introducing some *new staple*, or staples, to render the other crops in the rotation, besides cotton, rice, and tobacco, *remunerative*—and various other con siderations having a strong bearing on this whole question—I quote the following statements from *Southern*, as well as highly authoritative sources.

The Committee on Agriculture of the House of Representatives of South Carolina, through their Chairman, Hon. R. W. Roper, made a Report to that body, Dec. 14, 1842, from which the following are extracts :

" Let us now turn our consideration to one other great staple, cotton, of which the statis tics are so exact that we can ascertain by calculation what our prospects are as regards competition in that article. The United States produce at present 578,012,473 lbs.—more than one-half the crop of the whole world. South Carolina grows of this 43,927,171 lbs., or 1-13 part of the quantity ; but from this source of profit her palmy days are past. Every year opens new lands in the West, where congeniality of soil and climate to this commodity in creases the product per acre far beyond what can be reared at home, and consequently re duces the value infinitely below the costly prices which formerly enriched Carolina. These new lands produce, on an average, 2,500 lbs. of cotton per hand, while the lands in Carolina yield but 1,200 lbs., and the expenses of a laborer being about equal in either place, reduces the Carolina cotton to half its intrinsic value. We have also the declaration of Mr. Dixon H. Lewis, in a recent speech in Congress, that cotton, divested of Government embarrassments, might be grown in Alabama for three cents a pound.

" Your Committee will avail itself of the lucid calculations of a distinguished and talented a dividual,* to present another view of the subject, startling in its details, and bearing strong 'y on the propriety of summing up all our resources. The crop of the world amounts to 1,000,000,000 lbs., which would require, at the rate of 250 lbs. per acre, 4,000,000 of acres to grow this quantity. Now, the four States bordering on the coast of the Gulf of Mexico— viz., Louisiana, Mississippi, Alabama and Florida—contain 130,000,000 of acres; proving hat, if only one acre in 32 were found capable of producing 250 lbs. to the acre, these four States could, alone, supply the demand of all the markets in the world. In this calculation, the produce of Georgia, South Carolina, North Carolina and Virginia, with portions of other States, besides 150,000,000 acres in Texas, are entirely excluded. The lands of the Gulf States, therefore, and Texas, are sufficient to supply the demands of the world in all time to come. Where, then, is the hope or prospect of South Carolina in the competition ? . .

" South Carolina comprises within her borders 16,000,000 acres of land, of which only 1,300,000 are cultivated. Of this, cotton occupies 175,700 acres ; rice, 80,000 ; Indian corn 500,000 ; potatoes, 22,612 ; wheat, 24,079—making an aggregate of about 800,000 acres ; the balance of 500,000 are taken up in oats, rye, barley, hay, tobacco, and a limited portion of other articles necessary to the supplies of life. To what use, then, is the balance of our ter

returned quite as good or better profits, than on the wheat lands. He thought, taken as a whole, the graz ing farmers were doing better than the wheat farmers. The latter. though ostensibly making an equal and frequently better per centage, were *wasting their capital.* The grazing lands and the wheat lands were rap idly approaching each other in market value, by the rise of the former and the deterioration of the latter May this not afford a parallel to what will one day be witnessed in the Southern States ?

It is difficult for me to pass by the name of this accomplished writer—this pure, upright and philanthropic man—without throwing one stone on the cairn of his well-merited fame. He felt himself, from his infancy, cut off from the companionship of his kind, by disease and deformity ; but, notwithstanding the body was " ugly," he " carried a precious jewel in his head." Triumphing over constant physical sufferings which would have prostrated most men, he made attainments in general knowledge possessed by few of his con temporaries. His range of reading and study was remarkable. In his beautiful and sparkling letters to me, every subject and almost every science is touched upon by him in a manner that shows that he at least had mastered their general principles ; and, in the *abandon* of private intercourse, they seem to have been to him as the flowerets of a garden, among which his spirit could roam with that playful and joyous activ ey which was denied to his poor, frail body, among the objects of the outer and physical world.

Freely, unassumingly, and without an aspiration but for the good of his fellow men, his mind poured out its stores on a variety of topics in the publications of the day. Fortunately, he gave his principal attention to the subject of Agriculture, and, if not a *discoverer* (which he never claimed to be), he investigated and collated with an industriousness of research, discrimination and perspicacity, which brought the truth from all the different sources where discovery or experience had left its *disjecta membra*, into essays, so well com pacted, so clearly arranged, that men of the most ordinary parts could not only understand his separate sen tences and positions, but their connection and aggregate bearing, and thus master the whole subject. Peace to his ashes !

* Gov Hammond

ritory, of 14,000,000 of acres, to be appropriated? Are we forever to be supplied with stock from the West, bread-stuffs from the Middle States, and manufactures from the North? Is all that we can realize from our labor to be expended abroad? Nothing to be left for our own improvements or our luxury? As one means of correcting this evil, your Committee propose an Agricultural Survey of the State, to determine our natural advantages, develop our facilities of improvement, exhibit our profits and expenditures, and awaken our citizens to the importance of *vying with the rest of the human family in all the improvements of which our location is susceptible*

"The exposition which your Committee has given, showing the great competition of foreign rice with our own, and that South Carolina cannot compete with the West in the cheap production of cotton, and that she must, ere long, be driven from the market, demonstrates the necessity of looking abroad and around us *for other sources of advancement and profits than those we possess.* . . .

"We cannot expect that *accident* is continually to supply new staples suited to our soil and climate, and place us beyond the reach of contingent circumstances. We must resort to science to improve our Agriculture, and to machinery to enlarge and prepare present articles of culture, *or transplant and acclimate new products*, which will again, like those we have lost and will lose, lead off-for a period in the employment of capital, amassing of wealth and diffusion of human happiness."

The House and Senate agreed with the Report, the same day, and its principal recommendation, an Agricultural Survey of the State, was adopted.

The Committee appointed by the South Carolina State Agricultural Society to consider the scheme of Col. Davie to reduce the quantity of cotton grown, made a Report, through their Chairman, Judge Seabrook, at the winter meeting of the Society, 1845–6, from which the following are extracts :*

"Another cause of our distress is that, in a large portion of the southern country, cotton is cultivated, when its production does not now, and never can, at all compensate the planter for the labor bestowed. *There it is desirable for every one that other branches of industry should be pursued.* . . . We do not intend to encourage the cultivation of cotton to the neglect of the other products necessary to support or comfort. Every planter *should promptly render himself independent in reference to those articles which could be produced on his plantation.* In this way he would profitably curtail the quantity of land devoted to the cotton crop. An abandonment of the present extremely defective mode of culture, and the substitution of a better, would insure a larger quantity of cotton than would be lost *by diversifying the products of industry.* In other words, his cotton crop would be larger; his corn, wheat, rice, oats, barley, horses, mules, hogs, cattle, sheep, butter and vegetables, would be the produce of his farm.

"If, however, the cotton crop is to be given up one-half, after all the reductions of it which we have sanctioned, to what else can the planter of the South so profitably turn his attention? To grain? He already, in ordinary years, produces twice as much as the Middle States, and about one-eighth more than. the West. In Indian corn alone, the produce of the South, by her last census, was 300 million bushels. If the planter of cotton is engaged in an unprofitable business, much more is the grain raised. . . . *Millions of acres in South Carolina, including the lower country, are admirably adapted to the raising of rich grasses.* This might be added as another branch of industry, from which reasonable profits might be realized, and might very well be added to the cotton planter's income. The business of tanning and the manufactures of leather might be and ought to be enlarged. In this State, all the means of a successful pursuit of this branch of industry are at hand and within the reach of every one. Hides, lime, bark and mechanics (slaves) are abundant."

The remarks in both of the above extracts, though made exclusively in reference to South Carolina, will apply equally well, in many obvious particulars, to all the old cotton and tobacco growing States.

To a Northern man, accustomed from his childhood to see sheep husbandry blended, to a greater or less extent, in the operations of nearly *every farm*, and to live among farmers who regard it just as indispensable, and as much a matter of course, as the production of bread-stuffs, it seems singular enough that neither of the above able Committees, in looking for

---

* As has been before stated, the other members of the Committee were Judge O'Neall and W. J. Allston, Esq. Mr. A. did not concur with his colleagues in the proposition that there was not already an absolute over-production of cotton. He believed there was. In all other particulars, and consequently in all embraced in the extracts given, he concurred in the Report.

"other sources of advancement"—"new products"—"other branches of industry"—both to bring into use millions of acres of unproductive territory "admirably adapted to the raising of rich grasses," and to render profitable and preserve the fertility of the tillage lands of the State, should not have thought of wool growing—or only thought of it, as it were, incidentally—at the very heel of a catalogue of farm products, and in reference solely to supplying the home want!

Indeed, the estimate which has been set upon sheep husbandry generally, and by all classes of agriculturists, South, is a source of unmixed surprise to one acquainted with this pursuit, and with the resources of that region for sustaining it. There appears among many, if I may credit your own writers,* to be even a *prejudice* against *sheep* and sheep husbandry, *per se!* Is this because these animals bear a staple, and give employment to manufactories, which have claimed the "protection" of Government, to the prejudice, in the opinion of Southern politicians, of Southern interests?† Is any portion of it due to the scornful denunciations of the brilliant, but eccentric and cynical, statesman of Roanoke, who "would at any time go out of his way to kick a sheep"? Or is it owing to the, in most respects, justly popular writings of Col. Taylor, of Virginia? Hon. Andrew Stevenson, of the same State, in a letter to John S. Skinner, Esq., says:‡

"The prejudice which the late Col. John Taylor, of Caroline (who, by-the-by, did more for Agriculture than any man in America), had against sheep, has been the means of rendering this description of stock unpopular in many parts of the southern country. . . . If this distinguished patriot and statesman had lived at this day, he would have changed his opinion."

The impropriety and inexpediency of giving all the labor and prime land of the country to the exclusive cultivation of one or two crops, even leaving the deterioration of the lands, consequent on such a course, out of the question, is forcibly set forth in the Reports above quoted from. But that deterioration is an infinitely more fatal evil, both to individuals and States. An injudicious course of cropping can be easily changed; but, if the land is entirely impoverished, the change comes too late, until labor and capital have been employed on its restoration. The tendency, nay, the absolute connection as *cause* and *effect*, between the one-crop system and such deterioration, has been proved by too sad an experience at the South—is too universally recognized and conceded—to find a single questioner who possesses ordinary intelligence. Whether the consequent phenomena are solved by the excretionary theory of De Candolle, or the more ordinary one of the exhaustion of some of those substances which constitute the necessary food of plants, the facts presented are the same.‖ The soil yields constantly diminishing crops, until it becomes incapable of producing more than scattering and feeble plants; and the insect enemies of the latter, which would perish if deprived of their aliment by the substitution of some other plants, multiply in a constantly ascending rati o.§

* Hon. Andrew Stevenson, John S. Skinner. *et al*, in Monthly Journal of Agriculture, &c.
† If such protection has prejudiced the South, what stronger reason why she should remunerate herself by appropriating a share of it!
‡ Monthly Journal of Agriculture, July, 1845.
‖ The theory of M. De Candolle, apparently so strongly supported by the experiments of M. Macaire, has found many believers. But the statements of the latter have been contradicted by M. Braconnet, M. Mirbel, and finally are totally overthrown, in my judgment, by the experiments and investigations of Mr. Alfred Gyde, of Scotland. Mr. Gyde shows that the minute excretions of plants have the same composition with their *sap*; and he also watered plants with a solution of their excretions, not only without injury, but to their manifest benefit! For Mr. Gyde's able Prize Essay on this subject, see the Transactions of the Highland and Agricultural Society of Scotland (March, 1846). I am not aware that this essay has been republished in our country. It certainly should be.
§ Of the latter evil, the past year furnished a pregnant example. I saw it stated last winter, in the South Carolinian (published at Columbia, S. C.), on the authority of an United States Senator, that the falling off in the cotton crop would be enormous, by reason of the depredation of *worms*. This evil is constantly increasing, and must continue to, while the planter continues to provide aliment for each succeeding horde of destroyers, by continuing on the soil the plants on which they prey.

L

Experience has shown that if vegetables of different classes are made to follow each other, the soil will much longer retain its productiveness.— Even when "exhausted" of some one or more of those ingredients necessary for the healthy production of a particular plant, it is found to produce others luxuriantly which do not require the lacking ingredients, or but very minute portions of them And, by a most beautiful arrangement of physical causes and effects, when a plant is removed from the soil, and notwithstanding its place is occupied by others, a process of restoration at once commences to replace all that the absent plant has appropriated, and to prepare the kindly bosom of the earth again for its reception. Nature herself, in ministering to this beneficent end, becomes a great laboratory ; and in her most ordinary, as well as her most unusual operations, she is constantly producing those chemical changes, and furnishing those chemical ingredients, which restore what has been abstracted by man's cupidity, or lost by his improvidence. The gentle rain brings down ammonia and carbon to plants. The frost rives the solid rocks, to disengage their fertilizing constituents. The sun, in his flaming path, looks down not only to warm and give us light, but to perform functions in the vegetable economy without which all herbage, except a few miserable fungi, would perish; and to all he imparts their varied and beautiful coloring. The thunder which shakes the walls of cities, and strikes man with awe, brings to our aid one of the most efficient promoters of vegetation. Even the bursting volcano converts its fiery crater into a crucible and retort, and gives off that gas which forms so large a portion of all the vegetable and animal productions of the globe : and the wild winds, which strand navies in their course, equally diffuse it over the earth.

It follows from the above positions that naturally good lands* which are more or less exhausted will be gradually resuscitated by "rest," or an entire exemption from tillage ; and hence the absurd idea that lands require physical "rest," in the same sense in which the tired animal muscle requires it, after continuous exertion. But, apart from the theory, the practice of "resting" lands is inexpedient, for the following reasons: If a plant is not continued on a soil until it consumes any of those inorganic constituents necessary to its production—if, on the other hand, it is succeeded by a plant which makes its heaviest drafts on those inorganic substances which its predecessor required the least of, and vice versa—the natural recuperative process above adverted to, aided by means which lose to us none of the value of the crops, will repair the waste made by each plant, before it again occupies the soil, in a judicious rotation. Hence, by a rotation of crops, fertility can be indefinitely sustained, and the earth each year return its increase. Thus the ends of "rest" are attained, without its great and unprofitable sacrifices.

To sustain the fertility of the soil, some portion of the crops of every rotation must be converted into manure. These are the "aiding means" above alluded to. They may be converted into green or animal manure If the former, the whole crop is plowed under. If the latter, the crop is first partly converted into animal manure, by animals depastured on it, and then this animal manure, with the remaining vegetation, is plowed under. The last is always the most economical method, on good lands,† because the crop is worth almost as much for manure, after passing through the

* I say "naturally good lands," for those entirely deficient in several of the necessary constituents of a sterile soil might require ages of rest to obtain these constituents—if, indeed, they ever would, by merely natural causes.
† I have limited the assertion to "good lands," because a crop of green manure, turned under at the proper stage of its growth, will undoubtedly make rather more manure than in any other way ; and it may be expedient many times to give poor lands all. This is especially true in the reclamation of barren lands

bodies of animals, as it would be turned under green; and then we have all the profit made on or by the animals—meat, wool, &c.—without any additional cost. Sheep, being the best manurers, and otherwise the most profitable animals, will (with enough other animals to supply all the home demand for the necessaries furnished by them) best sustain a profitable rotation.

Here, perhaps, the discussion of this topic in connection with the subject matter of these letters should terminate; but I am unwilling to abandon it, without making a few practical suggestions as to the rotation which would be found most profitable at the South—more particularly on the valuable cotton lands, which are suffering most for the want of it. It is manifestly impossible to lay down any rule or rules on this subject, which can or should be rigidly acted upon, in all instances. Leading principles can only be declared, and, if correct, the intelligent man can always vary their application so as to meet the exigencies of his particular case.

First, I should consider it indispensable on all cotton (or tobacco) lands,* under all circumstances, to keep at least one-third of them in pasturage, to insure the proper amount of manure, over and above cotton seed, and such occasional supplies of swamp mud and marl as might be obtained at spare intervals—and all other incidental manures. Another third, 1 believe, should be generally devoted to grain for bread stuffs, for fattening the necessary amount of bacon, and for the winter forage of horses, mules, swine, &c. Unless the horses and mules, and, perhaps I should add, the cows, were wintered entirely, or in great part, on grain and the offal of the grain crops, one-third of the *cultivated land* in grass, would not support animals enough to produce the manure requisite for two-thirds in cotton and grain. But in making the above division, I spoke only of the arable lands fit for the growth of cotton. Most plantations have poor, or swampy or rough lands, which would most profitably be kept permanently in grass, and these would supply the deficit. The remaining third of the arable lands might be devoted to cotton, or, in the tobacco region, to tobacco.

By the course above proposed, the cotton (or tobacco) and wool would be made the salable products. The grain, grass, dairy products, bacon, &c., would be consumed on the plantation. This is as it should be. European famine has given a stir to the latter products this year, (and it may for a year more,) in the Southern markets; but with the ordinary European demand, the old Southern Atlantic States cannot, as we have seen, compete at a profit with these commodities, which debouch through the Mississippi, the St. Lawrence, and the northern canals. With the *two wools*, as they are sometimes called, the "vegetable and animal," these States can undoubtedly sustain themselves against the pressure of any outward competition.

Such a division of crops as the one above proposed, could be effected by a six-course system of rotation. Let us suppose the land of the plantation fit to grow corn and cotton, divided into six equal fields. I then propose the following rotation:

1st year, Grass depastured.	1st year, Grass depastured.	1st year, Cotton.
2d .. do. do.	2d .. Cotton.	2d .. Cotton with yard manure, &c.
3d .. Cotton.	3d .. Cotton with yard manure, &c.	
4th .. Cotton with yard manure, &c.	4th .. Corn with peas.	3d .. Corn with peas.
5th .. Corn with peas.	5th .. Small grains with grass seed.	4th .. Small grains with grass seed.
6th .. Small grains with grass seed.	6th .. Grass depastured.	5th .. Grass depastured.
		6th .. do. do.

* I have not included the rice lands, because being deep beds of alluvial deposits, composed in a great measure of organic matter, and being susceptible of irrigation, they will not wear out like ordinary soils, and stand less in need of rotation in their crops.

1st year, Cotton with yard ma-		1st year, Corn with peas.		1st year, Small grains with grass
nure, &c.	2d .. Small grains with grass			seed.
2d .. Corn with peas.		seeds.		2d .. Grass depastured.
3d .. Small grains with grass	3d .. Grass depastured.			3d .. do. do.
seed	4th .. do. do.			4th .. Cotton.
4th .. Grass depastured.	5th .. Cotton.			5th .. Cotton with yard ma-
5th .. do. do.	6th .. Cotton with yard ma-			nure, &c.
6th .. Cotton.	nure, &c.			6th .. Corn with peas.

Supposing each of these fields to contain 50 acres, this would give 100 acres of grass, 100 of cotton, and 100 of grain (50 of corn and 50 of small grains) annually.

By this course all the hauled* manure, each year, would be given to one-sixth of the land, and consequently the same field would not receive it but once in six years—yet *every crop* would be *adequately manured.* The first cotton crop would receive an ample amount from the grass roots and the droppings of animals for two years; the second, from the hauled manure; the corn, from the manure left by the previous crop, and, if needed, by a small amount of cotton seed, ashes, (or some other mineral fertilizer,) in the hill; the small grain crop would be amply manured by the peas sown with the preceding corn ; and the land would go back into grass in excellent " heart," and, if the previous tillage was what it should be, entirely free from weeds. The corn might intervene between the two cotton crops, and thus remove the objection which exists against taking two crops of the same kind in succession. But I placed cotton 4th, be-cause there should come a manured crop at this period of the rotation, and I thought it better to give the manure to the more valuable crop, and be-cause cotton, as the 5th crop, would not admit of the cultivation of the pea, to provide manure for the small grain succeeding. The rotation might be thus varied, however, if circumstances should seem to render it desirable.

I have put down no meadow in the rotation on the *arable* lands. But I believe the growth of hay to a certain extent, not only to supply any or-dinary deficiency in winter feed beyond the quantity furnished by the usual sources—but to guard against contingencies, would be good econo-my in all cases. All farm animals must be well wintered, to give a prof-itable return in summer ; and those occasional scarcities of fodder always liable to overtake the farmer, should be providently guarded against. It is never considered poor economy, in the North, to have a few tons of hay even to summer over. The necessary meadows for the plantation might be made on some of the less arable lands before referred to—and, when the tillage lands are in an uncommonly fertile state and pasturage plenty, it *would do* to mow one of the grass crops (the second one) of the above rotation, though, if avoidable, I should think the other course entirely pref-erable.

On poorer lands—the poorest class which can be profitably devoted to cotton growing—I would propose a five-shift course, as follows :

1st year, Grass depastured.   3d year, Cotton.
2d .. do. do.   4th .. Corn with peas.
    5th year, Small grains with grass seed.

The manure to be given to the third or fourth crop, according to circum stances, or divided between them.

On lands of a still inferior grade, but which it may be expedien to plow at intervals, I would propose the following :

1st year Grass depastured. 4th year. Grass depastured (or mown.)
2d .. do. do.   5th .. Corn with peas.
3d .. do. do.   6th .. Small grains with grass seed.

* I mean by this, the manure from *every* source which is carted upon the land in quantity, as contradis-tinguished from that which is dropped there by animals made by plowing under vegetables, or carried on in small quantities to drop in the hill, &c.

The number of years depastured to depend upon fertility—the poorer the land, the longer it should be kept in pasture.

The following is the rotation which was introduced by Col. Taylor, north of the cotton-growing region:

1st year, Corn.	3d year, Clover (and weeds) not mown nor
2d ‥ Wheat and clover sown—if too	grazed.
poor for wheat, left at rest and not grazed.	4th ‥ Clover not mown nor grazed.

Of this, Mr. John J. Thomas, one of the Editors of the Albany Culti vator, very justly remarks :

"It was materially opposed to the principles of good husbandry in several respects. It furnished vegetable manure only to the land. A large portion of the value of this vegetable growth was lost, by dissipation into the air, during its decay. The returns from the land were necessarily small, as only two years out of four produced crops for harvesting. And it greatly increased the labors of tillage, by the increase of noxious weeds."

Had this clover been fed off by sheep, a portion of the above objections would be inapplicable, and there would be no danger of the corn leaving the soil too impoverished for wheat, particularly if peas were sown with the former, to be plowed under. A crop of *weeds* is, of all others, the most to be avoided, as the seeds deposited by it will continue to sprout for years with the subsequent tillage crops, rendering them foul and difficult of cultivation.

I may be in a profound error, but I cannot but believe, after carefully studying Southern Agriculture, and the circumstances which invest it, that by adopting the six-shift system of rotation above recommended, or something analogous to it, on the cotton lands, the desideratum expressed in Judge Seabrook's Report will be attained. More cotton will ultimately, if not even now, be produced from less land : the other necessaries of life will become mainly the product of the plantation ; a new staple will be introduced to employ the surplus capital, as profitable at least in its acreable products as cotton, and tending to the constant reparation, as cotton tends to the constant waste of the fertility of the land.

I will not tire you, Sir, with a comparison between the relative profits of wool and cotton growing. On looking over the answers of Southern gentlemen to Mr. Walker's Treasury Circular, (1845,) I find that the stated profits on cotton in the Atlantic and Gulf States, west of Louisiana, range from 1 to 8 per cent. on capital invested—the average of all the statements being about $4\frac{1}{2}$ per cent. !

I may remark incidentally that in your own able replies to that Circular, you set down the profits of rice growing between 1842 and 1845, at $7\frac{1}{2}$ per cent.; for the ten preceding years, at "about 8 per cent."

A reference to Letter V. will show you how these profits compare with those of wool-growing. Admitting the accuracy of the data therein given, there is no *very great* difference in the cost of growing a pound of wool and a pound of cotton !

We come now to the *fourth* point of view in which we are to regard the profits of sheep husbandry in the Southern States—"whether independent of preceding considerations, and even if the staples furnished by sheep husbandry proved no more profitable, in direct returns on capital invested, than some of the present staples, it would not be better economy, on the whole, for the South to produce the raw material and manufacture domestic woolens, particularly for the apparel and bedding of slaves, than to be dependent for them on England and Massachusetts ? "

The woolen apparel and bedding of slaves, when no part of it is manu factured on the plantation, costs about $6 per head per annum. The blankets imported from England weigh about $4\frac{1}{2}$ lbs. and cost a little over

$3. The Welsh plains, imported from England, weigh usually not far from 13 ounces per yard, and cost from 65 to 70 cents; and the Chelmsfords, a heavy, coarse article, from Massachusetts, from 50 to 58 cents.

Now what is the cost of manufacturing (including wool and every other expense,) cloth of the same amount of stock, and better quality, than Welsh plains? To the present weight of the cloth per yard add one-third, and you have the weight of the wool in the fleece—as bought of the farmer.* If, then, the Welsh plains weigh 13 ounces per yard, they required 17½ ounces of fleece-wool as stock. Wool of the quality worked into "plain cloth" or "sheep's gray," in this State, (New-York,) many shades better in quality than the stock of Welsh plains, has averaged from June to December, 1846, from, say, 20 to 22 cents a pound†—or, if pulled from the pelts of slaughtered sheep, as is the case with large quantities of it worked into these cloths, it did not, during the same period, stand the purchaser-in to exceed 18 cents per pound. Assume the average to be 21 cents per pound, and the stock of a yard of these cloths (17½ ounces) would cost 22¾ cents. You are familiar with the character of the "sheep's grays" of New-York. They are worn almost universally by our farmers. Of the twenty-five thousand men you saw at the State Fair at Rochester, at least three-fourths of them ordinarily wear this quality of cloth for pantaloons, and say one-half of them for coats. Its ordinary weight is from that of the Welsh plain to 16 ounces per yard, and its style and expense of manufacture are superior to those of the former. It can be manufactured, including use of machinery, &c., and every process after the wool is received in the fleece, to fitting it for market, for *eleven cents per yard !* A merchant of this State owns a manufactory, employing say $25,000 or $30,000 of capital, which turns off from 500 to 600 yards of cloth per diem—the fleece-wool being converted into finished cloth in eight days. His whole expenses, including use of manufactory, averages, *according to his own statements*, not to exceed the above named price per yard. Add this sum to the cost of wool, and cloths containing an equal quantity and quality of stock with Welsh plains would cost 33¾ cents per yard ; and you therefore pay for this class of cloths about *one hundred per cent.* beyond the first cost, for transportation, duties, and manufacturer's profits. The latter, of course, absorbs most of the immense sum thus paid, or rather *thrown* away, annually by the Southern States. The Chelmsfords, and various other woolen goods imported by you, are probably manufactured at nearly equal profits.

Is it singular, then, that "*acres* of woolen manufactories" are now in the process of erection in the North? or that existing establishments are *declaring* dividends of from ten to fifteen per cent.?‡

But I have not done with the data of manufacturing. The manufacturer above alluded to has, to my certain knowledge, exchanged "sheep's grays" requiring a pound of stock per yard, for wool of the same quality as the stock, giving a yard of cloth for 1¾ lbs. of wool. Calling this wool

---

* After being washed in the ordinary manner on the back of the sheep.
† Wool has risen since December.
‡ I did contemplate an enumeration of the new woolen manufactories now building, or in contempla tion, within my knowledge, in this State and New-England , but will mention but a few of the most im portant ones. The Bay State Mills, now in process of erection in the new city of Lawrence, Mass., will work up 2,000,000 lbs. of wool per annum. One of the mills, 200 feet long and six stories high, will go into operation this summer. The machine-shop, wool-house, etc., (the mere offices,) will be, including wings, *thirteen hundred feet* in length, and three stories high. Their very *sewer* will cost $25,000 ! A splendid steam mill has just gone into operation in Utica, in this State, which will work up 1,000,000 lbs. of wool per annum. Another of the same size is in contemplation, in Utica; another in Syracuse; another in Auburn, &c. ! There never was a time when American manufactures stood on a firmer basis, or were making bet ter profits *with a prospect of having them continuous.* This is conceded by the *ablest of the manufacturers themselves* as I shall, in the proper place, show.

21 cents per pound, the cloth would thus cost the purchaser $36\frac{3}{4}$ cents per yard.

Any of the manufactories doing custom-work will manufacture these goods "at the halves," so that a yard requiring a pound of stock would cost two pounds of wool, or 42 cents. That as heavy as Welsh plains would thus cost $45\frac{1}{2}$ cents, it being from $19\frac{1}{2}$ to $24\frac{1}{2}$ cents per yard *less than you now pay.* Yet here the manufacturer of *custom-work* admits the suffi ciency of the profit, by asking no more.

Blankets are of still coarser wool, having the appearance of Smyrna, or inferior South American. They are not "sheared,"* which diminishes the waste. Neither do they need dyeing matter. But independent of these considerations, calling cost of stock per pound, and the waste from all causes the same, 6 lbs. of fleece-wool would make a blanket. To the wool costing 21 cents a pound add 11 cents per pound (of the stock) for manufacturing, and the actual cost of the blanket is $1 92. Have them manufactured by the halves, and they would cost you 12 lbs. of wool each, or $2 52.

I have in the previous estimates, based my calculations on the *market price* of the lower quality of medium wools.† But there is another and a most important view of the subject. It has already been shown that the South can produce wool, to any desirable extent, at a sum not exceeding 8 cents per pound—and, in favored localities, at a much lower rate. By the exchanging system (wool for cloth) you would get a yard of cloth equaling the Welsh plain in stock, and superior in quality, for 2 lbs. $2\frac{3}{4}$ oz. of wool, costing the producer just $17\frac{1}{4}$ cents! A blanket weighing $4\frac{1}{2}$ lbs. would be obtained for 12 lbs. of wool, costing 96 cents!

Does this sound a little like dreaming, Sir? I ask you to carefully examine the premises, and see if there is any escaping from these con clusions?

Will the South continue to slumber on, thus throwing away the fruits of her industry? Do you tell me that her people know nothing about manufacturing, and have no taste for it? The necessary knowledge is as readily acquired by a Southern as a Northern man; and when that is ob. tained, and there is a prospect of *profit* ahead, the *taste* will not long be wanting! You have the capital: you have natural facilities to an un- bounded extent both to propel the machinery and produce the staple. What more do you want? What more can you ask? A joint stock asso- ciation of planters, at any suitable point, might cause a manufactory to be erected worth say $25,000, under the direction of a skillful and experi- enced machinist. This would turn off, say, 500 yards of cloth per diem. If the machinery was in all respects good, and the water-power sufficient and unfailing, a competent and responsible Northern manufacturer could be obtained (if desired), to take the establishment, furnishing hands, &c., and work the wool furnished him into cloth of the kind before described— containing about the same stock with Welsh plains, and fitting it for mar

---

* After a sufficient number of fibres have been torn up from the threads by the teazles or cards of the "gig-mill" to form a sufficiently thick nap on the surface, these fibres are cropped or "sheared" by a ma- chine for that purpose; and in superfine cloths the process is several times repeated, each time cutting off an additional portion of fibre, which is called "flocks." A dishonest custom now prevails among some manufacturers of working these flocks again into the body of the cloth to give them weight, denseness, and apparent firmness. By this means the gigging and shearing process can be continued on thinnish cloths un- til a beautiful surface is obtained, without the additional thinness and lightness consequent thereon being apparent to any but an experienced eye. Sheep's grays and other coarse cloths are gigged and sheared but slightly. In some manufactories the former process is altogether omitted, and the cloth is simply "brushed" prior to shearing. Such cloths are stronger, but do not look as well.

† Say of the quality of common South-Down and Native and Long wools, with a sufficient dash of Me- rino blood in the last to make them carding-wools, and to bring them to about the same fineness with the first named

ket, for eight or nine cents a yard.* I know of a manufacturer, at no great distance from me, who thus takes a manufactory worth perhaps $8,000 or $10,000, and furnishes the cloth (of the above stamp,) fitted for market, for nine cents a yard, the owner furnishing the wool, the use of the manufactory, and the dyeing matter.† The supply of water at this establishment fails during two or three months each year; and one competent to judge informs me that seven cents would be better pay per yard, if the machinery could be kept in motion the year round. It is probable that it would cost rather more at the South to provide the necessary fixtures, obtain machinery, etc.; and it would also cost more, for a period, to carry on manufacturing, from the greater difficulty of obtaining operatives in case of losing any of those attached to the establishment. All these disadvantages, however, not of much importance at the first, will soon disappear. Slaves should, as rapidly as the nature of the case admits of, be converted into operatives, and when the number becomes once adequate to the end, it might be indefinitely multiplied, without those embarrassments which so commonly attend the attempt to mingle white and black labor.

It is cheaper to manufacture by hand,‡ (with the exception of carding, fulling, and dressing,) than to purchase your slave cloths at present prices, *if slave costs no more than free labor.*

On the average, 15 knots of *warp*, and 15 of *filling*, make one yard of flannel about 5 quarters wide. The ordinary shrinkage of this, in fulling it into cloth, is one quarter in length and width. It would therefore require 40 knots to make a yard of fulled cloth. The carding here in small parcels costs 3 cents per pound, and 18¾ cents per pound for fulling, dyeing and dressing. In considerable quantities, the carding can be hired done for 2 cents per pound, and the other processes for one shilling per yard. Spinning (by considerable quantities and for " cash-pay,"‖) can be hired done for 7 cents a *run* (20 knots) for warp, and 5 cents for filling—averaging 6 cents for both. Weaving can be hired done for 6 cents per yard (of flannel), which brings it, in the dressed cloth, to 8 cents per yard. The account would then stand thus:

	Small parcels.	Large parcels.
1 lb. of wool	21 cents.	21 cents.
Carding same...	3 "	2 "
Spinning	14 "	12 "
Weaving	10 "	8 "
Dyeing, fulling and dressing..........................	18¾ "	12½ "
Total...	66¾ cents.	55¼ cents.

Making 55½ cents the price of a yard of domestic cloth, estimating the wool at market price : estimating the latter at cost of production (8 cents) the price of the finished cloth would be 42¼ cents per yard, and it is a better article for wear than either the Welsh plains or Chelmsfords.§

* I have no doubt it could be done at a fair profit in the North for 7 cents per yard. I am understood, of course, to mean that the manufacturer pays no rent, insurance, nor for repairs. The stockholders furnish the wool, which is worked up by the former, at the stipulated price.
† Modern ingenuity has reduced the expense of this to a mere trifle. Most of the "sheep's grays," you have observed, are of ferruginous hue. Those of this color are dyed principally by *tan bark*—the bark of the hemlock (*Abies canadensis*), which is sold here at $1 75 to $2 a cord !
‡ I am aware that to "manufacture" *is* to *make by hand*, but I use the word in its popular and more general signification. It would have been better to have compounded a word from the Latin *machina* and *facio* (manchinfacture?) to signify *made by machinery*, and thus expressed the two ideas by properly derived and definitive words.
‖ This word "cash-pay" is one of mighty import in the regulation of prices in the interior, where a very general (but now decreasing) system of barter prevails, and under which Wealth too often dictates to Want *what* it shall receive for its labor, and also prescribes the *price* of the commodities in which it pays.
§ Home-made fabrics are usually stronger and wear better than those made by machinery, for, in other words, *manufactured* cloths outwear *machinfactured* ones!) but this is not *necessarily* so. The several processes *can* be done undoubtedly, and probably, generally *are* more perfectly by machinery than by hand. But in machine-made cloths the yarn is commonly spun finer, so there is less stock in a yard. And they are submitted to processes, described in a previous Note, which further impair their strength.

$1,500 will set up a carding and cloth-dressing factory, which, with three good hands, will turn off 50 yards of cloth per diem. By Table I. it appears that in 1839 there were but 114 of these factories south of the Potomac and west of the Mississippi, doing an annual business of $320,-938, while in the single State of New-York there were 323 factories, doing an annual business of $3,537,337 ! Of the 114 Southern factories 66 were in the States of Kentucky and Tennessee; 41 in Virginia; 3 in each o. the Carolinas; 1 in Georgia, and in the remaining four, *none* !

The number is decreasing in New-York, as manufactories of the com-mon fabrics, worn by farmers and other laboring men, are increasing in every direction—many of them doing custom-work either at the halves, or at a fixed sum per yard—and all of them exchanging cloth for wool. By either of these methods, the cloth can be obtained as cheaply, perhaps cheaper, than to manufacture it in families. But circumstanced as you are at the South, you can, as before asserted, manufacture more cheaply by hand (excepting carding, fulling and dressing), than to import your slave cloths at present prices, if provided with factories to perform the ex-cepted processes. Where the institution of slavery exists, and where spinning, weaving, etc., can be done in those intervals of bad weather when the time of laborers would otherwise be entirely thrown away, it is doubtful whether any extension of even the coarse cloth manufactories would, or ought to, in an economical point of view, banish the home-made article. If we count the slave labor thus *saved* one-half the value of free labor, and dispense with the fulling and dressing* (which we usually dis-pensed with in manufacturing *domestic* slave cloths, in the interior of the Carolinas, Georgia, etc.), the cloth would cost but 20 cents a yard, and the dyeing might carry it to 22 cents. Let one-half the fabric be made of cot-ton, and the cost would be still farther reduced.†

Since the above was written, I have received the samples of Welsh plains, Chelmsford plains, and slave blankets forwarded by you. None of these goods *exceed* in quantity the estimate I have put upon them in my preceding remarks.

The Welsh plain which you state cost 65 cents per yard by the piece, (32 inches wide,) is about the thickness of rather heavy—but not the heaviest—sheep's gray. It is not, however, by many shades, so close and firm a cloth, for the want of equal fulling; and perhaps even this would not give it equal firmness, by reason of the *loose twist* of the yarn. The yarn is considerably coarser, (larger in diameter,) than that ordinarily em-ployed in sheep's gray—but it derives no inconsiderable portion of its bulk (which gives the cloth its thickness) from the loose and imperfect man ner in which it was twisted in spinning. This is particularly the case with the *filling*, which you can scarcely detach from even so open a web, without its breaking in pieces. Accordingly, the cloth tears very easily *lengthwise*, for that presenting such an *apparent* amount of stock.

With a sufficient amount of fulling, dyeing, (it is white,) and a little gigging and shearing—or simply brushing—it would become identical in

* But still you want carding-machines, to card the wool; for, by hand, it is a slow and expensive process.
† I was shown a new article of satinets a day or two since. It was double or broadcloth width, black, and the cotton warp dyed black, and could only be distinguished from a very fair piece of black broad-cloth by examining the cut edge. The manufacturer stated that the cotton warp weighed but 3 oz. per yard; but I do not credit the assertion. One is strongly inclined to suspect that a cloth of this character could not have been "got up" for any very legitimate purpose, but that it belongs in the wooden-nutmeg and horn-flint category !
The ordinary satinet, when well made, is a profitable, cheap cloth.

appearance with heavy sheep's gray, excepting in the quality of the wool. That is inferior to any I ever saw in a single piece of the former. It appears to be of two qualities, the finest about like the Asia Minor or African ("Smyrna" or "Mogadore") wools; and this intermixed with occasional still coarser *sharp pointed hairs*, which could come only from an animal not many removes from the wild Argali.* In both, there is a peculiarly dry, harsh, wiry feeling, not found in North American wools, and which is more indicative of an inferior staple—of brittleness, and want of felting properties—than even their coarseness. The staple is not apparently a very long one. I conjecture that it is Iceland wool—or that, mixed with Orkney, or some of the coarsest short or medium staple wools of Scotland.

The Chelmsfords, (31 inches wide,) twilled, undyed,† cost, you inform me, 58 cents per yard. The plain article, (*i. e.* untwilled,) 28 inches wide, costs 50 cents per yard. The sample of the twilled, forwarded by you, is a thicker, decidedly stronger cloth, with larger and far more tightly twisted yarn, than the sample of Welsh plains. The wool is of about the same quality, though at first view it strikes you as decidedly coarser, as the longer nap shows more of the coarse fibres on the surface, and these are rendered more conspicuous still by their variety of color. But on resolving portions of each cloth back into unmanufactured wool, I can detect little or no difference in its fineness, unless it be that the stock of the Chelmsford plains possesses none of those peculiarly coarse fibres or hairs which characterize the other. The wool used in the Chelmsfords is apparently of a longer staple. It is probably South American, though it may be Smyrna or Mogadore, as it bears a strong resemblance to the wool of the broad-tailed sheep of Asia and Africa. You state that the Welsh is generally thought to outwear the Chelmsford plain. This may be true of the ordinary articles, but I think it cannot be of the samples forwarded. Of these, the latter possesses nearly double the strength of the former and is much the heaviest cloth.

The slave blanket, 6 feet 11 inches long, by 6 feet 5 inches wide, weighing 4½ lbs., you state cost about $3 12½ by the piece (a piece containing 16 blankets costs $50). It is manufactured of a very coarse and a long stapled wool—not much fulled—with a long nap raised on both surfaces. The wool in quality resembles that used in the Chelmsfords.

On the receipt of these samples, I forwarded a specimen of the Welsh plains to two manufacturers of experience and perfect pecuniary responsibility, asking them at what price per yard they would contract to furnish me 100,000 yards of cloth of the same style and equal quality with the sample. The question was put to both of these gentlemen and received by them, as purely a commercial one—the opening of a commercial negotiation. Each stood ready to enter immediately on the fulfillment of a contract, based on his offer.

The following is the answer of one of the above named gentlemen :

HENRY S. RANDALL, Esq.   MORRISVILLE, N. Y., April 20, 1847.

Dear Sir : Yours of the 13th is at hand and duly noticed. I have no wool of the quality of the sample sent, and do not wish to work foreign wool. I would like to make for you 100,000 yards like the sample, out of our American or domestic wool. I would make it as thick and tight as the sample sent, 32 inches wide, at 40 cents per yard. I could not say how much less it would cost to get up the article from the same kind of wool with that used in the sample. I do not know what that kind of wool is now worth in market. I have not worked any of it for two years past.
                                        Yours, truly,   C. TILLINGHAST.

* Many of the unimproved breeds have, as is common with wild animals, a coating of hair over a finer pelage beneath, and it is difficult to perfectly separate them.
† A small portion of the wool employed in the filling is black, giving the cloth a dirty drab or ash color. But this I take to be the natural color of the wool

The first answer of the other manufacturer, S. Newton Dexter, Esq. of Whitestown, Oneida Co., N. Y., (head of the Oriskany Manufacturing Company,) it is not necessary to transcribe entire. Mr. Dexter informed me that his machinery is calculated for the manufacture of fine cloth ; that the carding of coarse wool would injure his cards; that its manufacture would throw him out of his regular course of business ; that he had no wool of the quality used in the sample on hand ; that he should be com pelled to use domestic wool; and that for these reasons and some other named by him, he could not undertake to fill the contract at less than 42 cents per yard—which he knew would be considered a high price.

Mr. Dexter being a gentleman equally distinguished for his correct and able business character, and for that capacity and range of information which give value to his opinions on all the topics connected with this investigation, I addressed him a second communication, asking him what he could manufacture the cloth for, giving him time to procure stock of the same quality used in the sample. I also inclosed him proof-sheets of the preceding part of this letter, asking him his opinion of the correctness of my statements, in relation to the general cost of manufacturing, &c. The following extracts from his reply will be read with interest :

Col. HENRY S. RANDALL :   WHITESTOWN, April 24, 1847.

Dear Sir: Yours reached me on Wednesday. There is no doubt at all but what if I felt certain that wool could be procured of the quality of which your sample was made, at a price proportionably low, I could have afforded to have manufactured the cloth at 37 cents per yard, as well as at 42, and use our coarse native wool, at a probable cost of 25 cents. . . . . There has been an advance of more than 70 per cent. in the price of lard oil. The price a short time since was 55 cents. The last I bought cost 95 cents in New-York. Five quarts of this oil are wanted to every 80 yards of these cloths. . . . . I cannot imagine where the wool was from out of which the sample was made, probably from Iceland—for I recollect some twenty years ago the Oriskany Manufacturing Company obtained just such wool somewhere, when American wool was deemed too high, and manufactured it into miserable satinets, by which they lost a great deal of money The wool was said to have been imported from Iceland. I was one of the Directors of the mill then, but had nothing to do with "operating" it. . . . .

You request my opinion as to the correctness of your statements of the probable cost of Welsh plains, &c., and generally of the statements put forth by you on the subject of woollen manufactories. I am not *very good* authority as to the cost of manufacturing coarse woolens, never having done much in that way. I am free to say, however, that your estimates may generally be relied on. Certainly you have allowed liberally for what would have been the cost of such wool by the pound last year ; but I think your estimate of 17½ oz. of wool in the fleece, out of which to manufacture one yard of cloth 32 inches wide, similar to the sample inclosed in your letter, too low. I should think it would certainly take 20 oz., or 1¼ pounds. The allowance of 11 cents for manufacturing will, I am inclined to think, pay charges, but it will not afford any profit, nor interest on capital, nor leave anything for keeping machinery in repair. It is a very close calculation, when fuller's soap, lard-oil, &c., are so high.

The sheep's gray cloths that you speak of, you will observe, are generally not quite ¾ wide—say 26 inches—while the sample you sent me was 32 inches. One pound of well washed fleece wool *will* make a yard of sheep's gray of medium quality ; but unless the goods are *flocked*, the calculation is a very close one indeed.

I am inclined to think that you overestimate the profit of manufacturing woolen goods, although I admit that in well-managed institutions, that have the most improved machinery, with an abundant capital, the profits have, at times, been very large indeed, and our friend Samuel Lawrence, of whom you speak, is the most prominent example of such a manufacturer within my knowledge. . . . . Every new manufactory erected, if built with judgment, has one advantage over those already in operation, and that is, they have availed themselves of all the improvements of those in operation. And as machinery is constantly being produced at cheaper rates, a factory of increased capacity will probably have cost less money. . . . .

The Oriskany Manufacturing Company is the oldest company now manufacturing woolen goods in the United States. They have made satinets which have sold readily at $3 50 per yard, and have made cloths which have as readily sold for $12 per yard. Satinets full, as good can now be bought at 75 cents, and handsomer, if not better cloths, for $3. What a change is here ! And yet the Oriskany Manufacturing Company was perhaps never doing better than now. This Company availed itself of the opportunities offered last year to obtain wool very low, to purchase a supply for nearly two years. This year the business will be good, that is, pay a profit of 10 per cent. on investments, even where wool is purchased at current rates ; but I do not believe it will pay more. I will furnish you with a brief estimate :

A mill with a capital of $100,000 will manufacture, say 90,000 yards of 6-4 cloth, which
    will bring in market an average of $1 50 per yard, or $235,000
To get these cloths into cash (for they are sold at 8 months, and are charged
    with commission of 5 per cent., and other charges equal, in all, including in-
    terest, boxing and transportation, to 12 per cent. $16,200
Cost of 225,000 lbs. of wool at 30 cents. 67,500
   ,,   3,300 gallons sperm and lard oil at $1 3,300
   ,,   Soap, soft and hard 3,500
   ,,   600,000 teazies 1,000
   ,,   Dyeing materials of all kinds 11,500
   ,,   Fuel. 1,000
   ,,   Paper, tape, twine, nails, lumber, cards, candles, &c. 3,000
   ,,   Labor, $5,000 per quarter, or 20,000
   ,,   Insurance 2,000
          Total. .... ——$122,000

If I were under oath, I do not believe I should alter any of these items—or, at least, I should
add as often as I diminished, I have no doubt. You may think $1 50 a low average for cloths,
but it must be a very fair cloth to bring that sum, I assure you. You may also think 12 per cent.
a high charge for getting these cloths into cash, &c., but it is scarcely what we pay. And the
records of our wool book will show that 30 cents is the cost of such wool as we work. And our
books will prove that it has taken, for many years past, 2¼ lbs. of wool to make a yard of broadcloth. There is 13 per cent. left for profits here, because I have not allowed one cent for repairs
or taxes, or for the agents' salaries, which will swell the expenses fully up to $124,500—within a
fraction of swallowing up all over 10 per cent. .... Well, I admit that 10 per cent. is a great
business; but you speak of 15, and that is going too far. ....

    Very respectfully your friend and obedient servant,   S. NEWTON DEXTER.

It will be seen from the foregoing letters:

1st. That where their machinery is adapted to it, manufacturers are willing to make and sell goods of the same amount of stock and style of
manufacture, with Welsh plains, out of *domestic wool* for 40 cents per
yard; and that manufacturers of perfect pecuniary responsibility are ready
to contract so to furnish it. This (apart from the small item of transportation) is *twenty-five cents* per yard, or about *thirty-nine per cent.* cheaper
than you now obtain these cloths: and an article manufactured from domestic wool would, by reason of the far superior *strength* and *felting property* of the stock, be much stronger and more durable than the foreign
goods.

2d. It will be farther seen that a skillful and responsible manufacturer
would furnish cloth, corresponding with Welsh plains, at 37 cents per yard,
could he procure the same quality of wool now employed in the manufacture of those cloths at a price proportionably low with domestic wools,
calling the latter 25 cents per pound.

Blankets are manufactured at equally exorbitant profits; and the
Chelmsfords, paying less transportation and no duties, approach the same
standard of profit—though, judging from your samples, I consider them
the cheapest goods.

I have given Mr. Dexter's undoubtedly fair and candid statements in
the premises—my object in these letters being, as I once before have
stated, to arrive at *truth*, and not to support a favorite hypothesis, or to
maintain, at all hazards, preconceived views.

My own estimates and those of Mr. Dexter, of the actual cost of manufacturing Welsh plains, it will be seen, differ—but not so materially as
would as first appear, when the advance of wool, soap, oil, &c., are taken
into consideration. I have no doubt that, in making his estimates, he had
his eye more on the better and more elaborate machinery of his own mills
—the more expensive and perfect performance of the various manufacturing processes common in that class of establishments, than on the cheaper
machinery and processes necessary in the manufacture of coarse goods.
My estimates, or rather statements of cost of manufacturing sheep's gray
you will recollect, were given on supposed *actual knowledge* of what a
manufacturer of these goods *had made them at.* To these Mr. D. seems
to take no exceptions.

In relation to the shrinkage of wool, Mr. Dexter undoubtedly bases his opinion mainly on his own experience in manufacturing broad and other cloths of fine quality. In these, the shrinkage of the wool from the fleece is concededly at least *half.* And the firm, well finished and honestly made Oriskany cloths, I have no doubt require the highest rate of shrinkage in the stock. But Mr. D. concedes that a " pound of fleece wool will make a yard of sheep's gray of medium quality." Now the Welsh plain, of the quality of the sample, weighs 13 oz. per yard. As I have already stated, " the ordinary weight of the sheep's gray is from the weight of the Welsh plain to 16 oz. per yard." Thus a yard of "medium" sheep's gray out-weighs a yard of the Welsh plain. If this is so, the former, of course, re-quires the greatest amount of stock, the mere *width* making no difference whatever. Mr. Dexter was led into this error, evidently, by overesti-mating the *weight* of the Welsh plains—and this arose from the smallness of the sample submitted for his inspection.

His statement of the cost of manufacturing broadcloths by the Oriskany Company is entitled, I have no doubt, to the fullest reliance. In conse-quence of his remarks on this topic I have changed a statement in the preceding part of this letter alluded to by him, for fear it might convey an erroneous idea. Where I spoke of " existing establishments declaring dividends of *fifteen* per cent.," I have changed it, so that it now reads "from *ten* to *fifteen* per cent.," these being the dividends, respectively, of the Oriskany and Middlesex* Companies last year, and exhibiting ab ut the *range*, probably, of well-managed companies.

* Mr. Lawrence's great establishment at Lowell, which works up 1,700,000 lbs. of wool per annum.

LETTER VIII.

PROSPECTS OF THE WOOL MARKET—FUTURE DEMAND AND SUPPLY.

Amount of Wool which may be grown in the Southern States...If the demand is already supplied, where is it to find a Market?...The cheaper Producer can drive his rival from the market, unless the disparity of Capital is greatly against him...In Individual Capital, the South possesses the advantage over the North...The South can produce Wool cheaper than New-York...North of latitude 40° there will be little difference in the cost of producing Wool...Cost of producing it in New-England—Pennsylvania—New-Jersey—Ohio...The Prairies—Their vast Extent—Their anticipated Advantages for Sheep Husbandry—Flocks driven on them—Anticipations blasted, so far as keeping Sheep economically on the Natural Grasses is concerned...Character of the Prairie Grasses—Flourish but during a short season, rendering the time of foddering longer than even in New-England...Another Difficulty—The Wild Grasses which the Sheep feed on rapidly become extirpated—Statements of the Editor of the Prairie Farmer confirmatory of this, and of the assertion in relation to the length of the time of foddering...His proposition to introduce Grasses which will grow in the Winter—Impracticability—Reasons...Burning over the Prairies—Objections...Indifferent quality of Prairie Hay...Principal Advantages of the Prairies for Sheep Husbandry narrowed down to two—Cheapness of Land—Privilege of Pasturing the Public Lands...The latter Advantage rapidly lessening...Cost of Preparing the Prairies for Sheep Husbandry—Materials for Fences, Buildings and Fuel entirely wanting on the interior of them...Coal for Fuel plenty, but not economically available...Fences—those of earth inadequate...Hedges—Require fences to protect them while growing—Their success then doubtful...Timber may be grown for all of the above purposes, but would raise the cost of the land above those of the Sheep Lands of New-York and New-England...The Shepherd System as a Substitute for Fences—When the Sheep become numerous, it would cost more to keep them in separate flocks than fences cost in the East...Pasturing in Common considered—The Sheep could not be separated for any ordinary purpose of Sheep Husbandry—There would be no protection against theft, promiscuous interbreeding, untimely impregnation—No way of effectually combating contagious disorders—Reasons...Natural and unremovable Objections to the Prairies—Want of Water—A Climate far more fickle and excessive than in the Eastern States...Shown by the record of the thermometrical observations kept at the Military Posts of the United States...These compared...Wool-growing in Mexico—In South America.

Dear Sir : In recommending the production of Wool on a scale so extensive in the Southern States, as I have done in my preceding Letters, the fact should not be lost sight of, that were these recommendations complied with, one of the great staples of commerce would be enormously increased. The Southern States—the ten* to which I have confined all my preceding remarks and estimates—to say nothing of those in the same latitudes *west* of the Mississippi—include an area of 450,000 square miles, or 288,000,000 square acres. Allow one-eighth of this region to be in a state of cultivation,† or in natural pastures, and we have 36,000,000 acres which could be more or less devoted to the growth of wool. Assuming that, on the average, every two acres would, under proper tillage, support one sheep, (which, it seems to me, they might do with no very material diminution of present staples,) and that the sheep average 3 lbs. per fleece, the annual product of wool would be 54,000,000 lbs. This amount might be indefinitely added to, by diminishing the production of present staples. How far this could be economically done, experience must determine.

If we concede the adequacy of the present supply of wool to the demand, taking the world together, it is apparent that an increase of 50, 75, or 100 millions of pounds, in one quarter, will produce an over-supply, (and thus greatly depress prices,) unless met by an increased demand, or a corresponding diminution in production, in some other quarter. I do not concede the adequacy of the present supply, but shall, however, waive that point.

The question now arises, where is the wool thus produced to find a market, if the South should, within the next ten, fifteen, or twenty years,

* Nine, besides that portion of Louisiana east of the Mississippi.
† Probably the amount in cultivation, including that in natural pasture, is set down pretty high. It may not exceed a tenth.

furnish such a surplus? Where is the present supply to be diminished, or the demand increased?

Where agricultural competition exists, as a matter of course the pro ducer who can supply the market with the least expense to himself, has an advantage which nothing but a disparity greatly against him in capital can overcome. Large capital, satisfied with less gains than small capital, will sometimes sustain competition with the latter, with the advantages of the cheapness of production *somewhat* against it. But where the differ- ence in first cost is considerable, the cheaper producer can always drive his rival from the market. The aggregate agricultural capital in a region of given size in New-York, probably would ordinarily exceed that of an equal territory in South Carolina or Georgia. But it is not so with indi- vidual or personal capital. While the agricultural territory and capabilities of the latter States are in a comparatively few hands, those of New-York and New-England are parceled out among a multitude of small holders, who must realize the first class of agricultural profits, to support them- selves and their families. The advantage of capital is therefore, in reality, on the side of the South.

But independent of this consideration, I have already attempted to show that the South can produce wool so much cheaper than New-York, that the latter will stand no chance whatever in competing with her more favored rival—so soon as that rival sees fit to avail herself of her advan- tages. *North of latitude 40° there will be but little disparity in the cost of producing wool;* and therefore if the South can drive New-York to relin- quish the production of this staple, she can do the same with all portions of the United States lying north of this parallel, unless on the shores of the Pacific, where the isothermal line is at least 5° north of its course east of the Missouri. I will now enter upon some specifications, and, where ne- cessary, proofs, to sustain this proposition.

New-England has, concededly, no advantages over New-York for the cheap production of wool. Northern Pennsylvania is higher, colder, and more sterile than most of southern New-York. South-eastern Pennsyl- vania, and the fertile portions of New-Jersey, are the natural producers of bread-stuffs for the less favored regions of those States, and of provisions of all kinds for the New-York City and Philadelphia markets. The high price which good lands bear in the vicinity of such markets, would prevent them from competing with cheap interior lands in wool-growing. There are sheep lands of good quality in western Pennsylvania; and in the southern section, the winters are perceptibly a little shorter than in New- York. This will render the production of wool upon them somewhat less expensive than in the latter State, but it will not reduce it low enough to allow them to compete with the cheaper lands and still shorter win- ters of the South. The same remarks will apply to the hilly region con stituting the south-eastern portion of Ohio.

Proceeding still farther west, we find a region extending to a vast distance whose topographical and geological features, flora, &c., taken in connec- tion, effectually distinguish it from the territory lying east of the Missis- sippi and Ohio. Vast plains, called *prairies*, (so named by the early French settlers from the French word signifying meadow,) which can be purchased of the Government in the natural state for $1 25 per acre, and which are usually covered with natural grasses—would seem, if these grasses are adapted to the summer and winter subsistence of sheep, and there are no counterbalancing disadvantages, to unite facilities for the cheap production of wool not possessed in any other region of our country. And such supe- riority has actually and often been claimed for them.

I propose to investigate this question at considerable length, because there are various considerations which, at first view, give great plausibility to this claim. And if the prairies *can* produce wool cheaper than the South, it is in vain for the latter to embark in the business—at least, beyond the extent of supplying the home demand—for so limitless is the extent of these natural pastures throughout the whole northern basin of the Mississippi, that they could, perhaps, supply the entire market demand of the United States for this staple, for an indefinite period, vast as that demand is destined to be.

But a very few years have elapsed since the most sanguine anticipations were indulged in, by large numbers of our Northern and Eastern flockmasters, in relation to the superior capabilities and advantages of the prairies over Eastern lands for sheep-walks; and large flocks were driven hundreds of miles, lands purchased, and establishments created, to realize these supposed advantages. It is not too much to say that these anticipations—so far at least, as keeping sheep on the *natural* herbage of the prairies is concerned, were briefly and summarily blasted. Many of the flocks driven there, actually perished in the midst of seeming plenty. On the whole, the experiment is generally conceded to have resulted in failure. Let us see whether this was occasioned by mismanagement—temporary and removable causes—or whether we must look for those causes in natural and unchangeable circumstances.

A *portion* of the wild prairie grasses are relished by sheep, and they thrive on them ; but these grasses, as well as all the other varieties growing there, flourish during but an unusually limited portion of the season. They begin to dry up and lose their nutritive qualities in midsummer, and long before the foddering season has commenced on the bleakest highlands of New-England, they are as unfit for the subsistence of sheep, as dry *brush !* Where the natural grasses are alone depended upon, the foddering season on the prairies, north of latitude 40°, will range from six to seven months —rarely, perhaps, fall short of six, on lands which have been *previously depastured*, provided the sheep are maintained in good condition.

And there is another material difficulty with the prairie grasses which sheep feed on. They soon—many of them even in a single season—become extirpated if kept fed down while growing. This is so singular a fact in vegetable physiology, that I chose to state it in the words of an intelligent resident of the prairie region—whose local pride and partialities would naturally prompt him to give as favorable a coloring to the agricultural advantages of his chosen home, as a regard for truth would admit of. From a communication of J. Ambrose Wight, Esq., Editor of the Prairie Farmer, to L. A. Morrel*—replete with useful information, and characterized by an admirable candor—I make the following extracts :

" Sheep or other stock, but more particularly the former, put upon a given piece of wild prairie, and confined to it, unless the range be very large, would not continue to keep fat one season after another, though they would at first ; but if allowed a new range each season, they would always keep fat. The reason is this : Sheep in such cases will go over their range and select such food as they prefer, and will keep at it until it is gone. Hence the wild bean and pea vine, and a few other kinds of plants, will obtain their constant attentions, and will be kept so short that they will, on a given piece of land, die out the first year. Therefore if turned out on the same grounds another season, the best food will be gone, and the poorer, with which they must then take up, and which itself gets continually poorer, will not sustain them in their first condition. A small flock of sheep will thus run over a large extent of ground.

Hence the *utter hollowness of a supposition* which appears to be common at the East, *that large flocks of sheep can be sustained on the wild grass of the prairies alone.* There are many places, it is true, where a farmer might keep a large flock on the wild prairies

* American Shepherd   pp. 138—145.

during the summer months with profit, provided he had not too many neighbors in the same business. But such flocks would continually lessen their own range, at the same time that it is lessening by immigration, settlement and extended culture. I have been in the country about nine years; having gone, at the first, into an entirely unsettled region, and have paid much attention to the matter; and it is my belief that the wild prairies are desirable for wool-growing to a very limited degree; but that the cultivated prairies are desirable for this purpose to an almost limitless extent."

The following fully sustains my preceding statements in relation to the time of foddering. In answer to Mr. Morrel's question, " what length of time is foddering necessary in Northern Illinois ?" Mr. Wight says :

" The seasons have been extremely variable since my residence here—now nearly nine years. The winter of 1842 and '43 was the severest one since the settlement of the State; and the foddering season lasted from the middle of October to the middle of April. The winter of 1843 and '44, and the present one (1844-5) would require foddering for a less time, by full two months. *This is on the supposition, however, that good artificial pasturage is provided.* If the wild prairies are relied on alone for pasture and hay, full two months must be added to the foddering season; and stock would barely get through at that; and I think that sheep, in multitudes of instances, would perish. In this latitude with Timothy Red-top and Clover pastures, the average time would be from 4½ to 5 months. If a good blue-grass pasture were provided, in such winters as the last and present, it might be reduced to two months, and I am told that some so provided for, *one hundred miles south of here,* have, the present winter, scarcely foddered at all. I apprehend, however, that our winter here will always be variable, and that it will be far more difficult to predict their length and intensity than in New-England."

In another place Mr. Wight says :

" If, however, the question is asked, ' Does not the pasture on the prairies fail early in autumn, so as to compel the removal of sheep to other pasture before it is time to go into winter quarters ?' I answer, yes—long before. In many sections the prairies afford no adequate pasture for dairy purposes after the first of September. . . . . The wild grasses are extremely vigorous while they last, but are all, without an exception, short-lived."

The great diminution of the foddering season, where the domestic or cultivated grasses are already made use of, which Mr. W. anticipates *may* result from the introduction of blue-grass, will be found utterly unattainable. Blue-grass (known as June or spear grass), is one of the commonest varieties in New-York and New-England. Peoria, in Illinois, is in about the same latitude with the City of New-York, and consequently that portion of Illinois north of Peoria, corresponds with a considerable portion of New-York, and all of Connecticut and Rhode Island. And the climate of the former is not less rigorous, and is far more variable, than in the latter named States, as I shall presently show. Now in no portion of New-York or New-England will the blue-grass reduce the foddering season to two months, or anything like it. It is true that small flocks will pick up a subsistence on this and other grasses in the winter, when the ground is not covered with snow, and if the pastures are not fed down in the fall. To suppose, however, that this or any other herbage will *continue to grow,* when the earth is frozen almost to the consistency of a solid rock, far below its lowest roots, is an obvious error. In New-York, the ground remains so frozen usually during the entire winter, and in Northern Illinois the cold is equally intense, and there is less snow to protect the earth from its effects. The ground, therefore, is frozen quite as solidly, and considerably deeper than in the former. Grass left standing for winter consumption, in either State, becomes, by freezing and thawing, tough and innutritious. In New-York, the larger flock-masters have long since ceased to make any provision of this kind, for winter-feeding—preferring to keep their sheep in yards, and entirely from grass.

As Mr. Wight himself very accurately remarks in another part of his communication, " It is found to be decidedly better to keep sheep up in small flocks, with very little ground to run over, while kept on hay, than

N

to let them run out a part of the time and get such grass as they can pick, while there is not enough to sustain them." But the reason for this given by Mr. W., that " they eat much dirt, are liable to be poisoned and lose their appetite for hay," is very far from being the correct one. Green grass always, in a great measure, deprives sheep of their appetite for dry hay. The grass thus left standing loses its nutritive qualities, so that it will but imperfectly sustain animals, and when the snow falls and covers it, sheep not only cannot obtain it, but they are left without appetite for other food. Open winters, *i. e.*, winters without snow, are always particularly fatal to sheep which are suffered to run on the pastures, in this climate, and for the reasons above assigned. They sometimes appear to be doing well enough up to toward the close of February ; but they are imperceptibly losing condition and strength, and when the trying month of March, with its stormy and fickle weather, sets in, they begin to drop off, and all sorts of diseases—grub in the head, " the distemper," etc.—are assigned as the causes.

It is in vain to attempt to shorten the foddering season north of latitude 40°, on this side of the Rocky Mountains, by seeking for any plant to *continue its growth and thus produce green feed in winter*, unless in limited districts, and on the margins of large bodies of water. No plant can draw its nutriment from solidly frozen ground.

Mr. Wight proposes burning over portions of the prairies at intervals, to cause the vegetation to start afresh, and thus prolong the grazing season on the prairies. Mr. Flower makes the same suggestion. In some localities, and under favorable circumstances, this might, temporarily, accomplish the desired object ; but as population increases, and buildings and inclosures are erected, it would constantly lead to those unfortunate accidents, which have already, I believe, led at least one of the Western States to prohibit by severe penal enactments, the setting fire to the dead grass of the prairies. Besides, we have Mr. Wight's own authority for stating that sheep actually *extirpate* those of the prairie grasses which they will feed on, so that burning over could not cause *these* to re-sprout the same season or afterward.

It requires but little knowledge of the habits of the sheep to know that grasses rejected by it in summer, will not constitute a proper aliment for it in winter, and that if confined to such food, it will not prosper. A few sheep with liberty to *pick and waste*, will live on very inferior herbage in either summer or winter, (and hence the sanguine and erroneous statements put forth by owners of small flocks on the prairies,) but confine flocks to the same food—flocks which are too numerous to be allowed the privilege of selection and rejection in their food, and the disastrous consequences will not be long in exhibiting themselves.

In reviewing the preceding facts, the principal advantages of the prairies for the production of wool seem to be narrowed down to two points . the cheapness and fertility of the lands, with a contingent right inuring to the settler to use, without paying for it, all the unappropriated public domain ! If we admit that the soil of the prairies is as well adapted to the artificial grasses as that of New-York or New-England, (a point which, to say the least of it, is doubtful, for experience has shown it to be otherwise in Michigan and some other portions of the West,) the only peculiar and exclusive advantages which the prairies have over the lands of the old Middle and Eastern States, is their cheapness and freedom from rent where unsettled. Emigration is rapidly abridging the latter privilege, however—more rapidly than can well be appreciated without a reference to the statistics of the several new North-western States. And it will be remembered that when a prairie is belted round by population, and de pastured by numerous flocks and herds, its better grasses—at least for sheep—would be soon exterminated, and, consequently, though there might be ten or fifty thousand acres of common and free pasturage, it would be of trifling avail to the flock-master.

But taking this privilege for what it is worth, and taking into account the difference in the price of lands—calling one $1 25, and the other $20 per acre—and then, in my judgment, the Eastern will prove cheaper, all things considered, than the prairie lands, for Sheep Husbandry. I speak, of course, of the prairies as wholes—not of that narrow margin of each, which is attached to the farms lying partly on the outer and wooded lands.

The prairies must first be plowed, undoubtedly,* to seed them down effectually with the cultivated grasses. It requires from four to six yoke of oxen, says Mr. Solon Robinson, to break up from one to one and a half acres per diem. Suppose we concede this expense to be paid for by the first grain crop used as a covering for the grass; then the prairies are to be fenced—adequate buildings and other fixtures provided, for the use of a family, the storage of hay, the shelter of animals, &c. Where are the materials for these things and for fuel to be found, on a plain wholly destitute of trees, unless on the occasional "islands"—and where stones are entirely wanting, excepting sparsely scattered bowlders, and, very rarely, rocky ridges or cliffs ? Conceding that all the wood on the margins of the prairies will not be wanted for the local supply—which, as a general thing, it undoubtedly will—what would be the cost of fences, buildings and fuel, where every stick was transported from three to fifteen miles† by land carriage ? Fuel, it has been said, can be obtained from the local depositions of coal. It is true that Illinois and south-western Indiana, at least, constitute one vast coal basin. But any one possessing the slightest practical acquaintance with the subject, knows that it requires associated, aggregate and corporate wealth, to carry on mining operations to an extent sufficient to steadily and efficiently supply a considerable market. Even in a level country where coal is covered with a deep superficial deposition of earth, individuals may, where the stratum is cut through or uncovered in ravines or the beds of streams, quarry their own coal; but such opportunities are rare. The idea that individuals would find it within the compass of their means to sink vertical shafts and raise coal—each one for himself—on the bosoms of the prairies, is utterly preposterous. Coal has never yet borne a price in our *cities*, which would justify even Companies in lifting it by *vertical* shafts. Let the coal, however, be as cheap as it may be, at the mere point of excavation, the mere cartage of it, for the wants of a five-months winter—where the thermometer frequently indicates a degree of cold from 5° to 30° below 0°—will be an onerous tax on agricultural industry. And canals can never furrow the bosoms of most of those vast dry plains ; and ages must elapse before railroads will so interlace them, as to bring coal cheaply within the reach of population scattered over their entire surfaces.

If we suppose that adequate buildings can be constructed, with sufficient economy, with transported timber, the question still remains, What resource is there for fences ? Fences of earth have been proposed, but these will not stand long enough to pay for building, unless their sides are constructed at such an angle as would be wholly inadequate to "turn" sheep. Hedges, besides the other considerable expense of cultivating them, would

* I have seen it stated that the seeds of the cultivated grasses would "catch" sown on the surface of the prairie sod! That they would do this effectually and generally, is an assertion which no practical farmer will credit.

† Prairies are from one to thirty miles in diameter.

require *fences* to protect *them* from animals, until they attained a considerable size; and it is exceedingly questionable whether any good hedgeplant can be found, which is capable of resisting the rigorous and fickle climate of the North-western States. The different thorns, and other plants used in England, have generally failed in all the Northern States.

Timber *may* be *grown*, both for fuel, houses and fences, by the proper planting, cultivation and protection of suitable trees—but the expense and delay attending this course would raise the prairies to, or above the price of New-York and New-England sheep lands.

It has been claimed that the shepherd system will render fences unnecessary, to any but a very limited extent, on the prairies. Now, while there is but here and there a settler on the margins of some of these great plains, and while a flock of sheep can constantly seek new pasturage, as the old fails, over a boundless range, without encountering another man's flock, sheep require so little looking after that the shepherd system is entirely feasible and economical, notwithstanding the high price of labor. Under such circumstances, one man, provided with a horse and a brace of dogs, can perhaps give the necessary attention to 1,000 sheep, and have some time for other occupations. But this state of things, terminated already on most of the prairies this side of the Mississippi, will soon be unknown even on those in the territories bordering on the Missouri and its western tributaries. When wool-growers become to any degree numerous on the borders of the prairies, (as they certainly soon will, if these regions do possess any peculiar advantages for this branch of husbandry,) how are sheep to be kept *separate*, without that multitude of shepherds which the same services require in Spain, Germany, or Australia ?—and whose labor and subsistence* would cost more, during a series of years, than the *fences* in regions where wood and stone are plenty.

If the sheep are not kept separate—if allowed to run promiscuously together, how could the property of each holder be separated out of the vast general flock on a prairie five, ten or fifteen miles in mean diameter, for the purposes of slaughter, sale, washing, shearing, folding, or any other incident of their husbandry ? What protection would there be against wholesale theft, when no man could count his scattered flock ? What would prevent promiscuous interbreeding—and what object would it be, therefore, to attempt to procure choice breeds, or improve those already possessed ? What security would there be against those vagabond rams which the carelessness of some individual is always sure to let loose on a neighborhood, to beget lambs on every poorly-fenced farm, to perish in the storms of February and March ?† Finally, how could contagious and—unless promptly checked—highly malignant and fatal diseases, like the scab and hoof-ail, be met with the proper vigor, and treated with the necessary skill and care, among a multitude of holders scattered over miles of surface; and supposing all the necessary vigor, skill and care brought into action, what would they all avail where it was impossible to separate the healthy from the diseased—the cured from the sick ?‡ Let either of these diseases break out among a flock of ten thousand sheep, running together without inclosures, and any one familiar with their diagnosis and treatment, knows that if it were possible to drive them from the flock—which is extremely doubtful—it would cost far more than the value of the

---

* Costing four or perhaps six times more in this than in the former countries.
† It is questionable whether in a flock running in common on a prairie, one ewe in ten would escape untimely impregnation.
‡ Both of these diseases are susceptible of being communicated from a diseased sheep to one but recently cured of them; consequently, separation is the only safe and economical method, in large flocks prevent constant reinoculation.

sheep. True, these diseases have not yet visited, so far I am aware, the Western States. The scab is, in fact, but little known at present in any part of the United States. It may at any time; however, reäppear.* The hoof-ail, after the fury of its first onset is over, assumes a milder form— one which does not lead to death, if remedies are applied but once or twice during a season—and for this reason, probably, it is allowed to linger in many flocks in the sheep-growing regions of the U. S. It is a strictly contagious disease, and one animal having it would rapidly innoc- late, in the hot weather of summer, by itself and others receiving the dis- ease from it, one or five hundred thousand sheep having access to each other. A few years since it was a stranger to *this* region. Like the small- pox when unchecked by vaccination, or any other contagious malady, it gradually progresses from neighborhood to neighborhood—from State to State. Good fences, confinement to the farm, and a rigorous system of exclusion of all strange sheep, may and do save many flocks from its vis itation, but accidents and acts of carelessness are constantly occurring— and so long as they continue to occur, this malady will continue its on ward march. I consider it just as certain that it will visit and sweep over the North-western States, as I do that flocks are scattered along between those States and the present seat of the disease. And when it does visit them, if it finds any great flocks congregated on the prairies, not in a situ- ation to be immediately divided into small flocks, I venture to predict that, with all the care and attention which the sheep *will* receive, the miserable animals, eaten while yet alive by maggots—and festering in loathsome rottenness, will perish in multitudes—by whole flocks.†

Another objection to pasturing in common, would arise in the difficulty, if not impracticability, of establishing and *enforcing* an equitable system of joint occupancy, over or around a large prairie, so as to compel each farmer to regulate the number of his flocks and herds by the amount of cul- tivated pasture possessed by him.

But if we concede all the preceding difficulties to be removable, or even removed ; if we suppose the great north-western plains to be amply sup- plied with materials for building, fences, and fuel—there are two other dif- ficulties in the way of their becoming the best class of sheep-walks, which, from their nature are fixed, and, in the main, unchangeable. I allude to the *scarcity of water*, and the *climate*.

On the " dry and rolling prairies "—those claimed to possess the greatest advantages for Sheep Husbandry—running water is scarce, frequently ex- tremely so. The occasional streams are shallow and sluggish. Washing wool on the back of the sheep, conduces, I think, to the health of the ani- mal. It causes the sheep to shear much more easily—brings the wool into a better marketable condition, and diminishes transportation. Streams of considerable depth and rapidity (where, what is better, falling sheets of wa- ter over mill dams, &c., cannot be found), are almost indispensable to an effectual performance of this process. Sheep, also, in many periods of weather, *require* water for *drink*. When they are confined to dry feed, it is indispensable, in the absence of that snow which is often, in the Eastern States, made a substitute for water. Neither are attainable during consid- erable periods each winter, on the prairies, without resort to a pump—a sorry—and, (including the time of working it, when large flocks are to be watered), an expensive and troublesome substitute for running water.

Finally, the climate of the Western and North-western States is *more*

---

* Since writing the above, I have found, to my utter surprise, that this disease is within three miles of my own farm, in a flock driven into the country last fall.
† A history of this disease and its gloomy diagnosis, when neglected, will be given in a subsequent Letter.

variable—exhibits *more sudden* and *greater extremes*, than the climates of
New-York and New-England. The weak and easily prostrated muscular
and vascular system of the sheep, will better endure great extremes of continuous heat or cold, than rapid and marked variations in temperature.
Subjected to the latter, catarrh not violent enough to kill in its inflamma-
ry stage, but assuming a chronic form—and followed by a slow and wasting debility, frequently attacks flocks. Sometimes it assumes an epizootic and malignant character—as during the past winter—and sweeps away
thousands of sheep.

The isothermal line (or line of equal mean heat), does not vary particularly between the same latitudes in New-York or Wisconsin—or between
Virginia and Missouri. But as we leave the ocean and other large bodies
of water, the isotheral and isocheimal lines are found to diverge more
and more from the isothermal one—and the range of the thermometer (the
extremes of heat and cold indicated by it), rapidly increases. The following Table of temperatures, kept by officers in the Army, for a series of nine
years, is from Doct. Forry's excellent work on the "Climate of the United
States, &c."* It strikingly illustrates the fact asserted. The four points
specified are in about the same latitude.

	Highest.	Lowest.	Annual Range
Fort Wolcott. Newport, Rhode Island........	85	+2	83
Fort Trumbull, New-London, Conn...........	87	+9	78
Fort Armstrong, Rock Island, Ill.............	96	—10	106
Council Bluffs, near the confluence of } the Platte and Missouri }	104	—16	120

Doct. Forry states that the mean annual range of the thermometer at the
following places, is as follows : at Fort Sullivan (Eastport, Me.) it is 104°,
while at Forts Snelling (confluence of the St. Peter's and Mississippi in
Iowa) and Howard, (Green Bay, Wisconsin,) in about the same latitude, it
is respectively 119°, and 123°.

At Fort Preble (Portland, Me.) Fort Niagara (near the mouth of the
Niagara River, N. Y.), Fort Constitution (Portsmouth, N. H.) it is 99°
92°, and 97° ; at Fort Crawford, (confluence of the Wisconsin and Mississippi Rivers in Wisconsin,) on the same parallel, it is 120°.

The above instances are not isolated ones. The same law is found—
other things being equal—to generally prevail throughout our own, and
perhaps all other countries.†

While the cold of the Northern, and particularly the North-western
States, so greatly exceeds that of the Southern States, few would be prepared for the proposition that the extremes of heat in the former, often
reach points unknown many degrees farther South! Yet such is the
fact !

Fort Snelling, in latitude 44° 53', and occupying a central position in that
vast territory lying between the Great Lakes and the Missouri, and between
the 41st and 49th parallels of latitude—and which may therefore be presumed, to a certain extent, to afford a type of the climate of that whole region—feels a maximum summer heat of 93°—*the same* with that of Wash
ington City, in latitude 38° 53', and Old Point Comfort, Va., in latitude
37° 2'. At Fort Johnston, on the Coast of North Carolina, in latitude 34°
the maximum heat is but 90° ; at Fort Moultrie, in Charleston Harbor
n latitude 32° 42', it is also 90° ; at Fort Marion. St. Augustine, Florida.

* See the above named work. p. 43. I am also indebted to Doct. Forry for all the records of thermometri
cal observations, at the U. S. military posts, which are subsequently quoted.
† Local exceptions exist, owing to the prevailing winds and other causes. For example, Fort Howard
is much nearer a large body of water than Fort Snelling. Altitude also exerts its influence.

in latitude 29° 50', it is 92°; at Fort Brooke, Tampa Bay, Florida, in latitude 27° 57', it is 92°; and at Key West, *the most southern possession of the United States*, it is 89°!

It will thus be seen that the summer heat rises higher at Fort Snelling than at points on the sea-board more than 20° farther South!

Now let us compare their winter temperature. The minimum tempera· ture of Fort Snelling is —26°.* That at Washington is + 9°; Old Point Comfort + 20°; Fort Johnston + 28°; Fort Moultrie + 21°; St. Augus- ine + 39°; Tampa Bay + 35°; Key West + 52°! So the greatest cold of Fort Snelling is 35° below that of Washington—the most northern and by far the coldest of these posts—and it is actually 78° below that of a post, (Key West), which its summer heat exceeds by *four* degrees!

At Fort Howard, latitude 44° 40', the seasons are even more violently contrasted. Its maximum heat is 98°, its minimum—25. At Rock Island, Ill, latitude 41° 28' we have already seen that the maximum is 96°, the minimum — 10°; and at Council Bluffs, latitude 41° 45', the maximum 104°, the minimum —16°! At Petite Quoquille, near New-Orleans, the maximum is but 94°, the minimum + 30°!

And an examination of the *monthly* variations in temperature, at our North-western posts, will show that these are as excessive, in proportion, as those of the year—and their suddenness can scarcely be credited by an inhabitant of southern regions—more particularly those bordering on the Atlantic and Gulf of Mexico.†

It cannot be said that Fort Snelling, or Rock Island, or Council Bluffs, have the summers of Italy or the South of France—for the weather is much hotter at intervals, and is subject to far more frequent, abrupt and violent changes than in the latter: nor have these posts winters as mild as those of Europe, many degrees farther north.‡ And their winter exhibits the same sudden and violent changes which characterize the summer climate.

These facts, in my judgment, fully explain the remarkable mortality in the flocks which have been carried on the prairies, and which is usual'y attributed to over-driving, poisoning, &c. The climate itself, though not always a rapid, will prove one of the surest of *poisons*, unless great care— much greater than is requisite even on the bleak and sterile hills of New· England—is taken to protect them from its deleterious influences.

Facts sufficient have been adduced, probably, to convince every South ern man how much he has to fear, ultimately, from prairie competition, in the production of wool. Having thus attempted to measure the capabili- ties of the various regions of our own country for the cheap production of this staple, it may be well to turn our eyes to the comparative advan tages of other countries and nations—and to ask the question whether there is any danger to the domestic producer from *foreign competition* This can be done but briefly and rapidly in the limits which I have assigned to myself.

It will not be necessary for the purposes of the present inquiry, to ex- amine the climate, flora, &c., of all portions of the world. The wool- producing countries—those which have natural advantages to enable them to produce wool cheaply enough, and in sufficient quantities, to stand any chance in the general competition, are mainly embraced in a belt or region

[* It will be understood that the sign — before the number of degrees, indicates that it is that number of degrees *below Zero*, and the sign + used here, in the preceding Table, and in the subsequent paragraph, tc avoid confusion, signifies *above Zero.* *Publisher.*|
† In the Report of the Fishing Creek Agricultural Society, of your State, 1843, the Committee actually complain of the variableness of the climate! Truly, 'we can only judge by comparison!'
‡ The mean winter temperature of North Cape in Norway, latitude 71°, is 23° 72—that of Fort Snelling is 95—that of Council Bluffs, 24° 47—that of Rock Island, 26° 86.

about 15° in width, on each side of and at varying distances from the
Equator. The variation corresponds with the variation of temperature;
in other words, the wool zone is bounded by isothermal instead of latitudinal lines. Commencing on the eastern side of each continent, in the
northern hemisphere, between about 30° and 45°, it bears northwardly,
and strikes their eastern shores, say between 40° and 55°. In the southern hemisphere, I am not aware that the isothermal deviations, in the
corresponding parallels, have been noted—nor are they important, so
small, comparatively, is the latitudinal area of the surfaces included between them.

Independent of minor deviations everywhere exhibiting themselves in the
isothermal lines, more important local exceptions exist in many places, owing
to elevation, proximity of bodies of water, prevailing winds, &c. Thus, south
of latitude 30° in North America, the elevations of the Cordilleras give the
mild weather of the temperate, and even the rigors of the frozen zone;
and the same is true of the Andes of South America—in Bolivia, Peru,
Ecuador and New-Grenada—in the same latitudes, where, at the eastern
foot of these declivities, the tropical sun burns up, as with fire, the verdure
of the vast *llanos* of Brazil and Venezuela, and exhales death from the
pestilent fens of Guiana, and the reptile-teeming marshes of the Amazon.
The same exceptions exist on the Eastern Continent, wherever mountain
chains rise to sufficient elevations to bring to bear this well known and uniform law for the depression of temperature, albeit in tropical or sub-
tropical regions. The steady and mild climate of the Atlantic Ocean, and
its continual and peculiar motion on the west of Europe, preventing the
ice, which the north wind wafts down from the Arctic seas, from lodging
itself, or even approaching* those shores, strongly influences the climates
of the British Islands and Norway, rendering them more temperate than
others many degrees farther south in the interior of Europe and Asia.
Eastern Prussia, and Polish Russia, are rendered disproportionably cold
by the prevailing wind, which sweeps without resistance from the bosom
of the Arctic Ocean to the Carpathian Mountains : and the north-east wind,
laden with the frosts of Siberia, and untempered by the southern winds,
from which it is cut off by the lofty Altay Mountains, carries a cold under
which men, nay whole caravans,† perish in Persia, in the same latitude
with Northern Africa, and the confines of the burning Sahara.‡　The Caspian and Black Seas—Mounts Caucasus and Taurus prevent Asiatic Turkey,
and Mount Hæmus, European Turkey—from experiencing similar cold.
The same wind entering Europe, reduces the temperature of its eastern
considerably below that of its western confines ; and its effects are felt more
or less westwardly, in proportion as its course is arrested by mountains.
The climate of Silesia and Saxony is far colder and more mutable than
than that of Bohemia, from which they are only separated by the Erzge-
birge and Riesengebirge. In Northern European Russia, in Finland and
the basin of the Dwina—in the same latitudes where Norway exhibits the

---

* Malte Brun's Geography—*Art.* Climate of Europe.　　† Sir Robert Kerr Porter.

‡ From the delightful Arabian Nights—from the not less delightful strains of Lalla Rookh—from a thousand other sources, remembered and unremembered—song, fiction and Oriental tale—Persia always rises
before fancy's eye a realm and clime of beauty :

"—— deep myrrh-thickets blowing round
The stately cedar, tamarisks,
Thick roseries of scented thorn,
Tall orient shrubs, and obelisks
Graven with emblems of the time,
In honor of the golden prime,
Of good Haroun Alraschid."

There are portions of Persia where the soil is rich and the climate delightful—but, as a whole, it is a bleak,
sterile, unfruitful country—large portions of it covered with rugged mountains or saline deserts—with a
climate remarkable for the rapidity and extent of its variations.

flora of Northern Germany—spirits freeze and quicksilver becomes malle-
able. But it is unnecessary to continue this enumeration.

Let us now take a rapid view of the wool-growing countries embraced
in the specified zones. And we will first complete the description of our
own continent.

Mexico—that portion of it north of latitude 30°—bears too close a resemblance to our Western Territories conterminous with it, to require
separate notice.

But a small proportion of the great peninsula of South America is included between the 30th and 45th parallels of latitude, and admitting,
what seems probable, that the contiguity of two great oceans would so affect the climate as to carry the northern line of the wool zone a little
nearer to the Equator, this zone would still embrace but, say, two-thirds
of Buenos Ayres, nearly all of Chili, the little State of Uraguay, a mere
point of Brazil, and the north of Patagonia.

The growing of wool has already been commenced on the vast *pampas**
of Buenos Ayres—though as yet to a but limited extent. In 1832, the export of wool to Great Britain was 32,052 arrobas ;† but the same year
the import of English woolens considerably exceeded it in value. The
United States Tariff on foreign wools costing 7 cents per pound or under
being then but 5 per cent. *ad valorem*, the importation of wool of that
quality from the Argentine Republic‡ into our country in the fiscal year
ending June 30, 1846, was 4,295,659 lbs., and of wool costing more than
7 cents (paying a duty of 30 per cent. *ad valorem*, and a specific duty of
3 cents per pound) the import was 43,831 lbs.||

The *pampas* resemble the North American *prairies*, being plains covered with wild grasses, and entirely destitute of timber. The land is divided by the Government into estates a league square (5,760 acres,) and
sold at 10 cents per acre. Until recently the pampas were depastured almost exclusively by horses and cattle, and so plenty and cheap were
they, that they were frequently killed for their hides alone. The herds-
men and shepherds live in miserable huts, and temporary folds are formed
of the trunks of peach-trees. Western or south-western winds called *pam-
peros* often sweep the country with destructive fury, and there are instances in which flocks of sheep have been forced by them into streams
and have perished.

The inhabitants of the pampas are, on the north, the Gauchos—descendants of Spaniards—who, living in the saddle, and content to subsist on
jerked beef and cold water—having few wants, and none which the *lasso*
will not supply—lead a life of wild and roving liberty. Tribes of mounted Indians, wild, predatory, and constantly at war with the Gauchos, occupy the southern pampas.

The facilities for producing wool here closely resemble those of the
North American prairies, though wood is wanting over much more extensive tracts. The price of land on the pampas is less, but they are more
remote from markets, as there is little or no manufacturing done in South
America. Besides the cost of transportation, wool must pay, before reach-
ng market, the duties levied by some foreign nation. The duty in the
United States, by the Tariff of 1846, is 30 per centum *ad valorem*, without regard to quality, thus discontinuing that great discrimination in favor
of the coarse article, which allowed a large proportion of the wools of

---

* This word, like *llanos* in the Northern States of South America, and *prairies* in the North Western
United States, is applied to extensive plains. Those in the North of Chili are called *pampas del sacramento*.
† McCulloch's Commercial Dictionary. An arroba is 101½ lbs. avoirdupois.
‡ Buenos Ayres is so known in all the official documents of the United States.
|| Report of the Register of the Treasury, Dec., 1846.

O

Buenos Ayres, Africa, Turkey, &c., to enter our ports under a merely nominal duty. The present Tariff raised the duty on these wools to *six times* the former rate, *i. e.*, on wools costing 7 cents, from 3½ mills to 2 cents and 1 mill per pound. This will make an important difference to the foreign grower and exporter. If these wools continue, as hitherto, to be imported in the grease and dirt, from which state they lose about half weight in being brought as clean as well washed United States wool, every pound of them so imported will actually pay a double duty, or 4 cents and 2 mills, half of this being paid for *dirt*. If, on the other hand, they are washed prior to exportation, a reduction of 50 per cent. in their weight will call for a corresponding advance in their price. Wool now costing 7 cents at Buenos Ayres or Smyrna, will cost 14 cents; and if this is exported into the United States, it must pay a duty of 30 per cent., or 4 cents and 2 mills per pound. It will be seen, therefore, that the lowest priced foreign wools cannot enter our country without paying about this duty (4 cents) per pound, unless under fraudulent invoices; and this, as has been already shown, is *half* the cost of producing wool throughout a region of the United States much greater in extent than all that portion of South America included within the wool-growing zone.

The English duty on wools costing less .than 24 cents is 1 cent per pound; over 24 cents, 2 cents per pound. The French duty is 22 per cent. *ad valorem*, without regard to cost.

The security of life and property is far less in Buenos Ayres than in the United States; the character of the agricultural population less industrious, less skillful, and less methodical. Capitalists from other countries may, on account of the cheapness of the lands, make it profitable to purchase large *estancias*, and raise vast flocks of sheep; and this has already been done by a few Europeans. But the pampas are subject to the same general objections* with the North American prairies, and when the contagious diseases, adverted to in speaking of the latter, once obtain a footing on them, it is not difficult to predict how those diseases will be encountered by the wild and, so far as agricultural labor is concerned, indolent Gaucho. The difficulty of encountering them, with the best skill and industry, under such circumstances—of preventing their unlimited spread, constant return and frightful mortality, on plains without inclosures, where flocks have access to each other, or straggling sheep from one flock are liable, by every-day casualties, to be thrown among those of another flock —has been stated.

It is not improbable that while land remains so low, and the sheep healthy, the actual cost of production in Buenos Ayres will be somewhat less than in the United States; but taking all things into consideration, and looking to the future, I would sooner advise any one, even in an exclusively economical point of view, to purchase the cheap lands of our own Southern States for the objects of Sheep Husbandry, than any part of South America. With the present duty and the cost of transportation against the latter, there is no fear that it can undersell, in *our* markets, the produce of the former. The 7-cent South American wools, washed, will cost 14 cents, and washing will add about 1 cent a pound to the cost.† Add another cent for agent's commission, and also the U. S. duty, and the wool is brought to 20 cents a pound, independent of freight and insurance. which wi l carry it, I should think, to about two shillings. The United States producer can furnish wool of much better quality than the coarse South American article, at this price, and realize a high profit.

But is it said that the 7-cent South American wool sold in our markets in 1845 and 1846, was not all *coarse*—that much of it was actually of a superior quality? This is true. Many of the bales were *partly* made up of an article ranging with American Merino and Saxony wools. But there is little doubt that, to say the least of it, in *very many* such cases, if the invoice of the wool was not fraudulent, nominally, it was rendered so, in reality, by a previous fraud. The *modus operandi* is said to have been as follows: A sends his agent B to Buenos Ayres with instructions to purchase the best lots of wool and pay their market price; and he farther gives him secret instructions to re-sell these wools to C (a second agent) for 7 cents per pound, ostensibly in the ordinary course of business. The second agent C is subsequently sent out to buy, with no *information* of the mission of his predecessor; if he *suspect* the fraud, he has no *direct knowledge* of it, and having purchased wool for 7 cents which cost B 15 cents, he can invoice it at the former rate and support the invoice by his oath.

I have no *direct proof* of an instance of this species of fraud. The commonness of such transactions, however, was claimed to be a matter of perfect notoriety, by individuals who had investigated the subject. Allegations of this kind have appeared again and again in the most public manner, and I have yet to listen to the first denial of them, public or private. Fraudulent invoices are no new thing in our commercial history,* and the great discrimination made by the Tariff of 1842, in the duties on wool, offered the strongest temptations to them. The same kind of fraud may be still practiced, but the inducement to risk seizure for undervaluation is less where the diminution of duty is merely *pro rata* with the diminution of cost, and where getting the latter invoiced at as low a rate as 7 cents, is not followed, as before, by escape from a specific duty and a sudden descent of *five-sixths* in the ad valorem one.

I am free to confess, however, that it has always seemed to me that a determination to vigorously and faithfully discharge their duty in the premises, with a competent *practical knowledge of the quality of the article*, in the proper Custom-House officials, would always, in an unmanufactured staple, and one so readily classified and valued as wool, be a sufficient safeguard against fraudulent undervaluation, to any extent, in the invoice. They might perhaps be undervalued one or two cents on the pound, without making a case strong and obvious enough to justify appraisers in legalizing a seizure; but it is not for gains like these that perjuries would be ventured upon, or double agents and other expensive arrangements for the perpetration of more roundabout frauds, be found profitable.

Not having room, within the limits of this letter, to discuss the capabilities of the Old World to compete with us in wool growing, I will reserve that subject for my next.

---

* If any one dreams they are, let him read a speech on the Tariff made by Mr. Buchanan in the U. S Senate in 1841—another by Mr. Webster on ad valorem duties, made in the same body July 25, 1846, &c.

LETTER IX.

PROSPECTS OF THE WOOL MARKET—FUTURE DEMAND AND SUPPLY.

The Imports and Exports of Trans-Atlantic Nations...Means of ascertaining their Comparative Production...Table of the Imports of England...Amount of Wool grown in the United Kingdom, Consumption, Export, Facilities, including Soils and Climate, for its Cheap Production, and Prospect of its Increase or Diminution—Same of France—Same of Spain—Same of Italy—Same of Turkey in Europe—Same of Germany, including Prussia and Austria, with the exception of Hungary—Same of Hungary—Same of Russia—Same of Asia Minor—Same of Persia—Same of Independent Tartary—Same of Afghanistan and Beloochistan—Same of Thibet, Little Bucharia, and the remainder of China—Same of the Cape of Good Hope—Same of Australia and Van Diemen's Land...Conclusions in regard to Comparative Facilities, etc., of above Nations and the United States...The Northern States can compete with the most favored of them—and of course the South can, to much greater advantage...The South might safely embark in Wool-Growing, relying on the European Market alone...Rapid Extension of that Market Past and Future...But the American Wool-Grower is not compelled to seek a Foreign Market...Our Production does not meet the Demand of our own Manufactories...Table of the Imports of Wool into the United States...Table showing whence we Import Wool...Letter from Samuel Lawrence, Esq., showing the increasing call for Manufactories—The Stability of existing ones—and their ability to compete with those of Foreign Countries.. Extent of our Consumption of Woolens above the Supply made by our Manufactories...Table of Imports of Woolens...Probable Increase of our Manufactories...Reflections on the Tariff...Rapidly Increasing Consumption of our Population—Amount Consumed per head...Table of Increase of our Population... Future Increase...The Amount of Wool Necessary at various Future Periods.

Dear Sir : Probably there are few men who now dream of any danger to the wool-grower of the United States, in the *home* market, from *trans-Atlantic* competition. But there is another point of view, in which a glance at the facilities of the eastern nations, for the production of this staple, may not be uninteresting. *May we not undersell them with the raw material, in their own markets !* He who carefully and intelligently examines all the facts involved in the solution of this question, will find, in spite of the vague popular impressions which prevail on the subject, that so far at least as those nations are concerned, which *now* produce the greatest amount of the wool which supplies the markets of the Old World, the United States *can*, if satisfied with equal profits, *easily undersell* them.

As an importer of the raw and exporter of the manufactured article, England occupies the first place. In these particulars, she probably exceeds, by fully one-half, *all* the other nations of the Old World. France ranks next, and largely takes precedence of the remaining nations. Holland, though shorn, by disastrous political revolutions, of much of her ancient importance in this class of manufactures, still maintains a trade of some magnitude. Several of the German and Prussian States export particular descriptions of woolens; Italy sends out some light cloths ; and Turkey the carpets of that name. A full exhibit of the exports of all the wool-producing nations, would not, of course, lead us to an accurate knowledge of the amount of their production—for there is no one which does not manufacture the raw material to some extent. But with what knowledge we can obtain of their manufactures, the former information would enable us to ascertain, approximately at least, the amount of their production. This is all that is necessary for our present purpose, for we do not now, in reality, so much seek their *actual* as their *comparative* production.

England, as I have before remarked, is the great importer and exporter. Her duties on imported wool are, as has been seen,[*] exceedingly low, and she makes no discrimination in this particular, in relation to bottoms, or the places of export.[†] The vastness and variety of her demand give a

[*] See Letter VIII.
[†] With the exception, of course, of her own Colonies, from which it is exported free.

greater certainty to the exporter of prompt and favorable sales, in her markets, than in those of any other nation. France possesses the advantage of maritime contiguity, for securing the raw product of the nations bordering on the Mediterranean; and therefore, in some instances, as in the case of Turkey, she receives more of that product, in proportion to her manufacturing consumption, than England. But in one respect the latter has the advantage in securing the trade of the Levant. Between the natural products, and, of consequence, the exports of France and those of the other nations bordering on the Mediterranean, there exists a great similarity. She cannot send her wines to Hungary, nor these nor her silks to Italy, in exchange for wool. Her fruits, and indeed all of her natural products are the same with those of the whole south of Europe. England, the producer, and the great mart of the products of Northern Europe, can offer these in the Mediterranean on better terms than France; and in the manufacture of cotton goods, the main article of dress, and consequently one of the great ones of import throughout the whole Levant, the former possesses a decided superiority. All these natural and artificial circumstances have their weight, sometimes in favor of one, and sometimes the other of these nations, in determining the course of trade—and habit, ancient commercial associations, and even national predilections also throw their weight into the scale. In looking at the subject as a whole, however, all these facts, unless in a very few instances, so far offset each other, that in obtaining a view of the wool trade of England—her imports—we obtain a sufficiently accurate picture or index of the *proportionable* exports of all the nations of the Old World.

Before proceeding to ascertain the actual facilities of the several countries named in the Table, for the purposes of wool-growing, it may be well to briefly glance at that of England herself.

Mr. Luccock[*] estimated the produce of wool in England and Wales, in 1800, to be 393,236 packs,[†] or 94,376,640 lbs.; and in 1828, Mr. Hubbard[‡] placed it at 463,169 packs, or 111,160,560 lbs. According to a Table formed by order of a Committee of the House of Lords, the same year, the quantity produced on an average of years, in England, is 111,160,560 lbs. According to Mr. Luccock's estimate, (in 1800,) the number of sheep in England and Wales was 26,148,463. It is not thought to have varied much since. The Encyclopædia Americana,[||] (published 1835,) on the authority of the Edinburgh Encyclopædia, sets down the then present number of sheep in the United Kingdom as follows: in Scotland 3,500,000; in Ireland probably under 2,000,000; in England and Wales the same number as in the time of Mr. Luccock;—so that the aggregate number would be about 32,000,000. It will thus be seen that England and Wales, with an area much less than that of Virginia,[§] have almost 7,000,000 more sheep than the whole number in the United States in 1839!

Large as is the amount of wool produced in the United Kingdom, it does not meet, in the number of pounds, the amount required for woolens consumed in the United Kingdom alone.[¶] It is true that England has exported some combing wool, of her own growth, to meet the wants of a certain class of manufactories (of worsted) in France, which could not obtain stock of equal quality in any other quarter; and she has also exported considerable quantities of her own coarse short wools. Of the latter, I am ashamed to say, the United States *have* been considerable purchasers. The whole export of England, in 1824, amounted to but little over 18,000

---

[*] See Luccock on Wool, p. 311 and Table. [†] A pack of wool is 240 lbs.
[‡] Quoted by Mr. Bischoff—See vol. ii., Appendix. [||] Encyclopædia Americana—*art.* Sheep Raising
[§] The area of Virginia is 70,000 square miles, that of England and Wales 60,000.
[¶] See Bischoff, vol. ii., p. 171.

lbs. From that time it has gradually increased, and in 1838 it reache;
5,851,340 lbs.; in 1839, 4,603,799 lbs.; in 1840, 4,810,387 lbs.* Under the
last year of the late Tariff, we received from England, of wools not costing
to exceed 7 cents per pound, 1,188,800 lbs., and of those exceeding
cents, 28,406 lbs.; and from Scotland, of the cheaper class, 21,132 lbs.
This, however, only shows a surplus in *kind*, not in *quantity*. The Eng
lish short wools have, as has been abundantly shown by the testimony
of her most eminent manufacturers,‡ a *harshness* and *want of felting prop
erties* which render them unfit, unmixed with a better stamp of foreign
wools, for any but the very lowest description of cloths and stuffs, such a
blankets, baizes, army cloths, flushings or bearskins, &c. Nor will the
make *prime* articles, even of these low descriptions. England, therefore
after consuming such portions of these wools as she can, in the manufac
ture of the above-named and similar articles, and by mixing them, in the
nature of an *alloy*, with better foreign wools in a low class of fabrics, such
as flannels, livery and sergeant's cloth, etc., exports the balance to such
nations as are *foolish* enough to purchase it.||

The following Table, compiled from official sources, from Bischoff'
" Comprehensive History of the Woolen and Worsted Manufactures
&c.,"§ gives the imports of England every fifth year from 1810 to 1840

TABLE No. 8.

Countries fm.which Imp'ted	1810.	1815.	1820.	1825.	1830.	1835.	1840.
Russia	32,149	297,611	75,614	1,992,101	202.871	4,024,740	4,518,56
Norway	11,930	40,984			302		
Denmark	351,741	424,82..	13,527	554,213	179,717	366,444	605,521
Sweden	15,424	32,869		3,497	380	1,431	5,861
Prussia	123,037	105,073	107,101	131,100	713,246	256,147	24,646
Germany	778,835	3,137,438	5,113,442	28,799,661	26,073,282	23,794,196	22,812,00%
Holland	} 2,673	432,832	186,031	1,059,243	939,123	{ 301,555	46,247
Belgium						{ 231,22?	134,09?
France		756,427	230,909	436,072	45,093	104,535	48,80
Portugal	3,018.961	1,146,607	95,187	953,780	461,942	683,231	374,91?
Spain	5,952,407	3,929,579	3,536,229	8,206,427	1,643,515	1,602,752	1,266,90?
Gibraltar	349,033	12,891	3,651	19,250		476,737	242,734
Italy	21,554	97.679	2,815	227,453	9,461	1,051.005	1,608,541
Malta	40,040	55,804	5,050	72,131		39,913	2,206
Ionian Isles				25,983			121,110
Morea, &c						816,625	42,89?
Turkey		12,513	189,584	513,414		1,281,839	655,964
Syria							34,04?
Cape of Good Hope	29,717	23,363	13,869	27,619	33,407	191,624	751,741
Africn. other parts						5,102	337,90?
St. Helena							4,68?
East Indies	701		8,056			295,842	2,441,27?
New South Wales	167	73,171	99,415	323,995	{ 973,330	} 4,210,301	{ 6,215,32?
Van Diemen's Land					{ 993,979		{ 2,626,17?
Port Philip							785,39?
Swan River							42,74
South Australia							51,5?
British America	1,217		139	70		14	15,79
British West Indies	2,894	53	760		1,725	2,029	3,2?
United States of America		8,533	578	80,462	7,313	237,306	115,0?
Guatemala							3,00
Colombia							84
Brazil	43,014	4,311	4,277	37	1,148	18,760	9,18·
Rio de la Plata	73,159	41,527	68,759	331,265	19,441	962,900	616,7?
Chili			14,792	2			586,79
Peru				14,313	5,741		
Mexico						1,213,740	
Guernsey and Man	41,407	6,264	19,015	22,266	7,745	246	11.85?
Total Pounds weight.	10,914,117	13,640,375	9,789,020	43,795,281	32,313,059	42,174,532	46,224,76?

* Bischoff, Table 6th, Appendix.   † Report of the Secretary of the Treasury, 1846.
‡ See Bischoff, vol. ii., pp. 107, 153, 154, 163, 173, 175, 176, &c. The testimony here alluded to, or
abstract of it will be given in a subsequent Letter.
|| If these sound like *strong* expressions, I have to say that I shall be prepared to prove them, and sh
so do, in a subsequent Letter, from the testimony of the first manufacturers of England before a Committ
of the House of Lords. Nor were the facts *disputed* by an interest represented before the same Committ
who had every inducement to do so, if they could be sustained in it.
§ See Appendix of the above work, vol. ii. Misled by the title on the *cover*, I have nowhere befo
given the proper designation to Mr. Bischoff's work. Wherever the authority of this gentleman is giv
you will understand that it is derived from the work just named. Published London, 1842.

and therefore indicates, as well as the case admits of—home manufactures remaining the same—the rise or decline of wool-growing, in the several nations, for the period indicated.

It will be seen from the above, that Spain, (and we may include the whole Peninsula,) once so famous for her wools, has sunk to a fifth or sixth rate wool-producing country, and that her exports are still constantly declining; that Germany and Prussia have reached their climax, and are on the wane; that Russia, Italy, Australia and the East Indies are the most rapid increasers.

The high prices of land and provisions—nearly double those on the Continent* (far more than double those on many portions of it)—the onerous gen eral taxes and parochial assessments, will not allow wool to be grown in England for its own sake. The sheep must be reared, as a matter of pure necessity, to sustain her present system of convertible husbandry. A sheep fitted for that object, and to make the most meat in the shortest time, is the main desideratum. Wool is but a secondary consideration. None but the coarse, early maturing breeds will, therefore, ever be grown there. Unless some great revolution should take place in her Agriculture, these are not likely to ever materially increase or diminish from their present number. If any effect is produced on this husbandry by the abolition of the Corn-Laws, I think it will be to diminish rather than increase the num ber of sheep.

France, especially in some of her Southern Provinces, is admirably adapted to Sheep Husbandry. In 1825, the number of sheep in the Kingdom was estimated to exceed 30,000,000, but it is supposed to have materially diminished since that period, by reason of the division of landed property, and other causes.† With a population variously estimated from 163½ to 168 to the square mile,‡ a soil a fair portion of which is well adapted to the growth of bread-stuffs, and the remainder to the vine, fruits, the mulberry (for silk), etc., France finds it better economy to cultivate these, and draw a considerable portion of her supplies of wool from other countries—her fine wools from Germany and Spain, her coarse ones from the regions bordering on the northern shores of the Mediterranean. the Gulf of Venice, and the Black Sea. France exported 84,799 lbs. of wool, costing less than 7 cents a pound, to the United States in 1846.|| This small amount might have been of her own growth, or derived from her transit trade. By the statistical Tables appended to his description of France, by Malte Brun, it appears that of the 51,777,000 hectares§ which he estimates to comprise the surface, 22,818,000 are in arable land, while the entire extent of meadows and pastures (which are divided about evenly) but little exceeds 7,000,000 hectares.¶

Spain, it appears from the Table, now exports less wool to England than Italy or Russia! and is still (as late as 1840) on the decrease. This is not owing to the increase of her manufactures,** or by a diversion of her exports into other channels. The export to France would, undoubtedly, show a similar falling off. That to the United States is but nominal. In 1836 it was but 20,730 lbs.,†† and as this was wool costing less than 7 cents per pound, and came from the Mediterranean side of Spain, it was probably in her ports merely *in transitu.* The Gibraltar trade, given in the Table, I take to be exclusively or mainly a transit one. From the *Balan-*

* See Circular of John Maitland and others, Committee of the Woolen Trade in London—Bischoff, vol n, p. 33. † Bischoff, Youatt.
‡ Mitchell assumes the former, and Morse the latter to be the population.
|| Report of Secretary of the Treasury, 1846. § A hectare is 2 acres 1 rood and about 35·4 rods.
¶ Malte Brun, Am. ed. vol. iii., p. 1029.
** Spain is not estimated to manufacture more than one-twentieth of the woolens consumed by her Cyclopædia Amer., art. *Spain.* †† Report Secretary Treasury, 1846.

za Mercantil,[*] published by the Government, it appears that the exports of Spain of all kinds, in 1826, amounted to only £1,587,507. The exports of raw and manufactured silk and gut reached £243,390 ; lead, £215,360 ; wines, £189,340 ; wool, £161,650 ; fruits, £152,075 ; brandy, £107,715 ; barilla, £79,200, etc. This exhibits not only the smallness of the entire export of wool, but the diminished *comparative* importance of this once great national staple.

The number of sheep in Spain is still placed by many writers as high as 10,000,000 for the migratory flocks, and 8,000,000 for the stationary ones Even Mr. Youatt has fallen into this, as it strikes me, unquestionable error.[†] If Spain possesses 18,000,000 of sheep, what does she do with the wool, which should amount to at least 54,000,000 lbs.? Admitting—which probably exceeds the fact—that her export to France and other nations equals that to England, and that she manufactures a quantity equal to twice her whole export, the aggregate amount would be less than 8,000,000 lbs. The author of the article on Sheep Raising in the Encyclopædia Americana, places the number of the whole fine-wool sheep in Spain at 4,000,000. This I think high enough, and probably not far from the truth. This is a million less sheep than those of the State of New-York in 1839 !

The actual facilities for growing wool in Spain have already been alluded to in my fifth Letter. I should not consider it necessary to bestow farther examination on them, were it not for the fact that owing to various associations connected with the early history of the Merino sheep, and the lead once taken by Spain in the production of fine wool, her facilities have been, popularly, prodigiously overrated, and even the difficulties under which she has labored for this husbandry, magnified into advantages. Her northern mountains are high, broken, cold, and exposed to peculiarly piercing north winds,[‡] and the winter on them lasts, as I infer from Mr. Livingston, about six months. He says :[||]

"When the severe weather commences on the mountains, the shepherds prepare to depart, which is generally about the end of September and throughout the month of October to seek more temperate climates and fresher pastures In April or May, according as the season is late or early, they return to the mountains.

It *might* be practicable to prepare hay for winter use, in favorable positions, and particularly on the *parameras*, on these mountains, and thus the migratory sheep might become stationary on them. But the Spaniard is too much wedded to ancient customs, too little in love with change of any kind, and, most of all, a change bringing an addition of *labor*, to thus innovate on his own habits or those of his flocks.

The high basins of the Douro and Tagus (embracing the two Castiles and Leon) are too valuable for the cultivation of grain, vineyards, fruits, etc., to be profitably devoted to the pasturage of sheep. The wheat of Spain is among the best in Europe,[§] and it is stated in Mr. Jacob's *Tracts on the Corn Trade*, that she frequently does not raise enough for her own consumption.[¶] For the vine, olive, fig, mulberry, barilla, and various other products of equal profit both for home consumption and for export, she is not excelled probably by any country in Europe. A friend of mine who traveled in Spain in 1845, describes the valleys above alluded to, as almost exclusively devoted to tillage crops. In the Southern Provinces,

---

* Quoted by McCulloch—Com. Dic. art. *Cadiz.*
† See Youatt on the Sheep, Lond. ed., p. 147 *et supra.* Mr. Livingston in his day estimated the migratory sheep at 5,000,000, the stationary at 8,000,000. See Essay on Sheep, pp. 36, 39. Mr. L. was also undoubtedly in error.   ‡ Malte Brun. || Livingston on Sheep, p. 36.
§ Note by Percival to Am. ed. of Malte Brun : art. *Spain.*
¶ Quoted by McCulloch—Com. Dic.; art. *Odessa.*

where rain does not sometimes fall for months in the summer * the grass becomes entirely dried up, so that flocks, to be made stationary there, would require hay or other prepared food for several of the summer months

The *Transhumantes* or migratory flocks must still continue, then, to travel from the northern mountains to the warm basins of the Guadiana and the Guadalquiver for their winter quarters, and return to the mountains in the summer, or this branch of the husbandry would undoubtedly become extinct. The effect on the health and condition of the sheep, and the important item which it would form on the debit side of the account in Sheep Husbandry, to thus drive flocks a six weeks' journey twice a year, (consuming nearly *a quarter of the year* on the road,) can be estimated by any one acquainted with such matters.† The losses and expenses thus incurred would absorb all the profits of the husbandry, were it not for the extraordinary privileges conferred on the flockmasters (mainly consisting of the King, nobles and clergy) by the absurd and tyrannical regulations of the *Consejo de la Mesta.*‡ The abolition of the "Council of the Royal Troop," there cannot be a reasonable doubt, would be immediately followed by the downfall of the migratory Sheep Husbandry in Spain. That the day has gone by when this unfortunate and distracted country can ever again enjoy the blessings of permanent peace and settled institutions, under which this or any other branch of husbandry can increase or steadily flourish, until she reaches a point of political civilization entirely incompatible with the continuance of a relic of tyranny and barbarism so monstrous as the *Mesta,* I consider equally certain. I see, therefore, no possible, or at least probable contingency under which the migratory Sheep Husbandry of Spain is likely to be extended, or even to permanently maintain its present footing. Nor is there any probability of her again rising into importance as a wool-producing country, from her stationary flocks.

Italy, though too accessible to the dry, hot wind of Africa, (the *Solano,*' to exhibit the uniformity of deep-green verdure seen north of the Alps, is nevertheless—much of it—a country of fine pasturage. The great plain between the Alps and Appenines, the basin of the Po—including Lombardy, Sardinia, Parma, Modena, etc.—is one of the most productive in Europe, and its extraordinary facilities for irrigation allow five or six crops of hay to be mown in a single season. In Tuscany, the orange and lemon begin to make their appearance—the soil is alluvial and rich, and the mountainous districts are finely adapted to pasturage. The States of the Church are also highly fertile, and abound in good herbage; and on the deadly *Campagna di Roma,* and even the Pontine Marshes, flocks and herds find an abundant subsistence in winter, and are driven to the Appenines in summer. The same remarks apply to the northern portions of the Kingdom of Naples. The southern extremity of Italy is exposed to a burning climate, and exhibits the vegetation of Africa.

The whole superficial area of Italy does not exceed 122,000 square miles, and her population is 172 to the square mile. Scarcely raising bread-stuffs enough for her own consumption, taking one year with another,|| there is not the most remote prospect of her ever becoming an important wool-exporting country.

* See Hon. Wm. Jarvis's Letter to me on the subject of Merino Sheep, when I acted as Corr. Sec'y of the N. Y. State Agricultural Society—Transactions, 1841, p. 322.
† Since giving this as the distance from "the middle of Estremadura to the Cantabrian Mountains" (Let ter V.), I see it stated in the Encyclopædia Americana that "the whole journey from the mountains to the interior of Estremadura is reckoned at about 690 miles." Measurement on the map will show that it does not exceed 4 degrees or 277 miles, but the difference may be made by the circuitousness of the route, or the writer may refer to more eastern portions of the great Appenine Chain. I find it stated by several writers that each journey consumes six weeks.
‡ For a description of this odious tribunal see Livingston on Sheep, p. 35.
|| See McCulloch's Com. Dic.; art. *Odessa.*

P

Turkey both in Europe and Asia, it would appear from Table 8. is but a trifling exporter of wool. It should be remarked, however, that the wools of the Western Provinces, and of Greece, are generally exported from Trieste to France.[*] Under the late American Tariff, ("Tariff of 1842,") the export to the United States was becoming an important one—much greater than that to England. In 1846, it amounted, of wools costing less than 7 cents a pound, to 5,744,328 lbs.[†] European Turkey has a colder and less uniform climate than Italy, but still it is a fine one,[‡] and being a broken, mountainous country, well adapted to pasturage, and but sparsely populated, (55 to the square mile,) it is wonderful that so little attention has been paid to the culture of wool. But the proud and indolent Turk spurns all rural labor, or all interest in it, leaving it to his vassals—and these, destitute of any security to person or property, taxed, oppressed, liable to be compelled to make forced sales to bey or ayan—or, what is worse, their property seized outright—have little inducement to accumulate a species of property so easily pounced upon.[||]

Germany (including Prussia and Austria) is now the great producer of fine wools, supplying not only her own manufactories—which are estimated to consume half the whole product—but exporting the large surplus indicated in the Table. Nor is this all; for to France, the Netherlands, Switzerland, &c., she is supposed to export half as much as to England.[§] The whole region thus included—leaving out the Austrian States in Italy, which have already been considered—comprises a territory of 468,000 square miles, and a population of 58,800,000, or 130$\frac{2}{3}$ to the square mile. The country on the north is level, vast plains extending from the declivities of the mountains which occupy the center of Germany, to the North Sea and the Baltic. The center is mountainous, and its plains are ery elevated. The extreme South is covered with mountains. From the Little Carpathian or Jablunka Mountains, and from the eastern termination of the Styrian and Julian Alps, stretch away the vast Hungarian and Transylvanian plains to the confines of Turkey.

The great northern plain of Germany is low, sandy, flat, often consisting of naked silicious sands or those covered with lichens, interspersed with frequent marshes, and terminating in many places on the Baltic in vast morasses, or land redeemed from the sea by dikes. As a whole, the land, particularly in the maritime Provinces, is of an inferior quality, but some portions of it, as for example in Silesia and Saxony, is of a quality ranging from medium to good. The soil of Central and Southern Germany (including Austria) must, of course, exhibit many varieties. In general, however, it may be set down as productive in the valleys, and ordinary or poor on the high lands. The lower plains of Wirtemberg. Baden, the South of Bavaria, etc., are exceedingly fertile. The plains of Hungary on the south-east not uncommonly exhibit soils of remarkable richness, but they alternate with inferior ones, and with vast and unhealthy morasses. Taken together, the region which I have included under the designation of Germany, though not a sterile country, is not favored with soils naturally as productive as those of Italy or Spain; nor would it at all compare with that portion of the United States west of the Apalachians.

The climate of Germany is thus summed up by Malte Brun:[¶]

* Southey, quoted by Bischoff. vol. ii. p. 356.   † Report of the Secretary of the Treasury, 1846.
‡ For a picture of this as well as the other natural features of Turkey, both in Europe and Asia, Greece, and the Ionian Isles—as delicately accurate, as soft and rich as one of the scenes of Claude—see Childe Harold, Canto II., the opening of the Giaour, the Bride of Abydos, etc. Though this may be deemed a singular, it is the very best reference, which my reading enables me to make.
|| See Urquhart on Turkey and its Resources, p. 139.   § Encyclopædia Americana: art. Wool
¶ Am. ed., vol. ii., p. 594.

" The climate of Germany is greatly modified by the elevation and declivities of the coun
try; but independently of that cause, it does not admit, from its extent in latitude, of any
vague or general definition. It may be divided, however, into three great zones, and these
too, are susceptible of other subdivisions. The first is that of the northern plains, of which
the temperature is not so cold as it is humid and variable; they are exposed to every wind,
while fogs and tempests are conveyed to this region from two seas. The north-west plain is
subject, from its vicinity to the North Sea, to frequent rains and desolating hurricanes. The
influence of the Baltic on the north-east plain is less powerful; the climate, though colder,
is not so humid and variable.

The second general zone comprehends all the central part of Germany. . . . The mountains in that extensive region form a barrier against the effects of the maritime climate. The
sky is not obscured by mists, and the regular order of the seasons is not interrupted by
winds and tempests; but the elevation of the soil renders the climate colder than in other
countries in the same latitude nearer the level of the sea. . . . The third general zone is
that of the Alps. The lofty hights and rapid declivities connect very different climates;
thus the culture of the vine ceases in Bavaria and Upper Austria, and appears anew with
fresh vigor in the neighborhood of Vienna. The eternal glaciers of Tyrol and Salsburg are
contiguous to the valleys of Styria and Carniola, covered with fields of maize or vineyards,
and almost border on the olives of Trieste and the lemon-trees of Riva."

Contiguous mountains render the north of Hungary extremely cold.
Farther south, the climate rapidly becomes warmer, and on the lower
plains in the extreme south the heat is intense and the climate insalubrious

The *bauer* or farmer in those States of Germany where the feudal tenures have been abolished, and the land is held in fee simple, owns four or
five English acres of land. These men, says Mr. Jacob,

"although placed above the pressure of want, or possessing the bare necessaries of life,
have very little beyond them. Such as are industrious and frugal, by cultivating their small
portion of ground, may raise a sufficient quantity of potatoes for their own consumption, corn
for their bread, and provisions for two draught oxen. They all raise a small quantity of
flax, and some few *contrive to keep five or six sheep*. It is often no easy matter for those to
find occupation, who are desirous of other employment in addition to the cultivation of their
own land, for no agricultural labor can be carried on during the long and severe winters. . .
It is rare indeed that they can afford to have meat of any kind, and those only who are
more prosperous than their neighbors can keep a cow to provide themselves with milk."

The wool raised by these owners of five or six sheep, is annually
bought up by Jews and other traveling agents, who go from house to
house to collect it.

The following extracts from William Howitt's sprightly and interesting
" Rural and Domestic Life in Germany " will show under what circumstances a great portion of its wool is grown:

" Here you look in vain for anything like the green fields and hedge-rows of England. . . .
It is all one fenceless and plowed field. Long rows of trees on each side of the road are all
that divide them from the fields. . . . The keeping up of the cattle presents you a new
feature of rural life. As the quantity of land left for grass is very small, the grass is proportionably economized. The little patches of grass between woods and in the open parts of
the woods, the little strips along the river-banks and even in gardens and shrubberies, are
carefully preserved for this purpose. You see women in these places cutting grass with a
small hook or smooth-edged sickle, and carrying it away on their heads in baskets for their
cows. You see the grass on the lawns of good houses, on grass-plats, and in shrubberies,
very long and wild; and when you ask why it is not kept closer mown, the reply is that it
is given to the milk-woman, often for a consideration, who cuts it as she wants it. You see
other women picking the long grass out of the forests, or under the bushes on the hill-sides
where the slopes have been mown, for the same purpose. . . . The children may be seen
tanding in the stream in the villages carefully washing weeds before they are given to the
cattle. . . . Nettles, chervil, cow-parsnip, which in England are left to seed and rot, are
all here cut for the imprisoned cow. You go down to the river-side to fish, and a peasant is
soon with you, chattering and gesticulating, pointing to your feet and to the grass. It is to let
you know that you are not to angle there, because it treads down the grass; and accordingly,
in Germany, with rivers full of fish, you seldom see an angler; if you, he is pretty sure to be an
Englishman. . . . Not a sheep, a horse, or a cow is to be seen. . . . The mountain tops are
covered with wood. The slopes are covered with vineyards. You ask where the cattle are?
You are answered, in the stalls. Where are the sheep? Under the care of shepherds,
somewhere—Heaven knows where! you never come across them. It is only on the great

plains of the North that you afterward find large flocks and herds, under the care of keepers, kept close together; for as they have no fences, they are under the momentary peril of making ravages on their neighbor's crops."

Between Leipsic and Berlin, on the plains of Saxony, Mr. Howitt first saw flocks of sheep in the field, and he says:

"One thing which surprises an Englishman is to see what wretched creatures are the sheep which produce the famous Saxony wool. . . . . In fact, it is a prevailing idea that the leaner the sheep the finer the wool. It is the wool to which all the attention of the grower is devoted, and therefore, generally speaking, a more miserable assemblage of animals than a flock of German sheep is not to be seen. . . . . On the plains they wander under the care of a shepherd, and for the most part on fallows and stubbles, to pick up odds and ends, rather than to enjoy a regular pasture. You may see them penned on a blazing fallow, where not a trace of vegetable matter is to be seen, for the greater part of a summer day, which in this climate is pretty much like being roasted alive. . . . . For what purpose they are here, except to starve and melt them into leanness, I never could discover. . . . . The sheep, besides being lean, are generally dreadfully lame with that pestilent complaint the foot-rot, and their keepers, apparently, trouble themselves very little about it."

Mr. Howitt states that it is necessary to economize the land so closely, to sustain the population, in some parts of Germany, that the peasants actually convey earth up steep hill-sides in baskets, and cover the rocks with it, to thus add to the tillable soil!

In reviewing the preceding facts, you are struck with no one which would indicate particular natural advantages for sheep rearing in the States of Germany, Prussia, and—with an exception presently to be named—Austria. The climate of the North is humid, fickle and tempestuous; that of the middle cold with long winters. Neither possess any advantages over our own *Northern* States—and in some respects are decidedly inferior to them. This was the opinion of that eminent sheep-breeder and excellent man, Henry D. Grove, of this State, who was a native of Prussian Saxony, and who certainly would never be suspected by any one who knew him personally, of any want of partiality for anything pertaining to his *Fatherland!* In his letter to Benton and Barry on wool-growing, &c.. he says:

"Ten years' experience has fully satisfied me on this point. In some respects, we possess natural advantages over Germany."

In what particulars he awarded the preference to the United States, his letters and oral declarations to me, leave no uncertainty. It was both in *soil* and *climate*, and in instituting the comparison, he had his eye not on the most favored sections of our country, but on the hills of Rensselaer County in this State, where he resided.

If in *natural* advantages we surpass Germany, how much more we do in *artificial* ones, may be estimated from the preceding extracts from Messrs. Jacob and Howitt. To these general remarks portions of Hungary form an exception. In these, the climate is fine, the soil rich, and, the feudal tenures remaining unabolished, the land is yet held in those large estates so favorable to Sheep Husbandry. Prince Esterhazy, the former Austrian Ambassador to England, says Mr. Paget,* owns an estate of something more than 7,000 square miles, including 130 villages, 40 towns, and 34 castles. His sheep are said to amount to 3,000,000.† Other nobles own flocks of from ten to thirty thousand. The demi-savage Magyar serf, whose labor costs nothing, whose principal garment is a sheep-skin, and whose miserable and scanty food is more than half stolen,‡ makes a most *economical* shepherd! Hungary lacks facilities for internal communication, and her convenience to the Mediterranean markets—excepting Turkey—so as

* Paget's Hungary and Transylvania, vol. i, p. 46.   † Youatt.
‡ See Paget's Hungary, &c., p. 13 to 19.

to first throw her agricultural products into ports where the demand is good, is decidedly inferior to that of Italy, France and Spain. The Danube is the only natural outlet to her commerce—which, thanks to a liberality of policy on the part of Turkey,* contrasting most favorably with that of several *enlightened* nations † under similar circumstances, she enjoys without limitation. To reach Trieste, a long land carriage is indispensable. Her exports too, are embarrassed by the imposts and narrow restrictions of the Imperial Government. She cannot, therefore, export cheap heavy articles, such as provisions, to so great advantage as the Levantine nations : but every circumstance points to her as a country which should be one of the first on the Eastern Continent, for the production of wine, silk, wool, &c.

Separated from Hungary and Transylvania only by the Carpathian Mountains and Turkish Moldavia, lie the fertile provinces of South-eastern Russia; the basins of the Dniester, the Dnieper, and the Don. From the Carpathians to the Caspian, across the entire extent of the plains of ancient Scythia, not an elevation which could be properly dignified with the appellation of a mountain, breaks the immense expanse! The lower valley of the Dniester or Borysthenes, formerly known as the Ukraine, has been celebrated for centuries for its pasturage—for its horses ‡ and cattle : and recently flocks of Merino sheep have been introduced there and successfully crossed with the native variety. In 1839, Mr. Slade states that many of the colonists on the Steppe and in Bessarabia had 20,000 sheep. Merinos were introduced into Crimea or Taurida, by M. Rouvier, a French adventurer, in about 1802.‖ In this favored peninsula, which the learned Pallas describes as little less than an earthly Paradise, they have multiplied exceedingly, and extended to Cherson, Ekaterinoslav, Bessarabia and other provincial Governments.§ The export of wool from Odessa in 1829 was 3,402 lbs.; in 1830, 21,361 lbs.; in 1831, 35,058 lbs.; in 1832, 41,558 lbs.; in 1833, 66,457 lbs.; in 1834, 66,901 lbs.¶

In one respect Southern Russia has the advantage over Hungary. It is more sparsely populated, and land is perhaps in still lower estimation. As in the latter, the land, much of it, is fertile and well adapted to pasturage, and the price of labor is next to nothing. But for causes adverted to in the opening part of my eighth Letter, there is a wide disparity in the climates of the two countries, if we leave Crimea out of view. That of Russia, affected by the north and north-east winds—which the Carpathians exclude from Hungary—has a winter which for length and intensity is entirely unequaled in the latter, excepting in its northern mountainous regions. Sheep must be housed, and fed for some months on dry food, in Southern Russia. Taking into view the broad, level *steppes** and their luxuriant natural verdure—taking into view the climate, warm in summer, cold and exposed to winds of great severity in winter, it strikes me that there must be no inconsiderable resemblance between this portion of Russia and our own north-western prairies in corresponding latitudes (45° to 46°). But when the cost of land and labor is taken into consideration, wool can be produced cheaper, in my judgment, in South-western Russia than in Spain, France, Germany, Italy or any other portion of Europe, excepting Hungary. Were

---

* This power is remarkable for its liberality in all its regulations which affect the trade and commerce of other nations.

† *e. g.,* the policy of England in relation to the navigation of the St. Lawrence.

‡ This wild region and its horses have been rendered classic by Mazeppa. Who, that ever read, has forgot the description of the horse on which the Hetman performed his fiery and perilous ride !

‖ For an interesting account of the adventures of this fortunate French Jason, see Slade's " Travels in Germany and Russia," published London, 1840.

§ See Slade's Travels; also McCulloch's Com. Dic.—*art.* Odessa.

¶ McCulloch's Com. Dic.—*art.* Odessa.

** This Russian word has a similar signification to *prairie, pampas, llanos,* &c

European Turkey differently populated, and under different institutions, it might constitute another exception.

Central and Northern Russia, like the States north of Germany, are north of the wool zone. Their winters are too long and severe to allow them to compete with regions lying farther south, in wool-growing.

Asia Minor, or Turkey in Asia, and Persia have been alluded to—the former, much of it, a fine country with a most delightful climate, but its natural advantages all neutralized by its political systems and the character of its population—the latter, except in occasional favored positions, such as the valleys of Shiraz and Ispahan, a land of mountain and desert, of intense heat and intense cold.

Independent Tartary, lying immediately north of it, is less exposed to the hot winds of Arabia, but more so to the freezing ones of Siberia. Its vast dry plains are usually deserts, excepting on the borders of its exceedingly rare streams. Great Bucharia, however, in the south-east, on the head waters of the Amoo (Oxus)—from the Capital of which Timour ('Tamer lane) issued on his desolating path of conquest—is a country of great fertility. Its natural beauties constitute a favorite theme with the poets and geographers of Persia and Arabia. Since the opening of the navigation of the Indus, it has annually sent some wool to Bombay, which constitutes a part of that which is shipped thence to England, and is known in Table 8 as East Indian wool.

Afghanistan and Beloochistan, protected on the north from the Siberian winds by the lofty Hindoo Koosh mountains, and less exposed on the south to those of Arabia, exhibits a milder and less variable climate than that of the conterminous regions of Persia. Among the Highlands of the north, and those skirting the Indus on the east, there is much good pasturage. Sir Alexander Barnes states that four-fifths of the whole surface of Cabul, a Province of the former, is excellent pasture land. The wool of the broadtailed sheep of these countries also finds its way, by the Indus, to Bombay, and is classed as East India wool in the Table.

From the high, cold, mountain regions of Thibet, Little Bucharia, &c., some wools are exported, through the same channels, which come under the same classification. These countries also export shawl wool.* Most of China north of the great Desert of Cobi is a cold, mountainous country. The southern portion, or China Proper, is too densely populated and closely cultivated to be devoted to pasturage.

The wool trade which followed the opening of the Indus (the raw material being supplied by Afghanistan, Great Bucharia, Thibet and some of the Hindostanese Provinces) might doubtless be swelled into one of great importance, particularly by introducing finer breeds of sheep; but we can scarcely expect this, from what we know of the habits, agricultural and commercial, of the population. Among constant political changes wrought by the only Asiatic argument—the sword—the personal habits and occupations of the Asiatic remain ever the same, and are, perhaps, the best type of persistency to be found in anything short of immobile matter. Indeed, the stony features of the Sphinx have changed scarcely less through revolving generations, than have the ethnic ones of this great family of the human race!

Let us now pass to those regions of the Old World, south of the Equator, included in the wool-growing zone.

The southern extremity of Africa—the Cape of Good Hope—is included

---

* The table-land of Thibet is elevated 15,000 feet above the level of the sea. Mr. Trail remarks that every animal here, including Carnivora, produce that down under their hair which is known as shawl wool—though that manufactured comes mainly from a species of goat.

in the wool-growing zone. The following description of it is by Rev Robert Moffat, for twenty-three years a resident of it as the agent of the London Missionary Society :*

" The Colony extends from west to east about six hundred miles, its average breadth being about two hundred. . . . . Between the coast and the vast chain of mountains, beyond which lie the Karoo, the country is well watered, fertile and temperate. The other portions of the Colony, with few exceptions, and without a change in the seasons, appear to be doomed to perpetual sterility and drouth. The Karoo country, which is in the background of the Colony, is, as Lichstenstein correctly describes it, a parched and arid plain, stretching out to such an extent that the vast hills by which it is terminated, or rather which divide it from other plains, are lost in the distance. The beds of numberless little rivers, (in which water is rarely to be found) cross, like veins, in a thousand directions, this enormous space. The course of them might, in some places, be clearly distinguished by the dark green of the mimosas spreading along their banks. Excepting these, as far as the eye can reach, no tree or shrub is visible. . . . . But even on these hills and sunburnt plains thousands of sheep pasture on a thin sprinkling of verdure and esculents. . . . . The entire country, extending in some places hundreds of miles on each side of the Orange River, and from where it empties itself in the Atlantic, to beyond the 24th degree of east longitude, appears to have the curse of Gilboa resting upon it. It is rare that rains to any extent or quantity fall in those regions. Extreme drouth continues for years together. The fountains are exceedingly few, precarious, and latterly many of these have been dried up altogether."

According to Barrow, nearly seven-tenths of the Colony are destitute of vegetation during a greater part of the year. Sand drives before the winds, exercising an unfavorable influence on sheep and wool. Lions, tigers, wolves, hyenas, jackals, wild dogs, etc., are numerous on the very skirts of the settlements, making much vigilance necessary for the protection of the sheep; and they must be nightly driven into the settlements to be folded. But the natives have proved a vastly more destructive enemy than these.† The sheep introduced by the English colonists will probably eventually considerably increase beyond their present number in a country of so great extent, but we are scarcely authorized to believe that the Cape will ever take a high rank among the wool-producing countries of the world.

That great island, or continent, known as New South Wales, or Australia, has a superficial area equaling that of the United States. But a limited portion of it, however, is included in the wool zone. All of Van Diemen's Land, or Tasmania, is in that zone. The export of wool from these countries, as will be seen from the Table,‡ reached nearly ten million pounds in 1840—nearly half that of Germany, including Austria and Prussia, and almost *eight times* that of Spain ! Here, as at the Cape of Good Hope, there are no woolen manufactories, and being Colonies of England, their export to that country exhibits their *whole* production.

The soil, products, &c. of Australia are thus spoken of by Mr. McCulloch : ||

" The fertility of the soil in most parts of New-Holland that have been explored with any care, is very far indeed from corresponding with the glowing descriptions of some of its casual visitors, whose imaginations seem to have been dazzled by the magnificence of its botanical productions and the clearness and beauty of the climate. The truth is that the bad land bears a much greater proportion to the good in New-Holland than in almost any other country with which we are acquainted. . . . . Of course it is not to be supposed that in a country of such vast extent there must be some fertile districts ; but along the east coast, with which we are best acquainted, these seem to be much more confined than might have been expected ; and the little experience we have had on the west side, at Swan River and other places, does not seem to lead to any more favorable conclusions."

After stating that if the Government price of lands " is not a great deal

* Missionary Labors and Scenes in Southern Africa, pp. 23—24.   † See Letter V., and Note.
‡ Including Port Philip, Swan River, and South Australia, the exports of which are carried out separately at Table 8.
|| McCulloch's Com. Dic.—*Art.* Sydney.

above the mark in New-Holland, it must be a great deal below it in Upper Canada," Professor McCulloch continues :

"If the Americans exacted the same price for their public lands that we do, something might be found in favor of extending the principle to Canada. They, however, do nothing of the sort, but sell much better land at a decidedly lower rate. . . . . If slaves could be imported into a Colony of this sort, there might be some chance of its succeeding. But while land of the very best quality may be had in the Valley of the Mississippi for about a dollar an acre or less, we think better of the common sense of our countrymen than to suppose that any one able to carry himself across the Atlantic will resort to Australia."

Of the climate he says :

" The climate of such parts of New South Wales as have been explored by the English is particularly mild and salubrious. . . . . On the other hand, however, it has the serious defect of being too dry. It seems to be subject to the periodical recurrence of severe drouths. These prevail sometimes for 2, 3, or even 4 years together. The last ' great drouth' began in 1826, and did not terminate until 1829. Very little rain fell during the whole of this lengthened period, and for more than six months there was not a single shower. In consequence, the whole surface of the ground was so parched and withered that all minor vegetation ceased; and even culinary vegetables were raised with much difficulty. There was also a pretty severe drouth in 1835. This is the great drawback of the Colony; and were it more populous the drouths would expose it to still more serious difficulties."

Another drouth occurred in 1841, and Mr. Hood thus describes its effects on the sheep : *

" It will be scarcely believed in England that the estimated number of sheep which have died within the last twelve months in the Colony from catarrh and drouth is 70,000 !! that colonists are compelled in order to save the dam from starvation, to cut the throat of her lamb ; that no means are adopted for securing a stock of lambs for next year ; or that a stockholder would offer 8,000 sheep to any one that would remove them from his runs, and finding that no one could be prevailed upon to taint his own flocks by accepting so dangerous a present, had recourse to consuming them by fire, and had actually killed and burnt 2,000." . . . .

Of the country Mr. Hood remarks :

" The first object on the arrival of every settler should be to procure a good country for his flocks, and this, I have elsewhere said, is his grand difficulty. Let him be wary on this point. Almost every desirable or habitable spot in the old countries, as the early settled districts are called, is already occupied."

Some diseases seem to be peculiar to the country, or, rather, peculiarly inveterate in it. Mr. Youatt says :†

" The sheep frequently suffer from the wild and poachy nature of a considerable portion of the pasture. The foot-rot seems to assume a character of its own. . . . . If neglected, it speedily becomes inveterate and preys upon and destroys the animal. The losses occasioned by it in the early existence of the Colony were frightful."

The astringency of the water and other causes have produced severe epidemics. In some years, some of the flockmasters have lost half of their sheep.‡ The scab is a prevailing disease, and Doct. Lang says :‖

" When a convict shepherd has a pique against his master, or even against his overseer, it is often in his power to subject the whole of his master's flock to this obnoxious disease, merely by driving his own flock a few miles from their usual pasture, and bringing them into contact with a diseased flock. The chief source of the wealth and prosperity of the Colony is thus, in a great measure, at the mercy of the most worthless of men."

The cost of both land and labor is comparatively (*id est*, compared with the unoccupied lands of the United States) high. The Government minimum is 5s. ($1 15) per acre, but very little if any good land is sold at that price. Mr. Hood states that the portion of Capt. McArthur's immense estate which was obtained by purchase, cost, on the average, 7s. 6d. ($1 72½) per acre. Shepherds receive from £15 to £20 ($69 to $92) with

* Quoted by Spooner in " History, Diseases, &c., of the Sheep." London, 1844, p. 67.
† Youatt on Sheep, p. 189.   ‡ See Spooner, pp. 417-421.
‖ Lang—Historical and Statistical Account, vol. i., p. 351.

a *house and rations*, per annum ; overseers of a superior description £50 to £60 ($230 to $276),* also with a house and rations.†

The sheep are exposed to the depredations of various animals, but the wild dog is their most dangerous enemy, with the exception of the *runaway convict*. The sheep are therefore folded nightly, guarded by a watchman with his dogs, and with a fire to scare away the wild beasts.‡ One shepherd usually takes care of about 300 sheep, and " in the more sterile parts of the Colony, *where three acres of the uncultivated ground are scarcely sufficient for the support of one sheep*, the labor is very severe." ∥

Mr. Samuel Lawrence recently wrote me :

" I saw a gentleman from England a few months since who has an admirable flock in New South Wales, of twenty-five thousand sheep, and he assured me he had not received a penny of income from them since 1838."

Van Diemen's Land (containing 28,000 square miles) is claimed by Mr Youatt§ to be superior in several respects to Australia as a wool-growing country. Table 8 does not, however, show that its exports increase any more rapidly.

Both of these Islands, as colonies of Great Britain, send their wool to the latter duty free, and they save 1 cent per pound on wool costing less than 24 cents, and 2 cents on that exceeding that value. But this by no means offsets against the additional cost of freight, over that exported from the United States, Hungary, or the south of Russia. While it is only 3,375 miles from New-York to London, it is not less than 13,000 miles from Sydney or Hobart's Town to the latter place. Professor McCulloch states (art. *Sydney*) that the expense of conveying a passenger to Sydney is about three times that of conveying one to Quebec. I see no reason why a corresponding difference should not exist in the freights ; and in that case, freights from the United States would be two-thirds less than from Australia.

I pretend, Sir, to no power of vaticination on this subject, but the conclusions which *I* draw from a review of all the foregoing facts are as follows :

1. That wool-growing is never likely to permanently and importantly¶ increase in any of the countries of Europe, unless it be in Hungary, Turkey, and the south of Russia.

2. That it is more likely to decrease than increase in Great Britain, France, Portugal and Italy.

3. That such a decrease is next to certain in Spain and Germany, (including Prussia and Austria in the latter,) excepting Hungary and Transylvania ; that the decrease will be much more considerable in Germany ; that its rapidity and extent will be proportioned to the rapidity and extent with which the market is supplied from countries which can grow wool cheaper, such as North and South America, Hungary, Southern Russia, and Australia.

4. That wool-growing will undoubtedly largely increase in Hungary and Southern Russia—and that it *ought to* in European and Asiatic Turkey but will not, extensively, until the character of the people and their political institutions are changed.

5. That it will also increase at the Cape of Good Hope, Australia and Van Diemen's Land ; but that its economical extension in either of these countries is limited, especially if America becomes a competitor.

* Calling the English shilling 23 cents, according to Report of Director of U. S. Mint, 1827.
† Report of a Committee, &c., quoted by Mr. McCulloch—Com. Dic. ; art. *Sydney*.
‡ Cunningham's Two Years in New South Wales, vol. i., p. 254.
∥ Youatt on the Sheep, p. 188. § *Quæm vide*, p. 190.
¶ I say "importantly," because Sweden, Norway, Denmark, &c., in that spirit of rendering themselves in dependent of *foreign* supplies, which characterizes all nations, may, and probably will extend their wool culture ; but it will be too unprofitable a struggle against Nature, to be carried to a very great extent.

Q

6. That no part of the Eastern Continent or its islands, all things considered, possess equal advantages for wool-growing with some parts of the United States. 1. The climate of many portions of the latter (in the South) is not excelled by that of the most favored situations in Hungary or Australia; and in this respect it is decidedly superior to the south of Russia. 2. The soils of vast sections of the United States, with the above climate, are more *uniformly* fertile and adapted to pasturage than those of either Hungary or Southern Russia—and, as a whole, are entirely superior to those of Australia. 3. The regions alluded to in the United States, are better watered with running streams than either of the other named countries—have not the vast and unhealthy morasses of Hungary—and are not subject to the destructive drouths of Australia. 4. The land is cheaper in the United States than in Australia, and (my impression is) than in Hungary or Southern Russia; and, in the Southern States, labor costs no more than in the two latter, and far less than in the former. 5. In accessibility and nearness even to the great *European* wool market, the United States stand on equal terms, at least, with Hungary and Southern Russia, and the distance from Sydney (in Australia) to London is nearly *four* times the distance from New-York to London. 6. In *no respect* do either of these countries, the most favored in the Old World, excel, in my judgment, for the purposes of Sheep Husbandry, large portions of the United States; and I believe those portions of the United States can sell wool in the English market at a better profit on all the capital invested than either of the above countries, with the possible exception of the most favored portions of Hungary.

Our surplus wools can, therefore, at any time, be exported to England at a *reasonable* profit. This is true, even of wools grown in the Northern States. In 1845, the United States exported wool, (mainly to England,) to the value of $22,153; and in 1846, to the value of $203,996. This was a commercial experiment, and although it is not understood to have resulted in any profit to the exporters, the wool sold at an advance on the American prices current—and would have sold so as to have realized a handsome profit to the exporters, had it been properly sorted and otherwise prepared to meet the requisitions of the English market. Statements of this kind have been published by one of the most prominent of the exporters. It would seem, from Mr. Lawrence's statement, already quoted, that the prices of Australian wools have not yielded a profit over all expenses, during the same years. The quality and style of our wool have been praised by the English press, and are understood to have given high satisfaction to the English manufacturers. On the whole, then, we may regard this experiment as a successful one. The American prices current of those years were about 32 cents per pound. We have seen that the actual cost of wool (including all expenses, and 7 per cent. on price of land and sheep) in the Northern States may be set down at about 27 cents per pound.* These facts show that a *remunerating* price can be obtained for even *Northern* wool in England—if a profit on investment considerably exceeding the highest legal rate of interest (7 per centum) is to be considered "remunerating." And if this is true of the Northern wools of the United States, how much more so would it be of those of the South, the first cost of which has been estimated at less than *one-third* that of the former! †

I see not, therefore, a shadow of a reason why our Southern States might not embark, at once, with perfect safety, in an extensive production of wool, *if they had only the foreign market to look to.* I hesitate not to

* See Letter V.   † Ib

assert that they could drive all the European nations from the market, with the two or three exceptions heretofore specified; and with these, as well as the most favored Austro-Oriental regions, they could maintain a successful competition. The same remark is true of the Austro-Occidental regions of our own continent. And it is difficult to foresee the ultimate extent of this trans-Atlantic demand for wool. Vast portions of the Old World, in those zones where wool must eventually become the principal article of clothing, are but just stepping within the verge of civilization—just laying aside the skins and peltry of the pastoral nomad and the savage hunter, for garments of cloth. In 1771, England imported 1,829,772 lbs. of wool; in 1840, the import was 52,959,221 lbs.! In 1771, the export of woolens was £4,960,240. In 1840, the export of woolens was, £5,652,917, and of woolen and worsted yarn £3,796,644. Making all necessary allowance for the difference in prices, the increase in the export bears no comparison whatever to that in the import. What seems to be the unavoidable conclusion? It is that the *consumption* of a population of 27,000,000 (the population of Great Britain and Ireland) has thus enormously swelled within the period of *sixty-nine* years! This too in a country with a mild climate—which at the beginning of that period (1771) was as far advanced in social and political civilization, and the mass of whose people were as well clothed and better fed, than those of any nation on the Eastern Continent! It is not necessary to follow up this idea. Progress is an inseparable condition of humanity,* and civilization is its fruit. With the latter, new wants—a demand for greater comforts and luxuries—steadily keep pace; and with these again keeps pace the increase of population.† Both the latter causes conspire to swell the demand for cloths; and both these causes are at work in this Nineteenth Century, in a velocity of ratio which would fill a Malthus and Ricardo with consternation—if, indeed, it did not convince them of the fallacy of their gloomy theories. I dare to predict that the time will come when the present Russian Empire will consume a greater amount of woolens than the whole Eastern Continent now does! This may not come to pass in a day or a century—but unless retarded by unnatural, not to say *unusual* causes, our posterity in the third or fourth remove will be likely to witness it! Away, then, with those fallacious fears of over-production of cotton, bread-stuffs, etc.—the opposite extreme of Malthusianism—which have disturbed the repose of producers who are not content to let the great natural currents of demand and supply regulate each other; or rather, who are not content with those fair and just profits which they would receive under such an order of things.‡

But the American wool-grower is not compelled to look to the European market, unless he enormously increases his own production—and continues to increase it with the increase of the population. The Census of 1840 shows that the number of sheep in the United States, in 1839, was nearly 20,000,000. These have been steadily increasing, and probably now greatly exceed that number. Yet these have never supplied the demand of our

---

* This may not be thought to accord with preceding statements in relation to the unchangeability of Asiatic character and customs. Particular families or races of mankind have always advanced slowly, but the course of the world, as a whole, is onward. The circle of civilization widens, and races which come in contact with it, receive it, or are conquered and absorbed by the civilized races.

† When I speak of *luxuries* promoting the increase of population, I do not use the word in its invidious sense. I mean by it those things which, though not strictly speaking, *necessaries*, tend to promote human comfort.

‡ I mean this remark in no ultra spirit. Governments must be supported and resources raised. Incidental protection may be justly afforded to the products of agricultural or mechanical skill, under certain circumstances. But the fewer of these restrictions that are found necessary, the more rapidly, as a general rule, the wealth and comfort of mankind and nations are advanced.

own manufactories alone. The following Table* will show the value of the imports of wool into the U. S. from 1837 to 1847:

TABLE No. 9.

	Average imports of 1837, 1838 & 1839.	Average imports of 1840, 1841 & 1842.	Import of 1843.	Import of 1844.	Import of 1845.	Import of 1846.
Wool not costing to exceed 7 cts. a lb............	$558,458	$759,646	$190,352	$754,441	$1,553,789	$1,107 305
Exc'ding 7 cts.a lb	801,087	1,004,312	54 695	97,019	136,005	26,921
Total......	$1,359,545	$1,763,958	$245,017	$851,460	$1,689,794	$1,134,226

It may be a matter of interest to know from what countries these wools were imported. The following Table ‡ will give this information for the last fiscal year, (1846,) and will also give a general idea of our wool trade

TABLE No. 10.

WHENCE IMPORTED.	Wools not exceeding 7 cents per pound.		Wools exceeding 7 cents per pound.	
	Quantity.	Value.	Quantity.	Value.
	Pounds.	*Dollars*	*Pounds.*	*Pounds*
Russia............................	955,163	60,678		
Hanse Towns.....................	6,966	330	13,820	8,433
Holland.........................			170	93
Dutch West Indies..............	10,774	556		
Belgium........................	7,177	248	1,407	775
England........................	1,188,800	35,944	29,406	6,069
Scotland.......................	21,132	1,382		
Gibraltar.......................	207,006	12,330		
Cape of Good Hope.............	83,662	6,810		
British West Indies.............	8,694	537	522	70
British American Colonies......	168,520	9,543	39,346	4,562
France.........................	84,799	5,424	396	40
Spain..........................	20,730	1,425		
Italy..........................	81,156	4,720		
Trieste (Austria)...............	111,981	8,151		
Turkey.........................	5,744,398	398,822		
Morocco (Africa)...............	72,816	4,554		
Mexico.........................	425,148	26,984		
Brazil..........................	45, 215	3,083		
Argentine Republic.............	4,295,659	327,572	43,831	6 311
Chili...........................	1.819,772	130,837		
Peru...........................	122,686	8,588		
Asia, generally.................	945,729	58,778	2,397	269
Total.....................	16,427,952	1,107,305	130,295	26,921

That the course of trade indicated by the above Table, will, as has been already intimated, be materially affected by the New Tariff, I think there can be but little doubt. That of several of the places enumerated, too, has been, heretofore, merely a transit one.

To the following letter from the most extensive, and concededly leading American woolen manufacturer, I would call your particular attention Several of its declarations, placed in italics, by me, are highly significant

LOWELL, Mass., Feb. 10, 1847.

HENRY S. RANDALL, Esq., Cortland Village, N. Y

My Dear Sir : Your very kind and interesting favor of the 27th ult. duly came to hand and should, if practicable, have received an earlier reply. The business of wool-growing in this country is destined to be of immense importance, and I am firm in the belief that *within twenty-five years we shall produce a greater quantity than any other nation.*

* Compiled by me from Reports of the Secretary of the Treasury.
† The fiscal year 1842 ended on the 30th of September. Since then, the returns of imports and exports have been made up to the 30th of June. This year, therefore, embraces the imports of nine months only ending on June 30, 1843; and subsequent years end 30th of June, 1844, 1845, and so on.
‡ Report of the Secretary of the Treasury, 1846.

You ask, "Is the present home demand supplied?" There is not enough annually raised in the country by 10,000,000 lbs. to meet the demand of the manufactories.

You ask, "What countries we can export wool to, &c.?" This country will not export wool regularly for fifteen years, *for the reason that the consumption will increase as rapidly as the production. I can point out articles made of wool now imported, which will require thirty millions of pounds of that of a medium and fine quality, to supply the consumption.* . . . . .

The business of manufacturing wool in this country is on a better basis than ever before, inasmuch as the character, skill and capital engaged in it are such that FOREIGN COMPETI-TION IS DEFIED. *A very few years and all articles of wool used here will be of home manufacture.*

Now I beg of you to keep the wool-growers steady to the mark. Let them aim to excel in the blood and condition of their flocks, and the day is not distant when they will be amply remunerated. I shall always have great pleasure in hearing from you, and remain

Yours most truly, SAM. LAWRENCE.

Mr. Lawrence has certainly got the annual deficit of home wools low enough. Table 10 shows that it was upward of 16,000,000 lbs. during the last fiscal year, 1846. This, of itself, is something of a *margin* for the South, or some other new domestic producer, to fill!

Hitherto we have simply considered the amount of wool necessary to supply our *manufactories.* But these establishments fall very far short of working up all the wool consumed in the United States, even exclusive of home-made fabrics. The following Table* will show the value of the woolens imported for twenty-five years, up to and including 1845:

TABLE No. 11.

1821..	$7,437,737	1826..	$8,431,974	1831.	$12,627,229	1836.	$21,080,003	1841.. $11,001,939
1822...	12,185,904	1827...	8,742,701	1832...	9,992,424	1837...	8,500,292	1842... 8,375,725
1823...	8,268,038	1828...	8,679,505	1833...	13,262,509	1838...	11,512,920	1843... 2,472,154
1824...	8,386,597	1829...	6,881,489	1834...	11,879,328	1839...	18,575,945	1844... 9,475,762
1825...	11,392,264	1830...	5,776,396	1835...	17,834,424	1840...	9,071,184	1845...10,666,176

Here is another and still broader *"margin"* for both the American *Wool-Grower* and the American *Manufacturer* to fill!

With a country well adapted to the production of wool as any the sun shines on—which, all things considered, can produce it *more cheaply* than any extended portion of any trans-Atlantic country—shall we continue to import raw wool?

Whether we should continue to import woolens is sufficiently answered by the last paragraph but one of Mr. Lawrence's letter, fully sustained as the facts therein set forth are by those infallible tests—the *dividends* of our manufacturing establishments. The minimum of these, in *well managed* establishments, has already been stated to be about ten per centum per annum,† and in Mr. Lawrence's own great establishment the dividend of 1846 was *fifteen* per cent. Does any one suppose that the manufacturers of England, with all the advantage they can derive from cheaper labor‡— (but with vastly higher prices for suitable sites and buildings—land taxes, parochial taxes, income taxes—freights and duties on imported wools, etc. etc.)—do or can make dividends touching even the lowest rate above stated? They *cannot.*‖

* Report of the Secretary of the Treasury, 1845. † See Letter VII.
‡ Though not directly advised on the point, I take it for granted that the cost of machinery, also, is some-what less in England.
‖ It may be said that the two last-named expenses fall on the consumer. They doubtless would, but the English manufacturer has to compete with those of France and the United States, a much larger proportion of whose stock is of home growth—the latter entirely, in fine fabrics. The abrogation of the Corn-Laws will be of immense advantage to the English manufacturer, and enable him to better compete with other countries. But while the Bank of England ordinarily discounts paper at from 3 to 4 per cent., and while this is the common rate of interest in that country, it could not be expected that manufacturing capital would be allowed to draw 8 or 10, and much less 15 per cent. Such dividends, in a country whose uninvested capital, or that drawing so low a rate of interest, is so superabundant, would at once invite a competition which would speedily bring the profits of manufacturing capital down to a level with those of other commercial capital. We may, therefore conclude that no such dividends are made.

Is it said that our manufacturing companies have often been compelled to suspend, or break up, even under laws as favorable to them as those now in operation? The reason for this is too pointedly and pertinently stated by Mr. Lawrence to require any addition at my hands, in the following extract from a letter to me, bearing date April 13, 1847; and it will be seen in the concluding sentence that the bold and manly declarations of his preceding letter were not the result of a casual or momentary confidence, but are deliberately reasserted:

"The manufacture of wool has often been disastrous to parties who have embarked in it for many reasons, two of which are sufficient—*a want of capital and a want of skill.* These difficulties are being obviated. Capitalists are more ready to embark under certain auspices, and the amount of skill is very fast increasing, so that *this branch is on a footing not to be moved.*"

Undisturbed by those changes of vacillating legislation, or those movements in the National Legislature pointing to such changes—at one time enormously pampering the manufacturing interest, and leading to overaction and rash adventure—at another, threatening it with disaster and utter subversion—our manufacturers will steadily, nay, rapidly advance. If NOW LET ALONE, they will soon not only "*defy foreign competition*" in the home market, but there is not a single good reason to prevent them from *defying* it in the great and opening market of South America, and even *in the Old World*. Some evils or errors in commercial legislation are less to be deprecated than constant changes. The present Tariff, so far as it affects wool and woolens, is the result of a compromise of interests. It may not be perfect in principle or detail. But it does not seem to flagrantly favor or oppress any interest. I speak not in the spirit of a politician, or of the representative of an interest or section, when I express the hope that *no change will be made or attempted in this portion of the Tariff, until the lapse of years shall bring about other changes requiring it, or until ample experience shall clearly call for a revision of the system.*

I have spoken of two "margins" to be filled by the American woolgrower—the *present* deficit in supplying our own manufactories, and secondly, the *prospective* one, as our manufactures increase, so as to overtake and then keep pace with the consumption of an increasing population. The demands of our manufactories will advance *pari passu* with the *production*, Mr. Lawrence predicts, for at least fifteen years. Why not for fifty or a hundred! Let us glance at the prospective *consumption*, and see if, independent of exportations, it is likely to require any curbs or limits to be placed on *production* or *manufacture*.

In the debates in Congress on the Tariff in 1828–9, Mr. Mallary estimated the consumption of woolens in our country at $72,000,000 per ann.;—$10,000,000 imported; $22,000,00) manufactured; $40,000,000 home-made. The Committee of the "Friends of Domestic Industry," who met in New-York in 1831, reported that the proportion between the amount of wool worked up in factories to that in families was as 3 to 2; that the entire annual product of wool and its manufactures in the U. S. was $40,000,000. These are the only accessible published estimates which now occur to me.

The Census of 1840 shows that the value of woolens made in our manufactories in 1839, was $20,696,999. The import of foreign woolens the same year was $18,575,945, and of raw wool* $1,359,445. It should be remarked, however, that the import of woolens is considerably higher than that of any year before or since. Taking the average of the same three

* Taking the average product of 1837-8-9, as in Table 9. The separate import of 1839 is not before ms.

years for which the import of the raw wool is given,* (1837-8-9,) it would reach but $12,863,051. If we suppose the consumption to.equal the supply, this would give $33,560,050 as the value of the *factory-made* woolens consumed in the United States in 1839. I confess I have no data other than conjectural ones, to determine the amount of the home-made manufactures for that or any other year ; nor do I know that any other person has, or can, have such information. The United States Census, singularly enough does not include this as a separate item. It strikes *me*, however, that Mr Mallary's estimate is too high, and that of the Report of the " Friends of Domestic Industry " too low. The proportion of home-made to factory woolens is, no doubt, annually decreasing, for reasons already stated ;† but as far back as 1839, it would perhaps be a fair estimate to set them down as even. This would give $67,120,100 as the value of the woolens consumed by a population of 17,069,453, or nearly $4 per head. Allowing that every dollar in the manufactured article would represent one pound of stock, or raw wool—and taking slave-cloths, blankets, carpets, coarse home-made fabrics, factory plains, etc., all into account, a dollar is an ample sum to offset against every pound of the raw material—it follows that our whole population annually consume four pounds of wool per head. Judge Beatty of Kentucky, in an estimate published originally in the American Agriculturist, which has been much quoted, sets down the consumption as about 6 lbs. per head. An ordinary Northern farmer or laborer, in comfortable circumstances, will consume about 20 lbs. per annum ;‡ the poorer one not far from 15 lbs.; a boy of 8 years old, full 4 lbs.; a girl of that age (in the country, where females are dressed in woolens,) something more than half of that amount. In the cities and villages there is a large class whose consumption for dress ranges from 30 to 40 and even 50 lbs., and, including carpets, much more. A Southern slave consumes from 3 to 10 lbs. Four pounds, therefore, would not seem to be a high estimate, per head, for our whole population.

Let us now take a glance at the increase of population in the United States. The six different Censuses give the following results :

TABLE 12.

1790, Population	3,929,827	1820, Population	9,638,191
1800, "	5,305,941	1830, "	12,866,020
1810, "	7,239,814	1840, "	17,069,453

It will thus be seen that our population increases at a compound ratio of about three per cent. per annum, which would double it—assuming three per cent. to be the precise rate of increase—in 23 years 164 days.

Cheap and abundant provisions—a supply of fertile lands for all who choose to occupy them, &c.—the causes which have conspired to give so rapid an increase, hitherto, still operate to as great an extent as ever, and will continue to, at all events, for half a century, after the Census of 1840. Suppose the rate of increase, then, decreases to two per cent., which would double the population, reckoning as before, once in about 38 years, and

---

* In Table 9.   † Letter VII.

‡ He will wear out, during a year, 1 coat, 4 yards; 1 pair pants, 3 yards; 1 vest, 1 yard; 1 pair flannel drawers, 2 yards; 1 flannel shirt, 2½ yards; 4 pair hose, mittens, &c, 1½ lbs., which, calling a yard a pound of wool, all round, would amount to 14 lbs. His extra or holiday suit, 8 yards, will last 3 years, and his overcoat, 6 yards, 4 years—making the annual consumption of both, 3 1-6 yards. Two flannel shirts, 10 yards, will last two persons say 3 years, making the annual consumption of one, 1 1-9 yards. No account is here made of coverlids, wool hats, carpets, still used by many, and the latter, more or less of it, to be found in the houses of nearly all farmers in "comfortable circumstances." It will be seen that 20 lbs. of wool per head is a moderate estimate. The above enumeration would not equal to exceed two-thirds, and in some cases *half* the clothing annually consumed by the smartly dressing young men who have *labored* on my farm !

that it doubles twice at this rate—and the following would be the result, and the amount of wool required by the population at the periods indicated:

TABLE No. 13.

Year.	Population.	Amount of Wool.	Year.	Population.	Amount of Wool.
1863-4........	34,138,906	136,555,624	1925........	136,555,624	546,222,496
1886-7	68,277,812	273.111.248	1963........	273.111.248	1,092,444,992

Thus in a little over one hundred years, our population is likely to exceed the present one of Europe, (which is 233,500,000,) and we have *now* a sufficient territory to sustain it! At 3 lbs. of wool per head the number of sheep requisite to supply the *home demand* in 1963, would be over 364,000,000!—far more than are now to be found on the whole globe!— Such are some of the *reasonable* expectations which may be formed of the future prospects of the Home wool market.

LETTER X.

BREEDS OF SHEEP IN THE UNITED STATES

Enumeration of Imported Breeds... No indigenous ones... " Native " Sheep—their Origin—Views of Mr Youatt—Mr. Livingston—their true Origin—their Early Increase in New-England.. Vanderdonk's description of the Sheep and their increase introduced from Holland into New-Netherland (New-York)... Characteristics of the Native Sheep... Account of the Introduction of Merinos into the United States... Their valuation at different periods... The Spanish sub-varieties—Merged in the United States... Purity of blood of the descendants of the Early Importations... Spurious Merinos... Weight of Fleece of the Spanish and French (Rambouillet) families... Description of the latter... American Families—their Characteristics.. Doctor Emmons's Measurements of the Fineness of Wool of individuals of the American, Spanish, and French families —also of other breeds.. The Characteristics of the Merino—its Crosses... The Saxon Sheep—its Origin— Varieties—Treatment in Germany... Introduction into the United States... Purity of blood in our present flocks—Weight of Fleece—Characteristics... The New Leicester or "Bakewell"—Origin—Character in England—Introduction into the United States—Valuation in the latter—Characteristics... South-Down Sheep—Origin—Characteristics—Introduction into the United States.. Mr. Ellman's description of a perfect animal.. Cotswold Sheep—Original Stock—Crossed—the improved variety—Characteristics of—Introduction into the United States.. Cheviot Sheep—Importation into the United States—Original Stock—Crossed— Improved variety—Characteristics... Broad-Tailed Sheep—Introduction into our Country—Characteristics.

Dear Sir : It is believed by those competent to judge, and who have investigated the subject, that our country now possesses every known *breed* of sheep which could be of particular benefit to its husbandry. In proceeding to give an account of the sheep of the United States, I do not deem it necessary to take up your time with a detailed *history* of each race. The zoölogist or breeder anxious to obtain this information, will find it given with great elaboration and accuracy, in the admirable work on Sheep by the late Mr. Youatt.[*]

The principal breeds in the United States are the " Native," (so called) ; the Spanish and Saxon Merinos, introduced from the countries whose names they bear; the New Leicester or Bakewell, the South-Down, the Cotswold, the Cheviot, and the Lincoln from England. The common sheep of Holland were early imported by the Dutch emigrants who originally colonized New-York, but have long since ceased to exist as a distinct variety. The Broad-Tailed Sheep of Asia and Africa have several times been introduced from Persia, Tunis, Asia Minor, etc.

Chancellor Livingston also speaks of two "races as 'indigenous to this country, which we have not enumerated, as it is not known to the Committee[†] that they are now bred in any portion of the United States, viz., the Otter and Smith's Island Sheep, breeds said to have been discovered on two islands on our Atlantic coast. An almost infinite variety of crosses have taken place between the Spanish, English, and ' native ' families. To so great an extent, indeed, has this been carried, that there are,

[*] Also in Mr. Bischoff's, Spooner's, etc., (English) works. and Mr. Morrel's "American Shepherd"—the *historical* parts of all of which are compiled mainly from Mr. Youatt.

[†] At the Annual Meeting of the New-York State Agricultural Society, 1837, a Committee was appointed to report at the next Annual Meeting of the Society, on the " Condition ab✠ Comparative Value of the Several Breeds of Sheep in the United States." The Committee consisted of Henry S. Randall of Cortland, Henry D. Grove of Rensselaer, John B. Duane of Schenectady, Francis Rotch of Otsego, and C. N Bement of Albany. These gentlemen were at the time breeders of all or nearly all the most important varieties, and it was expected that each would write that portion of the Report treating of the one or ones bred by himself The Committee, however, desired—or rather required me to write the whole Report, which I did, with the exception of quotations from authors. The Committee met in Albany, prior to the presentation of the Report, and the late Thomas Dunn and several other breeders were present by invitation. The Report was unanimously adopted by the Committee, and assented to by the breeders present. I do not now quote or adopt *all* the conclusions of that Report. Experience has compelled me to modify some of my opinions, and *actual changes* in the breeds have taken place. But I have mentioned the above facts, to show the authority on which the statements were made, that I *have* quoted, rest; and also because the Report has been often quoted from, sometimes without any credit, and sometimes erroneously credited.

[To save constant reference, it will be understood that all the matter quoted in this Letter from the Report will, unlike the cases where Mr. Randall quotes at any length from the writings of others, be printed in the same type with the body of the Letter, and simply marked with quotation points.   *Publisher.*]

R

comparatively speaking, few flocks in the United States that preserve entire the distinctive characteristics of any one breed, or that can lay claim to unmixed purity of blood."

NATIVE SHEEP.—"Although this name is popularly applied to the common coarse-wooled sheep of the country, which existed here previously to the importation of the improved breeds, there is, properly speaking, no race of sheep 'native' to North America. Mr. Livingston, in speaking of a race as 'indigenous,' only quoted the language of another,* and his informant was either mistaken as to the fact, or misapprehended the term. The only animal of the genus *Ovis Aries*, originally inhabiting this country, is the Argali,† known to our enterprising travelers and traders who have penetrated to the Rocky Mountains, where the animal is found, as the Big Horn.‡ Though the pelage of the Argali approximates but little to the wool of the domestic sheep, they are, as is well known, considered by naturalists to have belonged originally to the same species; and the changes which have taken place in the form, covering, and habits of the latter, are attributed to his domestication, and the care and skill of Man during a long succession of years.

"The common sheep of the United States were of foreign and mostly of English origin. The writer of the volume on Sheep in the 'Farmer's Series,' [Mr. Youatt,] speaks of them as 'although somewhat differing in various districts, consisting chiefly of a coarse kind of Leicester, originally of British breed.'|| Others have seen, or fancied they saw, in some of them, a strong resemblance to the South-Downs. Mr. Livingston was of this number.§ But it is far more probable that they can claim a common descent from no one stock. Our ancestors emigrated from different sections of the British Dominions, and some portion of them from other parts of Europe. They brought their implements of husbandry, and their domestic animals, to fertilize the wilderness. Each, it would be natural to suppose, made choice of the favorite breed of his own immediate district to transport to the New World, and the admixture of these various races formed the mongrel family now under consideration. Amid the perils of war, and the incursion of beasts of prey, they were preserved with sedulous care. As early as 1676, Mr. Edward Randolph, in a 'Narrative to the Lords of the Privy Seal,' speaks of New-England as 'abounding with sheep.' "¶

Vanderdonk, writing in 1790, thus speaks of the sheep introduced from Holland into New-Netherland (now New-York) by the Dutch emigrants:—

"Sheep are also kept in the New-Netherlands, but not as many as in New-England, where the weaving business is carried on, and where much more attention is paid to them than by the New-Netherlanders. The sheep, however, thrive well, and become fat enough. I have seen mutton there so exceedingly fat that it was too luscious and offensive. The sheep breed well and are healthy; they find good pasture in summer, and good hay in winter; but the flocks require to be guarded and tended on account of the wolves, for which purpose men cannot be spared. There is also a more important hindrance to the keeping of sheep, which are chiefly cultivated for their wool. New-Netherland is a woody country throughout, being almost everywhere beset with trees, stumps and brushwood, wherein the sheep pasture, and by which they lose most of their wool. This is not apparent until they are sheared, when the fleeces turn out very light."

"The common sheep yielded a wool only suited to the coarsest fabrics, averaging, in the hands of good farmers, from 3 to 3½ lbs of wool to the

* Livingston's Essay on Sheep. pp. 56, 60.   † Godman's American Natural History.
‡ The "wooly sheep" of the Rocky Mountains, the description of which is quoted by Mr Morrel, (American Shepherd, p 131,) from Capt. Bonneville, is a *goat*.   It will be found described in Godman's Natural History, vol ii. p, 326, *et supra*.
|| Vol. on Sheep p 134.   § Essay on Sheep, p 53.   ¶ Colonial papers of Massachusetts.

fleece. They were slow in arriving at maturity, compared with the improved English breeds, and yielded when fully grown, from 10 to 14 lbs of a middling quality of mutton to the quarter. They were usually long-legged, light in the fore-quarter, and narrow on the breast and back, although some rare instances might be found of flocks with the short legs, and some approximation to the general form of the improved breeds. The common sheep were excellent breeders, often rearing, almost entirely destitute of care, and without shelter, one hundred per cent. of lambs, and it small flocks a still larger proportion. These, too, were usually dropped in March or the earlier part of April. Restless in their disposition, their impatience of restraint almost equaled that of the untamed Argali, from which they were descended ; and in many sections of our country it was common to see from twenty to fifty of them roving, with little regard to inclosures, over the possessions of their owner and his neighbors, leaving a large portion of their wool adhering to bushes and thorns, and the remainder placed nearly beyond the possibility of carding by the Tory weed (*Cynoglossum officinale*) and Burdock (*Arctium lappa*) so common on new lands.

" The old common stock of sheep, as a distinct family, have nearly disappeared, having been universally crossed, to a greater or less extent, with the foreign breeds of later introduction. The first and second cross with the Merino, resulted in a decided improvement, and produced a variety exceedingly valuable for the farmer who rears wool only for domestic purposes. The fleeces are of uneven fineness, being hairy on the thighs, dewlap, &c.; but the general quality is much improved ; the quantity is considerably augmented ; the carcass is more compact and nearer the ground ; and they have lost their unquiet and roving propensities. The cross with the Saxon, for reasons which we shall hereafter allude to, has not been generally so successful. With the Leicester and Downs the improvement, so far as form, size, and a propensity to take on fat are concerned, is manifest."

MERINO RAM.
[*Defiance*,] months old, bred by and the property of Henry S. Randall.]

SPANISH MERINO.—" The history of this celebrated race of sheep, so far as it is known, has so often been brought before the public that it is deemed unnecessary here to recapitulate it. The first importation of them into the United States took place in 1801. Four were shipped by Mr. Delessert, a banker of Paris, three of which perished on the passage.* The fourth arrived in safety at Rosendale, a farm owned by that gentleman near Kingston, in this State. The same year Mr. Seth Adams, of Massachusetts, imported a pair from France. In 1802, two pairs were sent from France by Mr. Livingston, the American Minister, to his estate on the Hudson; and later the same year, Mr. Humphrys, our Spanish Minister, shipped two hundred, on his departure from that country, for the United States." Hon. William Jarvis, of Weathersfield, Vermont, then American Consul at Lisbon, sent home large and valuable flocks in 1809, 1810, and 1811. The particularly favorable circumstances for obtaining the choicest sheep of Spain, under which these were procured, you will find detailed in a letter to me from Mr. Jarvis, dated December, 1841, published in the Transactions of the New-York State Agricultural Society of that year. Various subsequent importations took place, which it is not important to particularize.

The Merinos "attracted little notice, until our difficulties with England led to a cessation of commercial intercourse with that power, in 1808 and 1809. The attention of the country being then directed toward manufacturing and wool-growing, the Merino rose into importance. So great, indeed, was the interest excited, that from a thousand to fourteen hundred dollars a head was paid for them." Unfortunately some of the later importations "arrived in the worst condition, bringing with them those scourges of the ovine race, the scab and foot-rot. These evils and the increased supply soon brought them down to less than a twentieth part of their former price; they could now be bought for $20 a head. When, however, it was established, by actual experiment, that their wool did not deteriorate, as had been feared by many, in this country, and that they became readily acclimated, they again rose into favor. But the prostration of our manufactories, which soon after ensued, rendered the Merino comparatively of little value, and brought ruin to numbers who had purchased them at their previous high prices. The rise which has since taken place in the value of fine wool, as well as the causes which led to it, are too recent and well understood to require particular notice. With the rise of wool, the valuation of the sheep which bear it, has of course kept pace.

" The Merino has been variously described. This arises from the fact that it is but the general appellation of a breed, comprising several varieties, presenting essential points of difference in size, form, quality and quantity of wool." And writers of high authority differ even in their descriptions of these families or varieties. M. Lasteyrie, so celebrated as a writer on sheep, and particularly on the Merino, and Mr. Jarvis directly contradict each other on several points.† It is scarcely necessary now to quote their conflicting statements, or inquire which is right—as the questions involved possess no practical importance. These families have, generally, been merged, by interbreeding, in the United States and other countries which have received the race from Spain. Purity of Merino blood, and actual excellence in the individual and its ancestors, has long since been the only standard which has guided sensible men in selecting sheep of this breed. Families have indeed sprung up, in this country, ex

    * Archives of Useful Knowledge.—Cultivator, vol. i. p. 183.
    † See Lasteyrie on Sheep—or, if not accessible—his statements quoted by Mr. Youatt, p. 356. For Mr Jarvis's statements, see his Letter to L. D. Gregory, quoted in American Shepherd, pp. 73, 74.

hibiting wider points of difference than did those of Spain. In some cases they doubtless owe it to particular courses of breeding—but more often, probably, to concealed or forgotten infusions of other blood.

The point has, indeed, been occasionally mooted, whether there are any Merinos in the United States, descendants of the early importations, of *unquestionable* purity of blood. That there are, has been recently defi nitely settled by a connected chain of undisputable and undisputed testimony,* not necessary here to be repeated. That, on the other hand, in the recent rush of speculation, a marvelous facility has been evinced, in some instances, in suddenly recollecting lost links in the chain of pedigree—or in forgetting others which it would not be expedient to remember, no one would require any proof who has seen some of the animals which have been hawked through the country as full-bloods.

" Taken collectively, the Spanish rams, according to Chancellor Livingston, yield about eight and a half pounds of wool, and the ewes five, which loses half in washing—making four pounds and a quarter the average weight of fleece of the rams, and two and a half the average of the ewes.† Some varieties considerably exceed this estimate, and probably it would fall short if applied to the prime sheep of any variety."

The fleeces of the Merinos at Rambouillet in France, it is stated in the Report of M. Gilbert, to the National Institute, quoted by Mr. Livingston,‡ weigh, in the rams, from twelve to thirteen pounds (unwashed) wool —taking rams and ewes together, it has "not quite attained to eight pounds, after deducting the tags and the wool of the belly, which are sold separately." Mr. Livingston remarks that the French pound is about onetwelfth heavier than the English; but on the other hand, that from the manner of folding and housing sheep and feeding them on fallows in France, they are very dirty, and lose 60 per cent. in washing."|| This would bring the average of the Rambouillet flock to about four pounds, exclusive of tag and belly wool.

M. Lasteyrie gives the following annual averages per head of the Ram bouillet flock : 1796, 6 lbs. 9 oz.; 1797, 8 lbs.; 1798, 7 lbs.; 1799, 8 lbs., 1800, 8 lbs.; 1801, 9 lbs. 1 oz.—This is *unwashed* wool, and will lose half in washing. Mr. Livingston's imported ewes averaged 5 lbs. 2 oz.; his rams 6 lbs. 7 oz., of unwashed wool.§ The later importations will, judging from the specimens I have seen, average much higher than the latter. They are a large sheep, with good, but not the best, quality of Merino wool—some of the larger stocks being rather coarse—and not very uniform, one with another, either in their appearance or fleeces—and are most remarkable for the loose pendulous skin which hangs about their necks, and lies in folds about their bodies. They are free from hair— their wool, which is of good style, opens with a creamy color, and rich lustre, on a fine rose-colored skin. Their wool is long on the back, shortish on the belly—thick, but not so thick as that of many of the American Merinos—very yolky, but destitute of concrete external gum.

The American Merino has, as already intimated, diverged into families or varieties presenting wide points of difference. The minor distinctions are numerous, but they may all, perhaps, be classed under three general heads. The *first*, is a large, short-legged, strong, exceedingly hardy sheep, carrying a heavy fleece, ranging from medium to fine—free from hair in properly bred flocks—somewhat inclined to *throatiness*, but not so much so as the Rambouillets—bred to exhibit external concrete gum in some

* This testimony will be found in a Letter from me to A. B. Allen, Esq., in the December No of the American Agriculturist, 1844, and in the Cultivator, I think, of the same date—if not, the succeeding No.
† Livingston's Essay on Sheep, p. 39.   ‡ Ibid., p. 49, *et supra.*
|| Livingston's Essay on Sheep, p. 51.   § Ibid., Appendix.

flocks, but not commonly so—their wool longish on both back and belly and exceedingly dense—wool whiter within than the Rambouillets—skin the same rich rose-color. The ram on page 131 is a good specimen of this variety, though his age is not sufficient to give him the substance and compactness of an older animal, and the apparent want in these particulars is heightened by recent shearing.* His first fleece of well-washed wool at thirteen months old, was 8 lbs.; was of beautiful quality, and entirely destitute of hair. At three years old he would have sheared from 10 to 12 lbs. of well-washed wool.†

MERINO EWE.

The *second* general class of American Merinos are smaller than the preceding—less hardy—wool as a general thing finer—covered with a black pitchy gum on its extremities—fleece about one-fourth lighter than in class first.

The *third* class, which have been bred mostly South, are still smaller and less hardy—and carry still finer and lighter fleeces. The fleece is destitute of external gum. The sheep and wool bear a close resemblance to the Saxon; and if not actually mixed with that blood,‡ they have been formed into a similar variety, by a similar course of breeding.

Class *first* are a larger and stronger sheep than those originally imported from Spain, carry much heavier fleeces, and in well selected flocks, or individuals, the fleece is of a decidedly better quality. The ewe from my flock—the portrait of which is given above—sheared 7 lbs. 10 oz. of well-washed wool.|| The fibre numbered 1. in fig. 1, in the succeeding measurements by Dr. Emmons, is from this fleece. The fleece is exceedingly even and entirely destitute of hair.

For the purpose of exhibiting the comparative quality of the wool of

* The portrait, on the whole, is strikingly accurate, but the skill of the artist does not compensate for his want of experience, in animal painting, in giving the anatomical details and *expression* of the countenance The same remark applies to the portrait of the ewe.
† This valuable animal died since the above portrait was painted, and prior to his second shearing.
‡ I am not aware what pedigree is claimed for them. They are usually spoken of as Merinos.
|| *i. e.* washed as clean as practicable in a brook, under a heavy sheet of falling water.

the American, Rambouillet, and early imported Spanish Merinos, 1 copy the following, from the pen of Ebenezer Emmons, M. D., State Geologist, in the American Quarterly Journal of Agriculture and Science, of which publication Dr. E. is the Editor.

"Having given you a pretty full report of the farm and stock of Mr. Randall, embracing many details also in the several branches of husbandry, I now propose adding a few words as an appendix to that report. I gave some intimation, when speaking of the fineness of the wool of Mr. R.'s sheep, that on my return home I would furnish something more'exact as a test for fineness than the naked eye. In fulfillment of this intimation, I have been engaged since I returned, in measuring the diameter of the different staples which I procured while at Cortlandville, and which I have compared with others obtained of our mutual friend, Luther Tucker, Esq., of the Cultivator.

"The different kinds are indicated by numbers. I have prepared a scale which is equal to 100 millimeters; a millimeter is equal to 0·039 of an inch. The hundredth of a millimeter,* and the fibres of wool, are all subjected to the same magnifying power of an excellent Chevalier's compound microscope. The comparison is both absolute and relative; but it is highly interesting to see the perceptible difference between the different fibres of wool. The microscope also reveals other differences; some of the fibres appeared rather uneven or flattened, and destitute of a clear and distinct pith or tube; and, in fact, I may remark that the microscope is really the best method of testing the real quality of wool." . . .

Fig. 1.

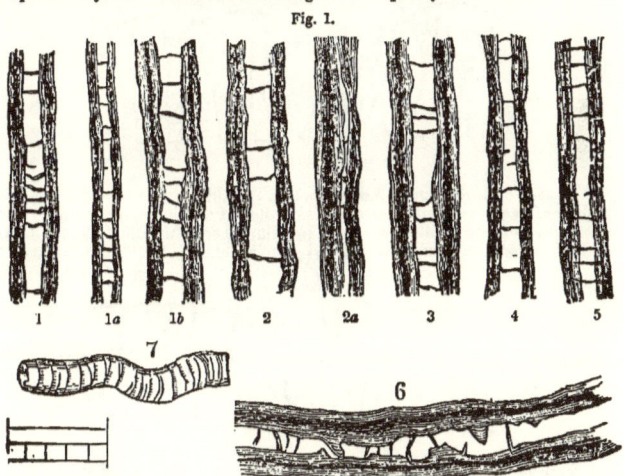

"No. 1, Mr. Randall's; No. 1a, fibre of Mr. Randall's prize Merino buck;† No. 1b, fibre from one of Mr. Randall's fleeces; No. 2 and 2a, fibres from Mr. Seth Adams's wool; No. 4, Remilles wool, Shoreham, Vt.; No. 5, fibre of S. O. Burchard's fine wool, Shoreham; No. 3, fibre of Charles L. Smith's wool, Shoreham; No. 6, fibre from Collins's Grandee. The last five were taken from wool left at the Cultivator office. In all the fibres examined there is a great uniformity in the parcels; only slight differences, in fact, could be detected in the several diameters. No. 7 shows the structure of wool as seen under the microscope. In the corner is the scale of measurement. The finest fibre as magnified in this cut is equal to about eighteen-hundredths of an inch in diameter.

"Another inquiry equally important with the preceding came up in this place: What is the strength of a single fibre of wool, and is the coarser comparatively stronger than the fine? I set about answering those inquiries at once, and now give you the result below:

"Mr. Randall's No. 1b, on three trials, supported on an average 62 grains; or, rather, broke when tried with the weight of 62 grains.

"Mr. R.'s No. 1a broke with 57·1 grains.

"The fibre from Collins's Grandee, on three trials, supported on an average 84·6 grains.

"Mr. Smith's specimen of Shoreham, Vt., on three trials, gave an average of 65·6 grains."

No. 1a is the wool of my ram "Premium," which received the first prize

---

* About 1-2500 of an inch.   † Taken from the animal by Doct. Emmons.

at the State Fair at Poughkeepsie, 1844,* and his fleece weighed 10 lbs. of well washed wool.

No. 2 and 2a, (Mr. Seth Adams's wool,) were from the sheep imported by that gentleman.

No. 6 was from Grandee, the best ram of Mr. Collins's Rambouillet importation.

It will be observed, first, that the American wool is the finest, and second, its strength is greatest in proportion to its diameter.

It will probably be as well to bring Doct. Emmons's subsequent measurements of the wool of other individuals and varieties together at this place, as to scatter them through the descriptions of the several breeds. It will render a comparison between them more convenient. I would remark that the cuts are copied from those of Doct. Emmons, with the strictest fidelity.† Indeed they are perfect *fac similes.*

Fig. 2.

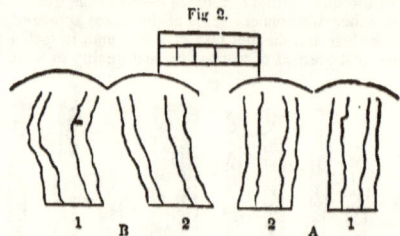

" Figure 2 (scale of mensurement same as in Fig. 1) exhibits the comparative diameters of the wool fibre of two premium Saxon sheep exhibited at the State Fair at Utica, 1845. A 1 is a fibre of wool from the shoulder of the 2d premium sheep (Mr. Church's); 2 do. from the ank. B 1, fibre from the shoulder of the first premium sheep (Mr. Crocker's); 2 do. flank.

Fig. 3.

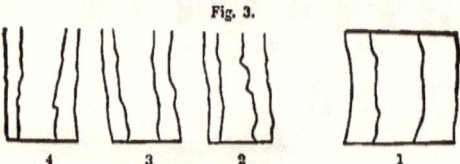

" Fig. 3, No. 1. fibre of Bakewell—about the average fineness of this kind of wool. No. 2, fibre from Merino ewe belonging to Col. Sherwood, 3 years old (Blakesley sheep.) No. 3 do. Mr. Bailey's ewe. No. 4 do. Mr. Atwood's.

Fig. 4.   .

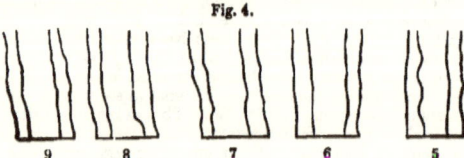

' Fig. 4.—No. 5, fibre of Mr. Ellis's ewe, fleece weighing 6 lbs. 13 oz. No. 6 do. Mr. Nettleton's yearling Merino buck. No. 7 do. a sample from the imported 5 per cent. South American wool, which is seen to be nearly as fine as the best of our flocks. No. 8 do. Col.

* This is the only time my sheep have ever been shown at a State Fair, and I first made arrangements for exhibiting, in the expectation of having the privilege of comparing my sheep with the imported Rambouillets of Mr. Collins. Mr. C., however, declined my invitation to show. I received the first prize on rams, and the first and second on ewes.

† Executed by William Howland, of New York, whom I take pleasure in recommending to all wishing to obtain wood engravings, as an accurate and most obliging artist.

Sherwood's three-year-old buck, sheared 8½ lbs. of wool. No. 9 do. finest Saxon wool is market.

Fig. 5.—No. 10, fine Ohio wool. No. 12, do. Saxon of the late Mr. Grove's excellent flock. No. 13, do. original imported Spanish wool by Seth Adams. No. 14, Mr. L. A. Morrell's Saxon

Fig. 5.

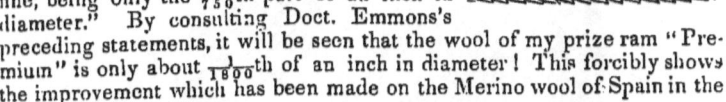

10   12   13   14

The following cut, copied from Youatt, exhibits a fibre of Merino wool viewed both as an opaque and transparent object, with a microscope manufactured by Mr. Powell, of London.

The serrations or "beards," which constitute the felting property of wool, are beautifully distinct and sharp. It was a picklock from a Negretti fleece, and Mr. Youatt says it is "very fine, being only the $\frac{1}{750}$th part of an inch in diameter." By consulting Doct. Emmons's preceding statements, it will be seen that the wool of my prize ram "Premium" is only about $\frac{1}{1800}$th of an inch in diameter! This forcibly shows the improvement which has been made on the Merino wool of Spain in the United States.

"The Merino, though the native of a warm climate, becomes readily in ured to the greatest extremes of cold, flourishing as far north as Sweden, without degenerating in fleece or form. It is a patient, docile animal, bearing much confinement without injury to health, and possesses none of that peculiar ' voraciousness of appetite,' ascribed to it by English writers.*— Accurately conducted experiments have shown that it consumes " a little over " two pounds of hay per diem, in winter; the Leicester consumes from three and a half to four; and the common wooled American sheep would not probably fall short of three. The mutton of the Merino, in spite of the prejudice which exists on the subject, is short grained and of good flavor, when killed at a proper age," and weighs from ten to fourteen pounds to to the quarter. "It is remarkable for its longevity, retaining its teeth and continuing to breed two or three years longer than the common sheep," and at least half a dozen years longer than the improved English Breeds; " but it should be remarked in connection with this fact, that it is corres, pondingly slow in arriving at maturity. It does not attain its full growth before three years old, and the ewes in the best managed flocks, are rarely permitted to breed before they reach that age."

The Merino is a far better breeder than any other fine-wooled sheep, and my experience goes to show that its lambs, when newly dropped, are hardier than the Bakewell, and equally so with the high bred South-Down. The ewe is not so good a nurse, however, as the latter, and will not usually do full justice to more than one lamb. Eighty or ninety per cent. is about the ordinary number of lambs usually reared, though it often reaches one hundred per cent. in carefully managed or small flocks.

"We have already adverted to the cross between the Merino and the native sheep. On the introduction of the Saxon family of the Merinos, they were universally engrafted on the parent stock, and the cross was continued until the Spanish blood was nearly bred out." When the admixture took place with judiciously selected Saxons, it resulted not unfavorably for certain purposes. But unfortunately these instances of judicious crossing were rare. Our country was flooded by eager speculators, with the feeblest and least hardy Merinos of Germany. Fineness of wool during

* Youatt, p. 149.   S

the period of this strange excitement, was made the only test of excellence, no matter how scanty its quantity, no matter how diminutive or miserable the carcass. Governed by such views, the holders of most of our Merino flocks purchased these over-delicate Saxons, and the consequence was as might have been foreseen—their flocks were ruined."

SAXON RAM

SAXONS.—" In the year 1765, Augustus Frederick, Elector of Saxony, obtained permission from the Spanish Court to import two hundred Merinos, selected from the choicest flocks of Spain.   They were chosen principally from the Escurial flock, and on their arrival in Saxony, were placed on a private estate belonging to the Elector, under the care of Spanish shepherds. So much importance was attached to the experiment, as it was then considered, that a commission was appointed to superintend the affairs of the establishment ; and it was made its duty to diffuse information in relation to the management of the new breed ; to dispose of the surplus rams at prices which would place them within the reach of all holders of sheep ; and finally, by explaining the superior value of the Merinos, to induce the Saxon farmers to cross them with their native breeds.   Popular prejudice, however, was strong against them, and this was hightened by the ravages of the scab, which had been introduced with them from Spain, and which proved very destructive before it was finally eradicated.   But when it became apparent that the Merino, so far from degenerating, had improved " in the quality of its wool, in Saxony, " the wise and patriotic efforts of the Elector began to reap their merited success, and a revolution took place in popular sentiment.   The call for rams became so great that the Government resolved on a new importation, to enable them more effectually to meet it, and to improve still farther the stock already obtained. For this purpose an individual, considered one of the best judges of sheep in Saxony, was dispatched to Spain in 1777, with orders to select three hundred.   For some reason, probably because he experienced difficulty in obtaining a greater number presenting all the qualifications he sought, he returned with but one hundred and ten. They were from nearly all the different flocks of Spain, but principally the Escurial—and were considered decidedly superior to the first importation. In addition to the establishment at Stolpen, already founded, others were now commenced at Rennersdorf, Lohmen, &c.; schools were established for the education of shepherds; publications were distributed by the commissioners to throw information on the subject before the people; and the Crown tenants, it is said, were each required to purchase a certain number of the sheep."

Mr. Spooner[*] states that there are two distinct breeds of the Saxon Merino sheep, the first "having stouter legs, stouter bodies, head and neck comparatively short and broad, body round. The wool grows most on the face and legs—the grease in the wool is almost pitchy." The other breed called Escurial have longer legs, with a long, spare neck and head, with very little wool on the latter, and a finer, shorter and softer character in its fleece, but less in quantity. The fleece in the Escurial averages from one and a half to two pounds in ewes, and two to three pounds in rams and wethers, while in the others it is from two and a quarter to three and a quarter in ewes, and from four to six pounds in ram and wethers. These varieties cannot be amalgamated successfully.

The preceding portrait is a favorable specimen of the Escurial Saxon, copied from a cut, after a drawing by Harvey, in Mr. Spooner's work.

That the German shepherds have sacrificed the hardiness of the Merino. and indeed almost everything else, for fineness of staple, there can be but little doubt. Their method of managing the sheep and its effects are thus described by Mr. Carr, a large sheep-owner of Germany :[†]

"They are always housed at night. even in summer, except in the very finest weather, when they are sometimes folded in the distant fallows, but never taken to pasture until the dew is off the grass. In the winter they are kept within doors altogether, and are fed with a small quantity of sound hay, and every variety of straw, which has not suffered from wet, and which is varied at each feed ; they pick it over carefully, eating the finer parts, and any grain that may have been left by the threshers. Abundance of good water to drink, and rocksalt in their cribs, are indispensables. . . . . They cannot thrive in a damp climate, and it is quite necessary that they should have a wide range of dry and hilly pasture of short and not over-nutritious herbage. If allowed to feed on swampy or marshy ground, even once or twice, in autumn, they are sure to die of liver-complaint in the following spring. If they are permitted to eat wet grass, or exposed frequently to rain, they disappear by hundreds with consumption. In these countries it is found the higher bred the sheep is, especially the Escurial, the more tender !"

Such are the common views of the sheep, and their treatment over Germany, Prussia, and Austria. Various statements of the methods adopted by Baron Geisler, Graf Hunyadi, and other eminent flockmasters, will be found in Dr. Bright's Travels in Lower Hungary, Paget's Travels in Hungary and Transylvania, Jacob's Travels in Germany, &c.

The qualities of the Saxons as breeders and nurses, may be inferred from the following regulations, for the management of his flock, by Baron Geisler.[‡]

"During the lambing period, a shepherd should be constantly day and night in the cote, in order that he may place the lamb, a soon as it is cleaned, together with its mother, in a separate pen, which has been before prepared. The ewes which have lambed should, during a week, be driven neither to water or pasture ; but low troughs of water for this purpose are to be introduced into each partition, in order that they may easily and at all times quench their thirst. It is also very useful to put a small quantity of barley-meal into the water, for by this means the quantity of the ewe's milk is much increased. When the lambs are so strong that they can eat, they are to be separated by degrees from their mothers, and fed with the best and finest oats, being suffered at first to go to them but three times a day, early in the morning, at mid-day, and in the evening, and so to continue till they can travel to pasture, and fully satisfy themselves."

The following history of the introduction of the Saxons into the United States, was compiled by me from written memoranda, and the oral statements of Mr. Grove, submitted to the Committee of New-York State Agricultural Society, already alluded to, of which I was Chairman, and was published in my Report, credited, of course, to Mr. Grove individually, as no other member of the Committee was conversant with the facts narrated. *

" The first importation of Saxony sheep into the United States was made by Mr. Samuel Henshaw, a merchant of Boston, at the instance of Col. James Shepherd, of Northampton. They were but six or seven in number. In 1824, Messrs. G. & T. Searle, of Boston, imported 77 Saxon sheep. They were selected and purchased by a Mr. Kretchman, a correspondent of the above firm, residing in Leipsic, and shipped at Bremen on board the American schooner Velocity. I was engaged to take charge of the sheep on the passage, and I also shipped six on my own account. I am sorry to say that as many as one-third of the sheep purchased by Kretchman, (who shared profit and loss in the undertaking,) were not pure-blooded sheep. The cargo were sold at auction at Brooklyn, as 'pure-blooded electoral Saxons,' and thus unfortunately in the very outset the pure and impure became irrevocably mixed. But I feel the greatest certainty that the Messrs. Searle intended to import none but the pure stock—the fault lay with Kretchman. In the fall of 1824, I entered into an arrangement with the Messrs. Searle to return to Saxony, and purchase in connection with Kretchman, from 160 to 200 Electoral sheep. I was detained at sea seven weeks, which gave rise to the belief that I was shipwrecked and lost. When I finally arrived, the sheep had been already bought by Kretchman. On being informed of what the purchase consisted, I protested against taking them to America, and insisted on a better selection, but to no purpose. A quarrel ensued between us, and Kretchman even went so far as to engage another to take charge of the sheep on their passage. My friends interposing, I was finally induced to take charge of them. The number shipped was 167, 15 of which perished on the passage. They were sold at Brighton, some of them going as high as from $400 to $450. A portion of this importation consisted of grade sheep, which sold as high as the pure-bloods, for the American purchasers could not know the difference. It may be readily imagined what an inducement the Brighton sale held out to speculation, both in this country and Saxony. The German newspapers teemed with advertisements of sheep for sale, headed ' Good for the American Market;' and these sheep, in many instances, were actually bought up for the American market at five, eight or ten dollars a head, when the pure-bloods could not be purchased at from less than $30 to $40. In 1826, Messrs. Searle imported three cargoes, amounting in the aggregate to 513 sheep. They were of about the same character with their prior importations, in the main good, but mixed with some grade sheep. On the same year a cargo of 221 arrived, on German account, Emil Bach, of Leipsic, supercargo. A few were good sheep and of pure blood ; but taken as a lot they were miserable. The owners sunk about $3,000. Next came a cargo of 210 on German account; Wasmuss and Multer, owners. The whole cost of these was about $1,125, in Germany. With the exception of a small number, procured to make a flourish on, in their advertisements of sale they were sheep having no pretensions to purity of blood. In 1827, the same individuals brought out another cargo. These were selected exclusively from grade flocks of low character. On the same year the Messrs. Searle made their last importation, consisting of 132 sheep. Of these I know little. My friends in Germany wrote me that they were like their other importations, a mixture of pure and impure blooded sheep. It is due, however, to the Messrs. Searle to say that, as a whole, their importations were much better than any other made into Boston.

" I will now turn your attention to the importations made into other ports. In 1825, 13 Saxons arrived in Portsmouth. They were miserable creatures. In 1826, 191 sheep arrived in New-York, per brig William, on German account. A portion of these were well descended and valuable animals, the rest were grade sheep. In June the same year, the brig Louisa brought out 173 on German account. Not more than one-third of them had the least pretensions to purity of blood. Next we find 158, shipped at Bremen, on German account.— Some were diseased before they left Bremen, and I am happy to state that twenty-two died before their arrival in New-York. All I intend to say of them is, that they were a most curious and motley mess of wretched animals. The next cargo imported arrived in the brig Maria Elizabeth, under my own care. They were 165 in number, belonging to myself and F. Gebhard, of New-York. These sheep cost me $65 a head when landed in New-York.— They sold at an average of $50 a head, thus sinking about $2,400 ! I need not say that they were exclusively of pure blood. A cargo of 81 arrived soon after, but I know nothing of their quality. The next importation consisted of 184, on German account, per brig Warren. With a few exceptions they were pure-blooded and good sheep. We next have an importation of 200 by the Bremen ship Louisa. They were commonly called the ' stop sale sheep.

* Mr. Morrel in his American Shepherd, quotes this as a "Report" drawn and read by Mr. Grove, (one is left to infer,) before the New-York State Agricultural Society. This is doubtless an inadvertance.

They were of the most miserable character, some of them being hardly half-grade sheep.—
The ship Phebe Ann brought 120 sheep, of which I know little; and 60 were landed at
Philadelphia, with the character of which I am unacquainted. Having determined to settle
in America, I returned to Saxony, and spent the winter of 1826–7 in visiting and examining
many flocks.   I selected 115 from the celebrated flock of Machern, embarked on board the
ship Albion, and landed in New-York June 27, 1827.   In 1828, I received 80 more from the
same flock, selected by a friend of mine, an excellent judge of sheep.   I first drove them to
Shaftsbury, adjoining the town of Hosic, where I now reside.   On their arrival they stood
me in $70 a head, and the lambs half that sum."

"It will be inferred· from the facts above stated that there are few Saxon flocks in the United States that have not been reduced to the quality
of grade sheep, by the promiscuous admixture of the pure and the impure
which were imported together, and *all* sold to our breeders as pure stock."

And independent of this, there are but exceedingly few flocks which
have not been *again* crossed with the Native or Merino sheep of our country, or both.   Those who early purchased the Merino, crossed them with
the Native ; and, when the Saxons arrived, these mongrels were bred to
Saxon rams.   This is the history of probably three-quarters of the " Saxon" flocks of the United States, and among them some, as *I know*, among
the most celebrated.

As these sheep have now so long been bred toward the Saxon that their
wool equals that of the pure-bloods, it is exceedingly problematical in my
mind whether they are any worse for the admixture : when crossed only
with the Merino, it is undoubtedly to their advantage.   Though I once
thought differently, experience has satisfied me that the American Saxon,
with these early crosses in its pedigree, is a hardier and more easily kept
animal than the pure Escurial or Electoral Saxon.   As with the Merino,
climate, feed, and other causes, have doubtless conspired to add to their
ize and vigor ; but, after all, I have not a doubt they usually owe more
of it to those early crosses.

The fleeces of the American Saxons weigh, on the average, from 2 or
2½ to 3 lbs.   They are, comparatively speaking, a tender sheep, requiring
regular supplies of good food, good shelter in winter, and protection in
cool weather from storms of all kinds ; but they are evidently hardier than
the parent German stock.   In docility and patience under confinement,
late maturity, and longevity, they resemble the Merinos, from which they
are descended ; though they do not mature so early as the Merino, nor
ordinarily live so long.   They are poorer nurses ; their lambs smaller, feebler, and far more likely to perish, unless sheltered and carefully watched.
They do not fatten so well, and, being considerably lighter, they consume
an amount of food correspondingly less.

Taken together, the American Saxons bear a much finer wool than the
American Merinos ; but Dr. Emmons's measurements show that this is
not always the case, and many breeders of Saxons are now crossing with
the Merino, in the expectation of increasing the weight of their fleeces
without deteriorating its quality.*   Though I am in possession of wool
from Saxons in Connecticut and Ohio, which compares well with the
higher grades of German wool,† and though there are doubtless other
flocks of equal quality in the country,‡ our Saxon wool, as a whole, falls
considerably below that of Germany ; and I never have seen a *single lock*
of the American equaling some samples, given me by a friend recently

* Mr. Lawrence believes this practicable, and Mr. Morrel and various other Saxon breeders have for
some time bred in this way.
† Fully equaling, and, I think, better than some German wool I recently saw, which, all expenses included, stood the purchaser in $1 60 per pound !
‡ Dr. Emmons stated, subsequently to his measurements above, that he had received wool from the
flock of Dr. Beekman, considerably finer than the Saxon wool figured.

from Europe, which came from Styria, south of Vienna, in Austria. The
inferiority of the American to the German wool is not due to climate or
other natural causes, nor is it owing to a want of skill on the part of our
breeders.   It is owing to the fact that but a very few of our manufactur
ers have ever felt willing to make that discrimination in prices which would
render it profitable to breed those small and delicate animals which pro
duce this exquisite quality of wool.   No American breeder thinks of hous-
ing his sheep from the *summer* rains and *dew*, or observing any of the *hothouse* regulations—at least in the summer—of Graf Hunyadi, or Baron
Geisler ! If he did, his wool would not probably pay *half* of its first cost.
When our manufacturers wish to find these wools in the *home* market,
they must learn to *pay* for them in the *home* market as liberally as they
are *compelled* to to obtain them in foreign ones !

THE NEW LEICESTER, OR BAKEWELL.

The portrait above is copied from one of a sheep of this variety, belonging to the Duke of Bedford, given in Mr. Youatt's work on Sheep.
  " The unimproved Leicester was a ' large, heavy, coarse-wooled breed'
of sheep, inhabiting the midland counties of England.   It is described also
as having been ' a slow feeder, and its flesh coarse-grained, and with little
flavor.'   The breeders of that period regarded only size and weight of
fleece.   The celebrated Mr. Bakewell, of Dishley, was the first who adopted a system more in accordance with the true principles of breeding.   He
selected from the flocks about him those sheep ' whose shape possessed
the peculiarities which he considered would produce the largest proportion of valuable meat, and offal,' and having observed that animals of medium size possess a greater aptitude to take on flesh, and consume less
food than those which are larger, and that prime fattening qualities are
rarely found in sheep carrying a great weight of wool, he gave the preference to those of smaller size, and was satisfied with lighter fleeces."   To
reach the wonderful results obtained by Mr. Bakewell, it was supposed
that he resorted to a cross with some other varieties, but it is believed by
some that he owed his success only to a judicious principle of selection.
and   teady adherence to certain principles of breeding.

It is exceedingly unfortunate that this eminent breeder has left us so much in the dark in relation to those principles of breeding, adopted by him, which led to such signal success in his efforts to improve both the cattle and sheep of the region in England in which he resided. All of his measures were veiled in impenetrable secrecy even from his most intimate friends, and he died without voluntarily throwing the least light on the subject. The whole inception and management of his famous "Dishley Society"* betrays selfishness the most intense, and, in plain English, *meanness* the most unalloyed. Should a man claiming to be a gentleman, in *this* country, make valuable discoveries in breeding, or in any other department of husbandry, and closely conceal them from the public, his conduct would meet with universal reprehension and contempt;† yet the thing seems to be considered a matter of course, or is at least passed over without censure, in Youatt, Spooner, Bischoff, and a host of earlier writers, all of whom laud Mr. Bakewell to the echo!

"The improved Leicester is of large size, but somewhat smaller than the original stock, and in this respect falls considerably below the coarser varieties of Cotswold, Lincoln, &c. Where there is a sufficiency of feed, the New Leicester is unrivaled for its fattening properties, but it will not bear hard stocking, nor must it be compelled to travel far in search of its food. It is, in fact, properly and exclusively a lowland sheep. In its appropriate situation, on the luxuriant herbage of the highly cultivated lands of England, it possesses unrivaled earliness of maturity; and its mutton, when not too fat, is of a good quality, but is usually coarse, and comparatively deficient in flavor, owing to that unnatural state of fatness which it so readily assumes, and which the breeder, to gain weight, so generally feeds for. The wethers, having reached their second year, are turned off in the succeeding February or March, and weigh at that age from thirty to thirty-five pounds to the quarter. The wool of the New Leicester is long—averaging, after the first shearing, about six inches; and the fleece of the American animal weighs about six pounds. It is of coarse quality, and little used in the manufacture of cloths, on account of its length, and that deficiency of felting properties common, in a greater or less extent, to all the English breeds. As a combing wool, however, it stands first, and is used in the manufacture of the finest worsteds, &c.

"The high bred Leicesters of Mr. Bakewell's stock became shy breeders and poor nurses, but crosses subsequently adopted " have, to some extent, obviated these defects. So far as my experience has extended in this country, however, the lambs are not very hardy, and require considerable attention at the time of yeaning, particularly if the weather is even moderately cold or stormy. Neither can the grown sheep be considered, in my opinion, *very* hardy. They are much affected by sudden changes in the weather, and a sudden change to cold is pretty sure to be registered on their noses by unmistakable indications of catarrh or 'snuffles.'

"In England, where mutton is generally eaten by the laboring classes, the meat of this variety is in very great demand; and the consequent return which a sheep possessing such fine feeding qualities is enabled to make, renders it a general favorite with the breeder. Instances are recorded of the most extraordinary prices having been paid for these ani-

---

* For the Regulations of this Society, see Youatt, p. 317.

† Of course I do not include in this category those nameless venders of recipes for killing Canada Thistles, rats, &c. &c.; and men who spend their time and property in inventing improved implements, etc., are entitled to the pay offered by the Patent laws. But, among our agriculturists of standing, who has ever known of a single instance of a valuable discovery in the operations of husbandry being concealed or withheld from the public? Who has known a breeder of rank wheedle a partner out of one-half of a valuable bull, and then refuse the quondam partner the services of that bull at any price, lest he should prove a dangerous rival in breeding? Yet, what English writer has expressed any contempt for such meanness? These things would not " go down" among us " *repudiators*" !

mals, and Mr. Bakewell's celebrated buck "Two Pounder" was let for the enormous price of four hundred guineas for a single season! The New Leicester has spread into all parts of the British Dominions, and been imported into the other countries of Europe and the United States. They were first introduced into our own country by the late Christopher Dunn, Esq., of Albany, about twenty-five years since.[*] Subsequent importations have been made by Mr. Powel, of Philadelphia, and various other gentlemen."

It is no more than justice to say that this breed has never proved a favorite with any large class of American farmers. Our long, cold winters, but more especially our dry, scorching summers, when it is often difficult to obtain the rich, green, tender feed in which the Leicester delights—the general want of green feed in the winter, robs it of its early maturity, and even of the ultimate size which it attains in England. Its mutton is too fat, and the fat and lean are too little intermixed, to suit American taste. Its wool is not very salable, from the much to be regretted dearth of worsted manufactories in our country. Its early decay and loss of wool constitute an objection to it, in a country where it is often so difficult to advantageously turn off sheep, particularly ewes. But, notwithstanding all these disadvantages, on rich lowland farms, in the vicinities of considerable markets, it will always probably make a profitable return.

The following description of what constitutes the desirable characteristics of this breed, is from the pen of Mr. Youatt :[†]

"The head should be hornless, long, small, tapering toward the muzzle, and projecting horizontally forward. The eyes prominent, but with a quiet expression. The ears thin, rather long, and directed backward. The neck full and broad at its base, where it proceeds from the chest, so that there is, with the slightest possible deviation, one continued horizontal line from the rump to the poll. The breast broad and full ; the shoulders also broad and round, and no uneven or angular formation where the shoulders join either the neck or the back—particularly no rising of the withers, or hollow behind the situation of these bones.— The arm fleshy through its whole extent, and even down to the knee. The bones of the leg small, standing wide apart; no looseness of skin about them, and comparatively bare of wool. The chest and barrel at once deep and round ; the ribs forming a considerable arch from the spine, so as in some cases, and especially when the animal is in good condition, to make the apparent width of the chest even greater than the depth. The barrel ribbed well home ; no irregularity of line on the back or belly, but on the sides ; the carcass very gradually diminishing in width toward the rump. The quarters long and full, and, as with the fore legs, the muscles extending down to the hock ; the thighs also wide and full. The legs of a moderate length ; the pelt also moderately thin, but soft and elastic, and covered with a good quantity of white wool—not so long as in some breeds, but considerably finer."

THE SOUTH-DOWN.—" This breed of sheep has existed for several centuries in England, on a range of chalky hills called the South Downs. They were, as recently as 1776, small in size, and of a form not superior to the common wooled sheep of the United States. Since that period, a course of judicious breeding, pursued by one man (Mr. Ellman, of Glynde, in Sussex), has mainly contributed to raise this variety to its present high degree of perfection, and that, too, without the admixture of the slightest degree of foreign blood. In our remarks on this breed of sheep, it will be understood that we speak of the pure improved family, as the original stock, presenting, with trifling modifications, the same characteristics which they exhibited sixty years since, are yet to be found in England—and as the middle space is occupied by a variety of grades, rising or falling in value, as they approximate to or recede from the improved blood.

" The South-Down is an upland sheep, of medium size, and its wool, which in point of length belongs to the middle class," has been estimated to rank with half-blood Merino, and was so estimated in my Report, quo-

---

tations from which constitute so large a portion of this Letter. But both subsequent experience, and information derived from other sources, have convinced me of the erroneousness of this opinion. South-Down wool is

SOUTH-DOWN RAM.

essentially different from Merino wool of any grade, though the fibre in some of the finest fleeces may be of the same apparent fineness with half or one-quarter blood Merino.

The following cut from Youatt,* gives the microscopic appearance, says that gentleman, of a "prime specimen of picklock South-Down wool," 1 being viewed as a transparent, and 2 as an opaque object." The fibre is $\frac{1}{800}$th part of an inch in diameter. 1

The cups or leaves of 2 "are roughened irregular, and some of the leaves have exceedingly short angles," but they are far sharper, more numerous and regular (the points which give wool its felting property) than in ordinary South-Down wool. In the latter, the cups are rounded and have a "rhomboidal" instead of that sharp and "hooked" character which distinguishes the Merino and Saxon.

South-Down wool is deficient in felting properties. It makes a "furzy, hairy" cloth, and is no longer used in England, unless largely admixed with foreign wool, even for the lowest class of cloths.

The following testimony was given by some of the most eminent manufacturers, wool-factors, staplers, and merchants of England, before the Committee of the House of Lords in 1828, several times previously alluded to :†

* Youatt, p. 236   † See Bischoff, vol. ii. pp. 145 to 155.

T

Mr. CHARLES BULL, wool agent, Lewes.—"Formerly it [South-Down wool] was used for clothing purposes; now it is impossible to sell it for that manufacture; . . . it is used for baizes and flannels in a very large way."

Mr. WILLIAM CUNNINGTON, wool-stapler, Wiltshire.—"The public will not wear the South-Down cloths, they are so very coarse."

Mr. JAMES FISON, wool dealer, Thetford.—"There has been deterioration in the quality of (South-Down) wool; the general weight of the fleece 20 years ago was 2 pounds to 2¼, and it is now 3 pounds to 3½, our wool used to be made into cloths, and returned into Norfolk, and used by myself and the agriculturists. We do not get the same cloth now; neither myself nor the farmer would wear it, because of the deterioration of quality."

Mr. JAMES HUBBARD, wool agent, Leeds.—South-Down wool is not "now employed for the purpose of making cloth; it has been forced down two or three steps in the scale of wool, and is now used for flannels and baize. . . . The wool gets more frothy and open, and in manufacturing it does not felt and improve so well; it works more flannely." . . . .

Mr. JOHN BROOKE, manufacturer, Howley.—"Manufacture principally blue cloths from 7s. to 24s. and 25s. per yard, and also narrow cloths. . . . Had the Duke of Norfolk's wool, Mr. Ellman, junior's, clip from 1817 and 1822, and Mr. Ellman, senior's, from 1817 to 1821. . . . Kept to English wool longer than any house in the neighborhood. . . . Ceased to manufacture it entirely in 1823 or 1824, . . . found our neighbors were sending out better cloths than we were, not only at the same price, but better manufactured cloths, and we lost our customers."

Mr. BENJAMIN GOTT, merchant and manufacturer, Leeds.—"I formerly used 150 packs of English wool weekly; the disuse of English wool was gradual, commencing about the year 1819, continuing to 1823 and 1824, about which time I began to manufacture exclusively from foreign wool. The disuse of English wool arose from the quality and the advantage of using foreign wool compared with our own. I could not now make an article which would be merchantable at all for the foreign market, (that remark applies equally to the home trade.) in certain descriptions of cloth, except of foreign wool." . . . These wools (the domestic and foreign,) "have different properties."

Mr. WILLIAM IRELAND, Blackwell Hall factor, London.—"We have been using English wool for second and livery cloths, but recently they have been so very much lowered in quality we have not been able to make use of them at all, and have been obliged to make use of low German and low Spanish wools for that purpose."

Mr. J. SUTCLIFFE, wool-stapler, Huddersfield.—"South-Down wool was formerly applied for making cloth for home consumption regularly, for the clothing of servants, &c. It was also used for army clothing. It is now no longer used for those purposes. It makes a furzy, soft, hairy piece; it has not that fastness in it that foreign wool has."

Many other individuals testify to the same effect, and the extremely low character of South-Down wool for carding purposes may be regarded as definitely settled. But as it has deteriorated it has increased in length of staple in England, and to such an extent that improved machinery enables it to be used as a combing wool—for the manufacture of worsteds. Where this has taken place it is quite as profitable, in England, as when it was finer and shorter. In the United States, where the demand for combing-wool is so small that it is easily met by a better article, perhaps thi would not be the case. And it may be problematical whether the proper combing length will be easily reached, or at least maintained in this country, in the absence of that high feeding system which has undoubtedly given the wool its increased length in England.*

The average weight of fleece in the hill-fed sheep is 3 lbs.; on rich lowlands a little more. Mr. John Ellman, Jr., testified before the Committee of the House of Lords that he was then "keeping his sheep better than formerly—fattening them, which rendered the fleece heavier—that they then averaged about 3 lbs. of wool."† "But the Down is cultivated more particularly for its mutton, which for quality takes precedence of all other" (from sheep of good size) "in the English markets. Its early maturity and extreme aptitude to lay on flesh, render it peculiarly valuable for this purpose. The Down is turned off at two years old, and its weight at that age is, in England, from 80 to 100 lbs. High fed wethers have reached

---

* Nearly or quite every individual who testifies to the deterioration and increased length of the South Down wool before the Lord's Committee, assign this as the cause of the change.

from 32 to even 40 lbs. a quarter! Notwithstanding its weight, the Down has, in the language of Mr. Youatt, a patience of occasional short keep, and an endurance of hard stocking, equal to any other sheep. This gives it a decided advantage over the bulkier Leicester, Lincolns, &c., as a mutton sheep, in hilly districts and those producing short and scanty herbage. It is hardy and healthy, though in common with the other English varieties much subject to the catarrh or "snuffles," and no sheep better withstands our American winters. The ewes are prolific breeders and good nursers. The Down is quiet and docile in its habits, and though an industrious feeder, exhibiting little disposition to rove." Like the Leicester, it is comparatively a short-lived animal, and the fleece continues to decrease in weight after it reaches maturity. It crosses better with short and middle wooled breeds than the Leicester. "A sheep possessing such qualities must of course be valuable in upland districts in the vicinity of markets. They have been introduced into every part of the British Dominions, and imported into various other countries. The Emperor of Russia paid Mr. Ellman three hundred guineas for two rams, and in 1800 'a ram belonging to the Duke of Bedford, was let for one season at eighty guineas, two others at forty guineas each, and four more at twenty-eight guineas each.' These valuable sheep were introduced into the United States a few years since by Col. J. H. Powell, of Philadelphia, and a small number was imported by one of the members of this Committee in 1834. The last were from the flock of Mr. Ellman, at a cost of $60 a head. Several other importations have since taken place."

The ram and ewe, the portraits of which are given, are the descendants of the importation of Francis Rotch, Esq., alluded to in the preceding paragraph. They are most spirited likenesses, and were kindly furnished me by that gentleman, to accompany this Letter. They are exceedingly

SOUTH-DOWN EWE.

characteristic of the Ellman stock. Not so large as the later importations of Mr. Rotch from the celebrated flock of Mr. Webb, they are, in the opinion of that gentleman, as well as in my own, a more beautifully formed and not less profitable animal. For compactness—great weight in a small compass—they are perhaps unrivaled.

The following is the description of the perfect South-Down by Mr. Ellman, the founder of the improved breed :

"The head small and hornless ; the face speckled or gray and neither too long nor too short ; the lips thin, and the space between the nose and the eyes narrow ; the under jaw or chap fine and thin ; the ears tolerably wide and well covered with wool, and the forehead also, and the whole space between the ears well protected by it, as a defence against the fly.

"The eye full and bright but not prominent. The orbits of the eye, the eye-cap or bone not too projecting, that it may not form a fatal obstacle in lambing.

" The neck of a medium length, thin toward the head, but enlarging toward the shoulders, where it should be broad and high and straight in its whole course above and below. The breast should be wide, deep, and projecting forward between the fore-legs, indicating a good constitution and a disposition to thrive. Corresponding with this, the shoulders should be on a level with the back, and not too wide above : they should bow outward from the top to the breast, indicating a springing rib beneath, and leaving room for it.

" The ribs coming out horizontally from the spine, and extending far backward, and the last rib projecting more than others, the back flat from the shoulders to the setting on of the tail ; the loin broad and flat ; the rump broad and the tail set on high, and nearly on a level with the spine. The hips wide ; the space between them and the last rib on either side as narrow as possible, and the ribs generally presenting a circular form like a barrel.

" The belly as straight as the back.

" The legs neither too long nor too short : the fore-legs straight from the breast to the foot : not bending inward at the knee, and standing far apart both before and behind ; the hock having a direction rather outward, and the twist, or the meeting of the thighs behind, being particularly full, the bones fine, yet having no appearance of weakness, and of a speckled or dark color.

The belly well defended with wool, and the wool coming down before and behind to the knee and to the hock ; the wool short, close, curled and fine, and free from spiry projecting fibres "

THE COTSWOLD SHEEP.

The above cut is copied from one in Mr. Spooner's work on Sheep— the original drawing being by Harvey.

The Cotswolds, until improved by modern crosses, were a very large, coarse, long-legged, flat-ribbed variety, light in the fore-quarter—shearing a long, heavy, coarse fleece of wool. They were hardy, prolific breeders and capital nurses. They were deficient in early maturity, and did not possess feeding properties equaling those of the Down or New Leicester. To a cross with the latter variety we owe the modern or improved Cotswold. Having had no personal experience with the breed,* I prefer quoting the descriptions of the later standard English writers, to the task of compilation.

The following is from Spooner : †

" The Cotswold is a large breed of sheep, with a long and abundant fleece, and the ewes are very prolific and good nurses. Formerly they were bred only on the hills, and fatted in the valleys, of the Severn and the Thames; but with the inclosure of the Cotswold Hills and the improvement of their cultivation they have been reared and fatted in the same district. They have been extensively crossed with the Leicester sheep, by which their size and fleece have been somewhat diminished, but their carcasses considerably improved, and their maturity rendered earlier. The wethers are now sometimes fattened at 14 months old, when they weigh from 15 lbs. to 24 lbs. per quarter, and at two years old increase to 20 lbs. or 30 lbs. The wool is strong, mellow, and of good color, though rather coarse, 6 to 8 inches in length, and from 7 lbs. to 8 lbs. per fleece. The superior hardihood of the improved Cotswold over the Leicester, and their adaptation to common treatment, together with the prolific nature of the ewes and their abundance of milk, have rendered them in many places rivals of the New Leicester, and have obtained for them, of late years, more attention to their selection and general treatment, under which management still farther improvement appears very probable. They have also been used in crossing other breeds, and, as before noticed, have been mixed with the Hampshire Downs. It is, indeed, the improved Cotswold that, under the term New or Improved Oxfordshire Sheep, are so frequently the successful candidates for prizes offered for the best long-wooled sheep at some of the principal agricultural meetings or shows in the Kingdom. The quality of the mutton is considered superior to that of the Leicester, the tallow being less abundant, with a larger development of muscle or flesh. We may, therefore, regard this breed as one of established reputation, and extending itself throughout every district of the Kingdom."

Of the method of crossing between the Cotswolds and Leicester, Mr. Youatt remarks : ‡

" The degree to which the cross may be carried must depend upon the nature of the old stock, and on the situation and character of the farm. In exposed situations, and somewhat scanty pasture, the old blood should decidedly prevail. On a more sheltered soil, and on land that will bear closer stocking, a greater use may be made of the Leicester. Another circumstance that will guide the farmer is the object that he principally has in view. If he expects to derive his chief profits from the wool, he will look to the primitive Cotswolds ; if he expects to gain more as a grazier, he will use the Leicester ram more freely."

Cotswold sheep of good quality have been imported into the United States by Messrs. Corning & Sotham, of Albany, and are now bred by the latter gentleman. I believe there were several earlier importations—but of their dates or particulars I am not advised.

THE CHEVIOT SHEEP.—Sheep of this breed have been imported into my immediate neighborhood, and were subject to my frequent inspection for two or three years. They had the appearance of small Leicesters, but were considerably inferior in correctness of proportions to high-bred animals of that variety. They perhaps more resemble a cross between the Leicester and the old " native " or common breed of the United States. Their fleeces were too coarse to furnish a good carding wool—too short for a good combing one. Mixed with a smaller lot of better wool, their this year's clip sold for 29 cents per pound, while my *heavier* Merino fleeces sold for 42 cents per pound. They attracted no notice, and might at any time have been bought of their owner for the price of common sheep of the same weight. I believe the flock was broken up and sold to butchers and others this spring, after shearing. They were certainly inferior to the description of the breed by Sir John Sinclair, even in 1792, quoted by Mr. Youatt,|| and

* With every breed previously described, I have had ample personal experience. I have merely seen Cotswold flocks.   † Q. v., p. 99.   ‡ Q. v., p. 340. || Q. v., pp. 265, 286.

had all the defects attributed to the original stock by Cully.* They might not, however, have been favorable specimens of the breed.

On the steep, storm-lashed Cheviot Hills, in the extreme North of Eng .and, this breed first attracted notice for their great hardiness in resisting

CHEVIOT EWE.

cold and feeding on coarse heathery herbage. A cross with the Leicester, pretty generally resorted to, constitutes the improved variety. The characteristics of the Leicester are quite evident in the portrait of the Cheviot Ewe, above, copied from Mr. Youatt.

Professor Low thus speaks of the result of this cross :

" The Cheviot breed amalgamates with the Leicester, and a system of breeding has been extensively introduced for producing the first cross of this descent. The rams employed are of the pure Leicester breed, and the progeny is superior in size, weight of wool, and tendency to fatten, to the native Cheviot. . . . The benefit, however, may be said to end with the first cross, and the progeny of this mixed descent is greatly inferior to the pure Leicester in form and fattening properties, and to the pure Cheviot in hardiness of constitution.

Of the improved Cheviot Mr. Spooner says :

" This breed has greatly extended itself throughout the mountains of Scotland, and in many instances supplanted the Black-faced breed ; but the change, though in many cases advantageous, has in some instances been otherwise, the latter being somewhat hardier, and more capable of subsisting on heathy pasturage. They are, however, a hardy race, well suited for their native pastures, bearing with comparative impunity the storms of winter, and thriving well on poor keep. Though less hardy than the black-faced sheep of Scotland, they are more profitable as respects their feeding, making more flesh on an equal quantity of food, and making it quicker. They have white faces and legs, open countenances, lively eyes, without horns. The ears are large, and somewhat singular, and there is much space between the ears and eyes. The carcass is long ; the back straight ; the shoulders rather light ; the ribs circular ; and the quarters good. The legs are small in the bone and covered with wool, as well as the body, with the exception of the face. The Cheviot wether is fit for the butcher at three years old, and averages from 12 lbs. to 18 lbs. per quarter—the mutton being of a good quality, though inferior to the South-Down, and of less flavor than the Black-faced. . . . . The Cheviot, though a mountain breed, is quiet and docile, and easily managed. The wool is *fine*, (?) closely covers the body, assisting much in preserving it from the effects of wet and cold ; the fleece averaging about 3½ lbs. Formerly the wool was extensively employed for making cloths, but having given place to the finer Saxony wools, it has sunk in price, and been confined to combing purposes. It has thus become altogether a secondary consideration." . . .

If Mr. Spooner is not made to say that the wool is "fine" by an omission of qualifying words, or some other misprint, his ideas of *fineness* must be singular indeed! The South-Down wool, rejected for carding pur poses, is several shades finer than the Cheviot! The latter is of about the quality of Leicester, the number of serrations about the same, and, says Mr. Youatt, speaking of the microscopic appearance of the wool, "the derivation of the breed (from the Leicester) is well illustrated by the formation of the fibre."

Mr. John Varley, manufacturer, of Stanningley, near Leeds, thus testified before the Lords' Committee :*

"I attribute the low price of Cheviot wool to deterioration ; it is deteriorated very much in point of hair ; it was formerly the fashion of the day for Cheviot wool to be worn as cloth ; it is not the fashion now. It is not fit to make fine cloths, as it was then. . . . . The wool is grown coarser and longer, and only fit to make low coatings and flushings."

This is confirmed by the testimony of other witnesses before the Committee; and Mr. Youatt on the same subject remarks,† "that the wool is inferior to the South-Down."

BROAD-TAILED ASIATIC AND AFRICAN SHEEP.—I allude to the Broad-tailed race of sheep, not from any high estimate which I place upon their value, but because they constitute one of the breeds now existing in a state of purity in the United States.

Some "Tunisian Mountain Sheep" were received by Col. Pickering when abroad, and were distributed by him in Pennsylvania.‡ They are highly spoken of by Col. Powell as a cross with the Dishley and South-Down. They have, I believe, long since become extinct.

It was Commodore Porter, I think, who, you informed me, sent home some of the Broad-tailed sheep of Asia, obtained from Smyrna, pure-blooded descendants of which yet exist in South Carolina.|| I have carefully examined the specimens of wool of the full blood and the grades of this variety forwarded by you. No. 3, taken from the skin of a full-blood, is 8 inches long, pure white, consisting of coarse hairs, uneven in their length and diameter—the same hair of uneven diameter in different parts of it, and the whole intermixed for about 4 inches from the roots, with a fine, downy or cottony wool. No. 2, about 3¾ inches long from the side of a three-fourths blood ram, is much evener in quality, with no hairs as coarse or wool as fine as in No. 3. It contains some jarr, or short, sharp-pointed hairs, and is a dry, and, I should judge, rather unworkable wool, not *well* adapted to either carding or combing. No. 1, from thigh of same animal, is 8 inches long, resembles No. 3, but not so great a distinction between the hair and the wool. No. 4, from a three-fourths blood 4-year-old ewe, is about 2 inches long, contains a few colored hairs, resembles No. 2, but is somewhat coarser. All these samples are destitute of yolk, and apparently come from loose, light, dry, open fleeces. They do not strike me as wools which could be as profitably cultivated as many others, for any objects or under any circumstances.

If the object is mutton instead of wool, it seems to me that a better selection can be made, from some of the English breeds—which intermingle

* Bischoff, vol. ii, p. 144. Mr. Youatt quotes the substance of the above, and fully sustains Mr. Varley's views.   † Q. v., p. 235.
‡ See Essay on Various Breeds of Sheep, by Col. John Hare Powell, published in the Memoirs of the Board of Agriculture of the State of New-York, vol. iii., p. 377, (1826.)
|| In Letter Vth I inadvertently spoke of these as a *large* breed of sheep. They are not above medium size, or rather, may be said to be a smallish race.

their fat and muscle in such a manner as to render both palatable, instead of depositing a greatly disproportioned share of the former in one luscious mass, forming an impediment to breeding, and an unsightly appendage in the eye of the breeder.

All the different varieties of the Broad-tailed and Fat-rumped sheep will be found described in Youatt, and I will not now consume your time with them.

LETTER XI.

THE MOST PROFITABLE BREED OF SHEEP FOR THE SOUTH.—PRINCIPLES OF BREEDING.

Breeds should be adapted to the circumstances of a Country...Circumstances requiring a Mutton Sheep ...Comparison between Mutton Sheep—The South-Downs, Leicesters and Cotswolds...How far the Feed Markets, &c., of the South demand such breeds...What breed of Sheep will give the greatest value of Wool from the feed of an acre?...Comparative Consumption and Wool Product of the Mutton breeds and the Merino—Other Expenses—Comparative Hardiness, &c...A pound of fine wool can be grown as cheaply as a pound of coarse—worth more for market or for consumption...The Mutton of the Merino and its Crosses...What sub-variety of the Merino best adapted to the wants of the South?...Review of the History of Wool-Growing and the Wool Markets since 1824...Tariffs and Prices...Injudicious course of the Manufacturers—Have discouraged the growth of fine wool and encouraged that of medium and coarse... A surplus of medium wools, and a bare or short supply of fine...Manufacturers now in the power of fine wool growers...Interest of the Manufacturers to encourage the growth of fine wools by paying better prices—are beginning to do so—will be compelled to continue this course...Will the North furnish the increasing demand?—No—Reasons...Fine wool in every point of view more profitable than coarse for cultivation in the South...Comparison between Merinos and Saxons...Crosses between them...Points which constitute excellence in a Merino—proper size—per centage of wool to live-weight—shape and general appearance—skin—wrinkles...The wool—what parts it should cover—its gum—length and weight of fleece—evenness—style—softness—scration—manner of opening, &c...Principles of breeding...In and-in breeding...Crossing...English Crosses with the Merino...Views of Mr. Livingston concerning the use of cross-bred rams—of the French breeders—of the author...Great importance of starting a flock with choice rams—with different strains of blood.

Dear Sir: No one breed of sheep combines the highest perfection in all those points which give value to this race of animals. One is remarkable for the weight, or early maturity, or excellent quality of its carcass, while it is deficient in quality or quantity of wool; and another which is valuable for wool, is comparatively deficient in carcass. Some varieties will flourish only under certain conditions of feed and climate, while others are much less affected by those conditions, and will subsist under the greatest variations of temperature, and on the most opposite qualities of verdure.

In selecting a breed for any given locality, we are to take into consideration *first*, the feed and climate, or the surrounding natural circumstances; and, *second*, the market facilities and demand. We should then make choice of that breed which, with the advantages possessed, and under all the circumstances, will yield the greatest net value of marketable product.

Rich lowland herbage, in a climate which allows it to remain green during a large portion of the year, is favorable to the production of large carcasses. If convenient to markets where mutton finds a prompt sale and good prices, then all the conditions are realized which call for a *mutton*, as contradistinguished from a *wool-producing* sheep. Under such circumstances, the choice should undoubtedly, in my judgment, rest between the improved English varieties—the South-Down, the New Leicester, and the improved Cotswold or New Oxfordshire sheep. In deciding between these, minor and more specific circumstances are to be taken into account. If we wish to keep large numbers, the Down will herd* much better than the two larger breeds; if our feed, though generally plentiful, is liable to be shortish during the drouths of summer, and we have not a *certain* supply of the most nutritious winter feed, the Down will better endure occasional short keep: if the market calls for a choice and high-flavored mutton, the Down possesses a decided superiority. If, on the other hand, we

* That is, remain thriving and healthy when kept together in large numbers.

U

wish to keep but few in the same inclosure, the large breeds will be as healthy as the Downs; if the pastures be wettish or marshy, the former will better subsist on the rank herbage which usually grows in such situations; if they do not afford so fine a quality of mutton, they, particularly the Leicester, possess an earlier maturity, and both give more meat for the amount of food consumed, and yield more tallow.

The next point of comparison between the Long and Middle wooled families, is the value of their wool. Though not the first or principal object aimed at in the culture of any of these breeds, it is, in this country, an important item or incident in determining their relative profitableness. The American Leicester* yields about 6 lbs. of long, coarse, combing wool; the Cotswold something more, but this perhaps counterbalanced by other considerations; the Down from 3 lbs. to 4 lbs. of a low quality of carding wool. None of these wools are very salable, at remunerating prices, in the American market. Both will become more so, as manufactures of worsted, and of flannels and baizes, increase. The difference in the weight of fleeces between the breeds is, *per se*, a less important consideration than would first appear, and for reasons which will be given when I speak of the connection between the amount of wool produced and the food consumed, by sheep.

Of the Cheviots I have taken no notice in this connection, as they are obviously inferior to the preceding breeds, except in a capacity to endure rigorous weather, and to subsist on heathy herbage. No part of the South has a climate too severe for the more valuable races, and its grasses and other esculents, wherever found, and as far as they go, are, making the proper allowances for wet and dry lands, highly palatable and nutritious to all the varieties which respectively feed in such situations.

Under the natural and artificial circumstances already alluded to, which surround Sheep Husbandry in many parts of England—where the fattest and grossest quality of mutton is consumed as almost the only animal food of the laboring classes—the heavy, early maturing New Leicester, and the still heavier New Oxfordshire sheep, seem exactly adapted to the wants of producer and consumer, and are of unrivaled value. To depasture poorer soils—sustain a folding system—and furnish the mutton which supplies the tables of the wealthy—the South-Down is an equal desideratum.

Have we any region in our Southern States, where analogous circumstances demand the introduction of similar breeds? The climate, so far as its effect on the *health* is concerned, is adapted to any, even the least hardy varieties; but not so its effects on the verdure on which they are to subsist. The long, scorching summers, so utterly unlike those of England, leave the grass on lands stocked heavily enough for profit, entirely too dry and short for the heavy, sluggish Long Wools. This is particularly true in the tide-water zone. Mutton, too, sheeted over externally with three or four inches of solid fat,† even if it could be made acceptable to the slave, in lieu of his ration of bacon—a thing more than doubtful—would never find any considerable market off from the plantation. So far as the supply of feed is concerned, the above remarks apply, though not equally, to the South-Down. It will live and thrive where the Long Wools would dwindle away, but it is a mistake to suppose that the heavy im-

---

* I use the word "American" Leicester, because it is notorious that this, as well as the Cotswold—and *all the other* heavy English varieties, soon lose in the weight of their fleeces when subjected to the climate and the (best ordinary) system of feeding in the United States. I should except, perhaps, a few highly pampered animals.

† Five and even six inches of solid fat, on the rib, is not uncommon in England. In the Cotswolds the fat and lean are more intermixed, and the mutton is of a better quality; but it would be considered entirely too luscious and tallowy by Americans.

proved South-Down will subsist, and attain its proper weight and fatness, on very poor or very scant herbage. The old *unimproved* variety would, like some other smallish and hardy races, obtain a living on keep as poor as that which grew on the lightest and thinnest soils of Sussex. Moulded by the hand of Ellman, and other breeders, to better fulfill the conditions of a mutton sheep, in size and other particulars, they demand that increased supply of food which the formation of additional fat and muscle require. Retaining some of the properties of the parent stock, they are less sluggish, and bear travel better than the Long Wools; but with them as with the latter, and *all other animals*, much or prolonged exercise in pursuit of food or otherwise, is unfavorable to obesity. Men, and particularly *owners*, in advocating the claims of this breed and that, seem not unfrequently to forget that the general physical laws which control in the development of all the animal tissues as well as functions, are uniform. Better organs will doubtless make a better appropriation of animal food; and they may be taught, so to speak, to appropriate it in particular directions—in one breed, more especially to the production of fat—in another, of muscle or lean meat—in another, wool. But, *cæteris paribus*, large animals will always require more food than small ones. Animals which are to be carried to a high state of fatness must have plentiful and nutritious food, and they must exercise but little in order to prevent the unnecessary "combustion" in the lungs, of that carbon which forms more than seven-tenths of their fat. No art of breeding can countervail these established laws of Nature.

Again, there are no facilities in the South for marketing large quantities of mutton—of a tithe of that which would be annually fitted for the shambles, were Sheep Husbandry introduced to anything like the extent I have recommended, and with the *mutton breeds* of sheep. With few cities and large villages—with a sparse population—with an agricultural population the greatest drawback on whose pecuniary prosperity is their inability to market their own surplus edibles—not a particle of rational doubt can exist on this point. True, I have expressed the opinion that, both as a matter of healthfulness and economy, mutton should be substituted for a moiety of the bacon used on the plantation; but with such a change, in a country so exclusively agricultural, each landholder would raise his own supply, and thus no market be created. It may then be regarded as a settled point that the production of wool is the primary, the great object of Southern Sheep Husbandry.

In instituting a comparison between breeds of sheep for wool-growing purposes, I will, in the outset, lay down the obviously incontrovertible proposition that the question is not what variety will shear the heaviest or even the most valuable fleeces, irrespective of the cost of production.— Cost of feed and care, and every other expense, must be deducted, to fairly test the profits of an animal. If a large sheep consume twice as much food as a small one, and give but once and a half as much wool, it is obviously more profitable, other things being equal, to keep *two* of the smaller sheep. The true question then is, *with the same expense in other particulars, From what breed will the verdure of an acre of land produce the greatest value of wool?*

Let us first proceed to ascertain the comparative amount of food consumed by the several breeds. There are no satisfactory experiments which show that *breed*, in itself considered, has any particular influence on the quantity of food consumed. It is found, with all varieties, that the consumption is in proportion to the live weight of the (grown) animal. Of course, this rule is not invariable in its individual application, but its gen-

eral soundness has been satisfactorily established. Spooner states that
grown sheep take up 3½ per cent. of their weight in what is equivalent to
dry hay per day, to keep in store condition. Veit places the consumption
at 2½ per cent. My experience would incline me to place it about midway
between the two. But whatever the precise amount of the consumption,
if it is proportioned to the weight, it follows that if an acre is capable of
sustaining three Merinos weighing 100 lbs. each, it will sustain but two
Leicesters weighing 150 lbs. each, and two and two-fifths South-Downs
weighing 125 lbs. each. Merinos of *this weight* often shear 5 lbs. per
fleece, taking flocks through. The herbage of an acre, then, would give
15 lbs. of Merino wool, and but 12 lbs. of Leicester, and but 9¾ lbs. of
South-Down (estimating the latter as high as 4 lbs. to the fleece)! Even
the finest and lightest fleeced sheep ordinarily known as Merinos, average
about 4 lbs. to the fleece, so that the feed of an acre would produce as
much of the highest quality of wool sold under the name of Merino, as it
would of New Leicester, and more than it would of South-Down! The
former would be worth from fifty to one hundred per cent. more per pound
than either of the latter! Nor does this indicate all the actual difference,
as I have, in the preceding estimate, placed the live-weight of the English
breeds low, and that of the Merino high. The live-weight of the four-
pound fine-fleeced Merino does not exceed 90 lbs. It ranges from 80 to
90 lbs., so that 300 lbs. of live-weight would give a still greater product
of wool to the acre.* I consider it perfectly safe to say that *the herbage
of an acre will uniformly give nearly double the value of Merino, that it will
of any of the English Long or Middle wools.*

The important question now remains, What are the *other* relative expenses of these breeds? I speak from experience when I say that the
Leicester† is in no respect a hardier sheep than the Merino—indeed, it is
my firm conviction that it is *less hardy*, under the most favorable circumstances. It is more subject to colds, and I think its constitution breaks up
more readily under disease. The lambs are more liable to perish from exposure to cold, when newly dropped. Under *unfavorable* circumstances—
herded in large flocks, pinched for feed, or subjected to long journeys—
its capacity to endure, and its ability to rally from the effects of such draw-
backs, do not compare with those of the Merino. The high-bred South-
Down, though considerably less hardy than the unimproved parent stock,
is still fairly entitled to the appellation of a hardy animal. In this respect
I consider it just about on a par with the Merino. I do not think, however, it will bear as hard stocking as the latter, without a rapid diminution
in size and quality. If the peculiar merits of the animal are to be taken
into account in determining the expenses—and I think they should be—
the superior fecundity of the South-Down is a point in its favor, as well
for a wool-producing as a mutton sheep. The South-Down ewe not only
frequently yeans twin lambs, as do both the Merino and Leicester, but she
possesses, unlike the latter, nursing properties to do justice by them. But
this advantage is fully counterbalanced by the superior longevity of the
Merino. All the English mutton breeds begin to rapidly deteriorate in
am unt of wool, capacity to fatten, and in general vigor, at about 5 years
old and their early maturity is no offset to this, in a sheep kept for wool-
growing purposes. This early decay would require earlier and more rapid
slaughter or sale than would always be economically convenient, or even
possible, in a region situated in all respects like the South. It is well, on

---

* It is understood that all of these live-weights refer to *ewes* in fair ordinary, or what is called store
condition.
† I speak of full-blood Leicesters. Some of its crosses are much hardier than the pure bred sheep.

properly stocked farms, to slaughter or turn off the Merino wether at four or five years old, to make room for the breeding stock; but he will not particularly deteriorate, and he will richly pay the way with his fleece, for several years longer. Breeding ewes are rarely turned off before *eight*, and are frequently kept until *ten* years old, at which period they exhibit no greater marks of age than do the Down and Leicester at *five* or *six*.— I have known instances of Merino ewes breeding uniformly until 15 years old! The Improved Cotswold is said to be hardier than the Leicester; but I have said less of this variety, throughout this entire Letter, as from their great size* and the consequent amount of food consumed by them, and the other necessary incidents connected with the breeding of so large animals, the idea of their being introduced as a *wool-growing sheep* anywhere, and particularly on lands grassed like those of the South, is, in my judgment, utterly preposterous. There is one advantage which all the coarse races of sheep have over the Merino. Either because their hoofs do not grow long and turn under from the sides, as do those of the Merino, and thus hold dirt and filth in constant contact with the foot, the coarse races are less subject to the visitations of the hoof-ail, and, when contracted, it spreads with less violence and malignity among them. Taking all the circumstances connected with the peculiar management of each race, and all the incidents, exigencies, and risks of the husbandry of each fairly into account, I am fully convinced that the expenses, other than those of feed, are not smaller *per capita*, or even in the number required to stock an acre, in either of the English breeds above referred to, than in the Merino. Nor should I be disposed to concede even equality, in these respects, to either of those English breeds, excepting the South-Down.

You write me, Sir, that many of the South Carolina planters are under the impression that coarse wools will be most profitably grown by them, *first*, because there is a greater deficit in the supply, and they are better protected from foreign competition; and, *secondly*, because they furnish the raw material for so great a portion of the woolens consumed in the South. Each of these premises is true, but are the conclusions legitimate? Notwithstanding the greater deficit and better protection, do the coarse wools bear as high a price as the fine ones? If not, they are not as profitable, for I have already shown that *it costs no more to raise a pound of coarse than a pound of fine wool.* Nay, a pound of *medium* Merino wool can be raised *more cheaply* than a pound of the South-Down, Leicester, or Cotswold! This I consider clearly established.

Grant that the South requires a much greater proportion of coarse than of fine wool, for her own consumption. If a man needing iron for his own consumption, wrought a mine to obtain it, in which he should happen to find gold equally accessible and plentiful, would it be economical in him to neglect the more precious metal because *he* wanted *to use* the iron? or should he dig the gold, obtain the iron by exchange, and pocket the difference in value? Would it be economical to grow surplus wool, wool for market, worth from 25 to 30 cents per pound, when it costs no more per pound to grow that worth from 40 to 45 cents? And even for the home want, for the uses of the plantation—for slave-cloths, &c.—*fine wool is worth more per pound than coarse for actual wear or use!* Is this proposition new and incredible to you? I challenge the fullest investigation of its truth, through the testimony of those familiar with the subject, or through the direct ordeal of experiment. It is true that a piece of fine broadcloth is not so strong, nor will it wear like a Chelmsford plain of treble thick-

* I saw two at the late N. Y. State Fair, at Saratoga, which weighed over 300 lbs. each!

ness. The threads of the former are spun to extreme fineness to economize the costly raw material. To give it that finish which is demanded
by fashion—to give it its beautiful nap—these threads are still farther reduced by "gigging" and "shearing." But spin fine wool into yarn as
coarse as that used in Chelmsfords, and manufacture it in the same way,
and it would make a far stronger and more durable cloth. The reasons
are obvious. Merino wool is decidedly stronger than the English coarse
Long and Middle wools—or any other coarse wools—in proportion to its
diameter or bulk. It felts far better, and there is therefore a greater cohesion between the different fibres of the same thread, and between the
different threads. It is also more pliable and elastic, and consequently less
subject to "breaking" and abrasion.

Unless the views I have advanced are singularly erroneous, it will be
seen that, for wool-growing purposes, the Merino possesses a marked and
decided superiority over the best breeds and families of coarse-wooled
sheep. As a mutton sheep, it is inferior to some of those breeds, but not
so much so as it is generally reputed to be. If required to consume the
fat and lean together, many who have never tasted Merino mutton, and
who have an unfavorable impression of it, would, I suspect, find it more
palatable than the luscious and over-fat New Leicester. The mutton of
the cross between the Merino and "Native" sheep would certainly be
preferred to the Leicester, by anybody but an English laborer used to the
latter. It is short-grained, tender, and of good flavor. The same is true
of the crosses with the English varieties. These will be, hereafter, more
particularly alluded to. Grade Merino wethers (say half-bloods) are favor-
ites with the Northern drover and butcher. They are of good size—ex-
traordinarily heavy for their apparent bulk*—make good mutton—tallow
well—and their pelts, from the greater weight of wool on them, command
an extra price. They would, in my opinion, furnish a mutton every way
suitable for plantation consumption, and one which would be well accepted in the Southern markets.

In speaking of the Merino in this connection, I have in all cases, unless
it is distinctly specified to the contrary, had no reference to the Saxons—
though they are, as it is well known, pure-blooded descendants of the
former.

Assuming it now as a settled point, that it is to the Merino race that the
wool-grower must look for the most profitable sheep, let us now proceed
to inquire which of the widely varying sub-varieties of this race are best
adapted to the wants and circumstances of the South. A brief glance at
the history of wool-growing, and of the wool markets, for the last few
years, will form an useful preliminary inquiry, and will assist us materially
in arriving at a correct conclusion.

On the introduction of the Saxons, about twenty-four years since, they
were sought with avidity by the holders of the fine-wooled flocks of the
country, consisting at that time of pure or grade Merinos. The Tariff of
1824 imposed a duty of 20 per cent. on wools costing above 10 cents per
pound, gradually rising to 30 per cent., and 15 per cent. on those costing
less than 10 cents. Foreign woolen cloths† were subject to an *ad valorem*
duty of 30 per cent. until June 30th, 1825, and after that it was raised to 33¼
per cent. The Tariff of 1828 immediately raised the duty on all wools to 40
per cent. *ad valorem* and 4 cents per pound specific duty, and 5 per cent.
was to be annually added to the *ad valorem* duty, until it should reach 50

---

* On account of the shortness of their wool, compared with the coarse breeds.
† Where I use the word "cloths" here and in the statements of the different Tariffs which follow, you
will understand that I do not include carpetings, blankets, worsted stuff goods. &c.

per cent. (in 1831.) The duty on woolen cloths was also raised (after June, 1829) to 45 per cent., and that exceeding $4 the square yard to 50 per cent. Under the decisive encouragement offered to both the wool-grower and manufacturer by this Act, a great impetus was given to the production of the finest wools, and the Saxons everywhere rapidly superseded, or bred out by crossing, the Spanish Merinos. The latter disappeared almost entirely from New-York and New-England. In the fine-wool mania which ensued, weight of fleece, constitution, and every-thing else, were sacrificed to the quality of the wool. The Tariff of 1832 imposed a 40 per cent. *ad valorem* and 4 cents per pound specific duty on wools costing over 8 cents ; and it raised the duty on all broadcloths to 50 per cent. It made wools costing less than 8 cents per pound free of duty. The "Compromise" Tariff of 1833 commenced a system of progressive reductions until the maximum rate of duties should not exceed 20 per cent. The following Table will give the duties of each year, on wool and cloths, under this Act, estimating the *ad valorem* and specific duties on wools exceeding 8 cents, together in an average per centage :*

TABLE 14.

	Per ct. ad val.	1833.	1835.	1837.	1839.	1841.	1842.
Wool costing less than 8 cents per pound at place of exportation............	free.	free.	free.	free.	free.	free.	20
Wool costing over 8 cents per pound......	54	50·60	47·20	43·80	40·40	30·20	20
Woolen cloths........................	50	47	44	41	38	29	20

The Tariff of 1841 struck out the 20 per cent. duty on the 8 cent wools. The Tariff of 1842 again imposed an *ad valorem* duty of 5 per cent. or wools costing *seven* cents or under, and raised it on the higher wools to 30 per cent. *ad valorem* and 3 cents per pound specific duty, and on cloths to 40 per cent. *ad valorem*. The Tariff of 1846 established an *ad valorem* duty of 30 per cent. on all wools, and on cloths. By referring to Table 7, Letter V., it will be seen that the prices of wool have not been controlled by the amount of the protection. They reached their maximum in 1836, and then fell off, not again to rally, (except during the single year 1839)— not again to reach 40 cents—until 1844. Why was this ? What produced the sudden depreciation of 1837 ? The Tariffs of 1828 and 1832 gave *too much protection* to both wool-grower and manufacturer. Their pursuits became the *El Dorado* of agricultural and mercantile speculators. Skill without capital, and capital without skill, and in some cases probably thirst of gain without either, rushed into these favored avocations. The bank inflations of the period fanned the fires of speculation, and taught some of the wisest commercial heads of the country to forget the providence that had hitherto distinguished them. The natural result followed. In the financial crisis of 1837, manufacturing, and all other monetary enterprises which had not been conducted with skill and providence, and which were not based on adequate and real capital, were involved in a common destruction, and even the solidest and best conducted institutions of the country were shaken by the fury of the explosion. Wool suddenly fell almost 50 per cent. (from 54 to 30 cents per pound.)† In 1838 it rallied a little, and in 1839 it again reached 50 cents, but it went down nearly to the minimum point in 1840. The grower began to be discouraged. He who bred the delicate Saxons, (and, as I have already said, they now comprised the flocks of nearly all the large wool-growers in the country,)

* The reduction of one-tenth of the excess over 20 per cent. took place Dec. 31st, each year, to 1841; then one-half of the residue of the excess; and on the 30th of June, 1842, the other half of said residue was deducted.

† The quality of the wools here alluded to will be found specified in a note on the second page of Letter V.

was not obtaining the actual first cost per pound of his wool. He clam
ored loudly for an increase of duties on the foreign article, as the reductions of the " Compromise " Act were now approaching their ultimate
standard—20 per cent.—and he attributed the low prices to this cause:
Saxon wool continued low, and did not pay its first cost in 1841 and 1842.
Was this due solely to the reduction of the Tariff? A reference to Table
11 (Letter IX.) will show that the import of foreign woolens was less from
1836 up to and including 1842, than for the six preceding years! Where
then was the foreign competition which was driving the manufacturer to
keep down the price of wools? The Tariff of 1842 raised the duty on
wool 10 per cent. and added a specific duty of 3 cents per pound; and it
raised the duty on cloths from 20 to 40 per cent. The import of foreign
woolens sunk, the succeeding year, to a lower point than it had touched
since 1821, and in 1844 and 1845 it did not reach the *average* of the six
years preceding the enactment of the Tariff of 1842. A reference to
Table 9 (Letter IX.) will show that the import of foreign *fine* wools also
largely fell off. This coincided with the expectations of the advocates of
a higher Tariff, but another and equally legitimate expectation entertained
by the great body of Northern wool-growers—that *they* were to *share* in
the benefits arising from the exclusion of foreign competition—was sig-
nally disappointed. The Tariff of 1842 was enacted on the 30th day of
August, and part of the clip of that year was sold under its operation.
Wool sold that year *lower* than it had for the five preceding years, viz., for
30 cents. The next year it advanced one penny! General discouragement now seized upon the growers of fine wool. The market was not
overstocked—foreign competition was light, but *still they could not* sell
their wool for its first cost! To add to their mortification, the manufacturer, by a most short-sighted policy, would scarcely make a discrimination of 6d. per pound between Saxon wool and medium Merino and grade
wools weighing nearly twice as much to the fleece. If the grower of medium wool got 25 cents per pound for fleeces weighing 4 lbs.—thus realizing $1 per fleece—the ordinary Saxon grower would get but 30 cents
per pound for fleeces weighing 2½ lbs., and thus realize but 75 cents! *
When the Saxon growers found that the Tariff of '42 brought them no
relief, they began to give up their costly and carefully nursed flocks. The
example, once set, became contagious, and there was a period when it
seemed as if all the Saxon sheep of the country would be sacrificed to
this reaction. Many abandoned wool-growing altogether, at a heavy sacrifice of their fixtures for rearing sheep. Others crossed with coarse-wooled
breeds, and rushing from one extreme to the other, some even crossed
with the English mutton breeds! Some more judiciously went back to
the parent Merino stock, but usually they selected the heaviest and
coarsest wooled Merinos, and thus materially deteriorated the character of
their wool. As the preceding period had been distinguished by its mania
for *fine* wool, this was, by its mania for *heavy fleeces*!† The English
crosses, however, were speedily abandoned.‡ The Merino regained his

* And though the larger, stronger sheep, bearing the medium wool, would eat more, it was far hardier,
required less protection and care of every kind, and would increase more rapidly—circumstances which
would far more than counterbalance its excess of consumption
† I make no claim of having possessed greater sagacity or foresight in these particulars than the mass of
breeders. I began with the Merino. These I crossed with the Saxon, and I also bred the pure-blood Saxons for several years. Unsatisfied with these, I made some experiments with the English mutton breeds,
both as pure bloods and crosses. Finding none of them equal to the Merino as a wool-producing sheep, I
returned to the latter, and I bred for *heavy fleeces* until the manufacturers saw fit to make a juster discrimination in the prices paid by them for the different qualities of wool.
‡ I mean by those who sought to improve their *fine-wooled* flocks by an English cross. English and all
other coarse-wooled sheep are immensely and rapidly improved, for wool-growing purposes, by a proper
fine wooled cross, as I have already and shall again have occasion to mention.

supremacy, lost for nearly twenty years, and again became the popular favorite. It was generally adopted by those who were commencing flocks in the new Western States, and gives its type to the sheep of those regions.

It will be seen from the preceding facts that the supply of fine wool* has proportionably decreased, and that of medium and coarse increased. This has driven the manufacturers to make a juster discrimination in prices. They now realize that their own short-sighted economy has been all but fatal to fine wool-growing in the United States. And they cannot but feel that in destroying this interest, *they destroy themselves.* Our manufacturers are not so miserably blind as to dream of drawing their raw material from foreign countries—of paying an import duty of 30 per cent. and then competing with the English manufacturer who pays an import duty not exceeding two pence per pound! It is doubtful, in my mind, whether the home supply will not fall considerably short of the home demand for fine wool *for this year !*† The point has been already reached where but a little more discouragement, or a little *longer continued* discouragement, would have banished these wools from the country! So far, the manufactories have not felt this evil, for they have not been compelled to import. Neither pampered nor persecuted by the Tariff of 1846—called for by the consumption of the country—with solid capital and greater experience and skill at their command—they are rapidly increasing, and rising on a solider basis than ever before. So, to *sustain our manufacturing interest,* (that engaged in the manufacture of fine cloths,) it is absolutely necessary that the diminution of *fine* wools be not only immediately arrested, but that the growth of them be immediately and largely increased. These facts now first beginning to be clearly appreciated by the manufacturer— will deter him from resorting to his former suicidal policy. Instances have recently come to my knowledge of manufacturers offering to contract with fine-wool growers for their entire clips, for a term of years, at an advance on present prices—prices, be it remembered, higher than they have been except for two years (1839 and 1844) since the overthrow of 1837. Should the manufacturer, however, again forget his own interest, the fine-wool grower has it in his power to teach it to him most effectually. Instead of being discouraged and driven from the business, he has but to withhold his wools for a season—say for a few months, to compel the former to import wools at a ruinous cost—stop his machinery, or pay fair prices at home! I believe in no combinations to control prices. Something far better than vague report, however, says that several of the large manufacturing establishments of New-England employed the *same agents,* last season, to buy much, if not all of their wools—and that these wools were subsequently divided by bidding or otherwise, among the parties to the transaction! *Is this denied?* I think it will not be denied. If this was so, what was it but a combination to control prices?‡ But whether

---

* To make myself clearly understood, I will, in the remarks which follow, classify wools as follows: *superfine,* the choicest quality of wool grown in the United States, and never grown here excepting in comparatively small quantities; *fine,* good ordinary Saxon; *good medium,* the highest quality of wool usually known in the market as Merino; *medium,* ordinary Merino; *ordinary,* grade Merino and perhaps selected South-Down fleeces; *coarse,* the English long wools, &c. This subdivision is not minute enough, by any means, to express fully the number of well-defined classes which exist in wool. A farther multiplication of them here, however, I have thought would only tend to confusion.

† The position has been all along taken that the general supply was under the demand, but the deficit hitherto has been principally in medium and coarse wools. See Table 9, Letter IX.

‡ And before leaving this point, I will ask another question: Why were most of the wools of New-York and New-England untouched and unlooked at by the agents of the manufacturers this year, contrary to all preceding customs for two or three months subsequently to shearing? These same agents flocked in droves to the Western States and bought up their entire clip immediately after shearing, while reports were constantly coming back that this manufactory and that had purchased its entire supply for a year, or perhaps two years? Was this because the Eastern growers demanded exorbitant prices? Was it because anything like an approach to a supply of fine wools could be found in the West? Or was it the result of a

X

so or not, when we compare the profits which have inure
and manufacturers of fine wool for the last few years, it
mer both to *speak* and *act* decidedly. Their interests ha
long enough! But it is to be hoped that the grower of
not be hereafter driven to the alternative of either suffer
defending himself by retaliatory measures. Some few
turers have always, I believe, taken a high and liberal
others, as already remarked, now see the necessity of a
prevent any combined or general effort to depress prices

Will the North again turn its attention to the growth
fine wools—again supply the demand, and keep up with
Not unless stimulated by the inducement of extraordir
certainly, against the competition of the South. The clir
or, beyond all dispute, north of 42°, is too severe for any
commonly known, which bear either of these classes of we
only such variety, in anything like general use, is the S
a delicate sheep, entirely incapable of safely withstand
winters, without good shelter, good and regularly admin
careful and skillful management in all other particulars.
is a little more than usually backward, so that grass does
the lambing season, it is difficult to raise the lambs of th
the young ewes will in many instances disown their lamb
them, not have a drop of milk for them; and if in such a
happens, a north-east or north-west storm comes drivin
snow or sleet on its wings, or there is a sudden depressic
ature from any cause, no care will save multitudes of la
ing.* And it will not do to defer the time of having the
cape these evils, or they will not attain size and strengt
safely through their first winter.† A few large sheephold
buildings, etc., have been arranged with exclusive referer
of these sheep, may continue to grow fine wool *until drii
competition of the South;* but many of these have re
Merino cross. The ordinary farmers, the small sheepho
aggregate, grow by far the largest portion of our Norther
bibed a deep-seated aversion—nay, a positive disgust—a
sheep. They have not the necessary fixtures for their
and they are entirely unwilling to bestow the necessary a
them. Besides, mutton and wool being about an equal c
this class of farmers, they want larger and earlier maturi
above all, they want a strong, hardy sheep, which demar
than their cattle. The strong, compact, medium-wooled
haps still more generally, its crosses with coarse varieti
wool which I have classified as ordinary—will be the ge
The same reasons will weigh still more strongly in the N
as I have shown, the climate is a still worse one for deli
these causes will tend to swell the amount of medium, ore

---

concerted movement to bring the Eastern grower into taking last year's prices ?
multitude of instances—or, he was contented to receive the slightest advance on t
true or nearly all who sold soon after the market opened in the East. If not the
combined movement, the *general* desertion of the Eastern and resort to the Weste
facturers was a most singular coincidence ! These manufacturers are now *fain* to
at a *considerable* advance from the prices of 1846—and, as already hinted, it is his
mind, whether they will not be compelled to *import* at a still higher advance, to ek
to be hoped that this will be the last *Act* in the *drama* of folly and suicide played b
   * Not even in close barns, and with constant attendance.
   † North of latitude 42°, it is necessary, as a general rule, that lambs be dropped i
give them this requisite size and strength Occasional cold storms come nearly
period, and not unfrequently up to the first of June. Mr. Grove was a decided ad
He used to say that "it was better to lose two of them in the spring than one in th

wools. Though the reäction has been but recent, the market demand for medium and ordinary wools is now better supplied—nearer being glutted, so far as I am enabled to judge—than that for fine and superfine. And should the market become glutted with either or both, it is important to remember that the *latter* will be far more profitable for *export* than the former.

Every consideration, then, in my judgment, points to wools ranging from good medium upward, instead of the lower classes, as the most profitable staples for cultivation in the South. The only question which now arises on this point is, from which variety, the Saxon or Merino, shall the South attempt to cultivate these wools?

It is generally supposed, and as a general thing it is true, that the Merino bears no better wool than that which I have classified as good medium. But the measurements of Dr. Emmons (given in Letter X.) show, by the infallible testimony of the microscope, that heavy-fleeced Merinos sometimes equal—nay, surpass Saxons, in fineness. The fact is more decisive, as the Saxon fibres there measured came not only from the most celebrated flocks—from the prize sheep at State Fairs—but it also came from samples, in most instances, given by the *owners* for *public exhibition.* I do not claim that Merinos like these are common. They are rather to be regarded in the light of those prodigies of excellence which occasionally appear, but which it is difficult to reproduce with anything like uniformity. Nor are lesser fleeced Merinos, bearing wool equal to ordinary Saxon, very common. During the *fine wool mania,* all, who sought fine wool, bred the Saxon sheep, or crossed with it; and the few who stood out, and clung to the Merino, generally aimed to distinguish it as widely, as possible from the former, by increasing the weight of its fleece, to the disregard of its fineness. This, too, was the general disposition during the *heavy-fleeced mania.* Of consequence, but very few of our breeders have ever, or until recently, sought a high degree of fineness in fleece in breeding the Merino. Recent experience has satisfied me that this is rapidly attainable. Mr. Lawrence, in a quotation already made by me (in Letter l.), says: "I believe a breed may be reared which will give four pounds of exquisitely fine wool to the fleece." I know by multiplied experiments that once interbreeding between an ewe bearing good medium wool (the fleece weighing, say, from 4½ lbs. to 5 lbs.), with a Merino ram of sufficiently high quality, will produce wool in the offspring equaling ordinary Saxon, and a fleece averaging 4 lbs., with none of its weight made up of *gum.* The result of *two* such interbreedings will bring the progeny of a heavy-fleeced medium ewe (provided her fleece is properly *even*) to the same point. The four-pound fine-fleeced Merino would be a far more profitable animal than the Saxon, other things being equal. But other things are not equal. The former is every way a hardier animal, and a better nurse. It is about 20 lbs. heavier, and therefore consumes more feed; but I consider this additional expense more than counterbalanced by the additional care and risk attending the husbandry of the Saxon. If required to keep the number good, and give the proper attention to the rearing of lambs, I would sooner engage to keep, at the same price, one thousand such Merinos for a year, than to keep the same number of Saxons.

It would be practicable, doubtless, to increase the Saxon's fleece to 4 lbs.; but any one, familiar with such experiments, knows that it is far easier to increase fineness of wool, by diminishing weight of fleece and carcass a little, than it is to increase weight of fleece and carcass without lowering the quality of the wool. And there is this additional objection to the latter system of breeding, so far as the Saxon is concerned. The breeder is not only called upon to increase the weight of its fleece and carcass, but to engraft on it hardiness of constitution, nursing properties, etc., which by no means follow, as a matter of course, its improvement in the former particulars. These, and particularly the latter, could only be attained, so as to be transmissible with a proper degree of certainty from parents to offspring, by years of breeding, accompanied by a rigorous course of selection. If, therefore, you were called upon to form a variety just suited to your wants, the Merino would present the most ductile and the safest materials. But the Southern agriculturist, just entering upon sheep-rearing, would not be prepared to conduct nice experiments in breeding. He wants a breed or variety already prepared to his hand. And for the same reasons, notwithstanding the fineness of his climate, he wants a hardy breed—one that demands no extra skill, no great experience, for its management. Merinos reaching or closely approaching the standard above specified are now to be found, while there is no corresponding variety of Saxons; and to incur the risks arising from inexperience, want of preparation, &c., the superior hardiness of the former would, of course, render them entirely preferable.

Some have recommended a cross between the Saxons and Merinos, as a cheap and ready method of obtaining a four-pound fine-fleeced sheep. A properly selected Saxon ram, crossed with good medium and medium-wooled Merino ewes, cutting from 5 lbs. to $5\frac{1}{2}$ lbs. of wool, will almost uniformly produce this result. And it is easier now to get the Saxon than the Merino, fine enough for this purpose. Or a flock may be bred up from Saxon ewes and a Merino ram. The objection to both courses is the same, though not equal to that which exists against breeding the full-blood Saxons—viz., the production of a feeble and a poor nursing sheep. The latter evil, especially, clings for generations to these cross-bred animals, so far as my experience and observation have extended. And unless Saxons are selected which do not possess the characteristic faults of the variety, the cross-breds are inferior to pure-blood Merinos in many *other and essential* particulars, notwithstanding the fleece may be all that we desire.

There is another important point where the pure-blood Merino possesses a marked advantage. Few Southern wool-growers will *commence* their flocks exclusively with high-bred animals of any kind. With a few of them to breed rams from, and to *gradually* grow up a full-blood flock, they will *mainly* depend upon grading up the common sheep of the country. With the long-legged, bare-bellied, open-wooled sheep common in the South (as it once was in the North), the Saxon makes an indifferent cross. Their faults run too much *in the same direction*, in all save the fineness of wool, for, however good its shape, the wool of the Saxon is comparatively short and open. It therefore shortens the wool of the common sheep, without adding much or any to its thickness, and thus the fleece remains a light one. Precisely all this is the reverse of what results from a cross between the Merino and the common sheep. The wool is but little shortened, unless the staple of the common sheep was very long; it is essentially thickened; it is made to extend over the belly; the fleece is, therefore, greatly increased in weight; the sheep is rendered more compact and "stocky," and it is brought nearer to the ground. Even the first cross, though its fleece is somewhat *uneven*, is a prime sheep for the wants of ordinary farmers, and among these it is, accordingly, a decided favorite, over the whole Northern States. A majority of them would, I think, give it preference over any other kind or variety of sheep. Two or three more proper Merino crosses raise it to the rank of a *first-rate wool-growing sheep*—scarcely inferior to the full-blood Merino in anything, save that *it does not transmit its good qualities with quite so much certainty to its offspring.**

Let us now proceed to inquire what are the points which constitute excellence, or mark a departure from it, in the class of Merino sheep which I have attempted to show form, in every point of view, the most suitable variety to commence wool-growing with in the South. What should be its size, weight of fleece, shape, general appearance, style of wool, &c. &c.?

Size, within extremes, is not, *per se*, a matter of much consequence.— There should, however, be *uniformity* in this particular, at least through the same flock, not only for their good appearance, but larger sheep are apt, by their superior strength, to crowd away small ones from the rack or trough. A sheep very small of *its breed and family*, is commonly less hardy. If very large, *it must travel farther to fill itself*; and, therefore, this would be an objection to it in a breed designed to graze on short and scant pasturage—for the extra exercise thus made necessary would cause it to waste (in the form of carbon, in the lungs) a considerable portion of the food, which would, under other circumstances, be converted into animal tissues. Very large, like very small animals, of the same species—and, I am inclined to think, the former more frequently—lack the robustness, vigor of muscle, capacity to endure unusual and protracted exercise, or privation of food, or any other unfavorable deviation from ordinary habits, possessed by compact medium-sized animals. This rule will be found to apply among all domestic animals. Lastly, I am not prepared to prove, but I *believe* that, with the *same breeding*, the woolly, like the osseous and muscular tissues of a large Merino sheep, will not be as fine as those of a smaller one. I do not found this opinion, so far as wool is concerned, upon, nor do I claim that it is supported by, any analogies. I state it as solely the result of individual observation. If it is a tendency which can be successfully resisted, I never have been fortunate enough to have a sufficient number of instances brought under my eye, in any one flock, to have them constitute anything more than sparse exceptions to what I deem a well established rule. I have never known a family of very large Merinos bearing anything better than medium wool; and the first step to any decided improvement in them immediately reduces their weight, for it can only be effected by interbreeding with finer and smaller families. Ewes weighing from 80 lbs. to 90 lbs. alive, in good fair store condition, are of about the proper size, in my judgment, where *fine wool* is the object.† Rams should weigh 40 lbs. or 50 lbs. more. Ewes of the large Merino families weigh from 100 lbs. to 110 lbs.—the rams 50 lbs. more; nor do even these equal the size of some of the late imported French Merinos.

A relation analogous to the preceding one, exists between the weight of the fleece and its quality. This point has already been sufficiently set forth on another page. The opinion is there expressed that the Merino may be easily bred, by judicious selection of sire and dam, to bear 4 lbs. of fine wool, or wool equaling ordinary Saxon. I would now add that, as a *general rule*, and in large flocks, I do not believe *more* than this can be obtained, without a depreciation in the quality, among ewes. The ram's fleece should in all cases, in a very superior animal, be about double that of the ewe. Five per cent. of the live-weight of the carcass, with ewes, is the maximum weight of *fine* wool, which we can, in the present state of breeding, look for with any uniform certainty. This would give a fleece of 4 lbs. to 80 lbs. of live-weight. As the fine-wool Merinos increase, and thus give a wider range and better selection of materials for nice experi-

* The latter point will be more particularly adverted to in a subsequent part of this Letter.
† Saxons weigh about 20 lbs. less.

ments, it is very possible that the per centage of the fleece
Mr. Lawrence, in speaking of attaining a four-pound flee
quality, undoubtedly alluded to the wool which I have cla
The four-pound fleeced *fine* Merino can undoubtedly be
by diminishing the weight of its fleece 10 or 12 oun
and even then it will be a hardier and better anima
class of Saxons which now produce this wool. But w
rence's standard can be fully attained, neither exper
vation enable me to decide. If it could, and the shee
four-pound *fine*-fleeced Merino in other respects, we sho
heep. Such wool has sold this year at upward of 60
which would bring the fleeces to $2 40 a piece! It ma
glance at the comparative worth of fleeces in the several
taking this year's prices, and taking the weights which
accompanying the several qualities, in prime ordinary floc
of 4 lbs., at 50 cents,* would be worth $2; good medi
lbs., at 40 cents, $1 80; medium, weighing 5 lbs., at 32 c
the consumption of feed rises with the diminution of qu
the daily consumption of hay for 150 days to be 3 per
weight, 100 fine Merinos, averaging 85 lbs. each, would c
tons of hay; and 100 medium Merinos, averaging 105
consume about 23½ tons—an important difference in
penses! The fine-wooled Merino does not, like the So
vantage in this particular by his inferior hardiness.

The shape and general appearance of the Merino shot
The head should be well carried up, and in the ewe ho
be better on many accounts to have the ram also hornless
ally characteristic of the Merino, many prefer to see
should be shortish, broad between the eyes, the nose po
ewe fine and free from wrinkles. The eye should be b
prominent, and gentle in its expression. The neck shoul
curving downward), short, round, stout—particularly so a
the shoulder, forward of the upper point of which it shou
the level of the back. The points of the shoulder shou
perceptible extent above the level of the back. The b
should be straight; the crops (that portion of the body i
of the shoulder-blades) full; the ribs well arched; the b
pacious; the flank well let down; the hind-quarters ful
flesh meeting well down between the thighs, (or in the
bosom should be broad and full; the legs short, well apar
ular, (*i. e.*, not drawn under the body toward each other v
standing.) Viewed as a whole, the Merino should preser
of a low, stout, plump, and—though differing essentially
mutton-sheep model—a highly symmetrical sheep.

The skin is an important point. It should be loose, si
of a rich, delicate pink color. A colorless skin, or one
proaching to a butternut hue, indicates bad breeding. O
wrinkles, there is a difference of opinion. Being rather
the Merino—like the black color in a Berkshire hog, or t
color in Durham cattle—these wrinkles have been more
ices, than those points which give actual value to the ani
breeders have not been slow to act upon this hint! Man
that more wool can be obtained from a wrinkled skin; ar

* This is not high for *fine* Merino wool. Though I sold my lot for 42 cents, I wo
fleeces of nearly all my later-bred sheep, if I would sell them separately.

of the case which has induced both the Spanish and French breeders to cultivate them—the latter to a monstrosity. I confess that I agree, to a considerable extent, with Mr. Joshua Kirby Trimmer,* that "this idea is as wild as that which some of our theorists have entertained, that, by laying lands in high ridges and low furrows, the surface of the earth and its produce is increased." Though I once entertained a different opinion, the steel-yards have satisfied me that an exceedingly wrinkled neck does not add but a little to the weight of the fleece—not enough to compensate for the deformity, and the great impediment which it places in the way of the shearer. I have owned rams, the labor of shearing six of which, in a nice and workmanlike manner—cutting the wool off short and smooth, on and among the multitude of folds and wrinkles—was fully equivalent to shearing fifteen ordinary Merino rams, or twenty-five ewes—that is to say, a day's work for one man. And none but a skillful shearer could, with *any* time given him, clip the wool short and smooth among the wrinkles, without frequently and severely cutting the skin. A smoothly drawn skin, and absence of all dewlap, on the other hand, would not, perhaps, be desirable.

The wool of the Merino should densely cover the whole body, where it can possibly grow, from a point between and a little below the eyes, and well up on the cheeks, to the knees and hocks. Short wool may show, particularly in young animals, on the legs, even below the knees and hocks—but long wool covering the legs, and on the nose below the eyes, is unsightly—without value—and on the faces it frequently impedes the sight of the animal, causing it to be in a state of perpetual alarm, and disqualifying it to escape real danger. Neither is this useless wool, as seems to be thought by some, the slightest indication of a heavy fleece. I have as often seen it on Saxons scarcely shearing 2 lbs. of wool, and on the very lightest fleeced Merinos.

The amount of *gum* which the wool should exhibit, is another of the mooted points. Here, as in many other particulars, experience has changed my earlier impressions. Merino wool should be yolky or "oily," prior to washing—though not to that extreme extent, giving it the appearance of being saturated with grease, occasionally witnessed. The extreme tips of the wool may exhibit a sufficient trace of gum to give the fleece a darkish cast—particularly in the ram—but a black, pitchy gum, resembling semi-hardened tar, extending an eighth or a quarter of an inch into the fleece, and which *cannot be removed in ordinary washing*, is, in my opinion, decidedly objectionable. There is a white or yellowish *concrete* gum, not removable by common washing, which appears in the *interior* of some fleeces, which is equally objectionable.

The weight of fleece remaining the same, medium length of staple, with compactness, is preferable to long, open wool, inasmuch as it constitutes a better safeguard from inclemencies of weather, and better protects the sheep from the bad effects of cold and drenching rains in spring and fall. The wool should be as nearly as possible of even length and thickness over the whole body. Shortness on the flank, and shortness or thinness on the belly, are serious defects.

"Evenness of fleece" is a point of the first importance. Many sheep exhibit good wool on the shoulder and side, while it is far coarser and even hairy on the thighs, dewlap, &c. Rams of this stamp should not be bred from by any one aiming to establish a superior fine-wooled flock, and all such ewes should be gradually excluded from those selected for breeding.

The "style of the wool" is a point of as much consequence as mere

* "Practical Observations on the Improvement of British Fine Wools, &c." by the above, 1828.

fineness. Some very fine wool is stiff and the fibres almost straight, like hair. It has a dry, *cottony* look. This is a poor, unsalable article, however fine the fibre. Softness of wool—a delicate, silky, highly elastic feel, between the fingers or on the lips, is the first thing to look after. This is usually an index, or inseparable attendant, of the other good qualities, so that an experienced judge can decide, with little difficulty, between the quality of two fleeces, in the *dark!* Wool should be finely serrated or crimped from one extremity to the other—*i. e.*, it should present a regular series of minute curves, and, generally, the greater the number of these curves in a given length, the higher the quality of wool in all other particulars. The wool should open on the back of the sheep in *connected masses*, instead of breaking up into little round spiral ringlets of the size of a pipe-stem, which indicate thinness of fleece; and when the wool is pressed open each way with the hands, it should be dense enough to conceal all but a delicate rose-colored line of skin. The interior of the wool should be a pure, glittering white, with a lustre and "liveliness" of look not surpassed in the best silk.

The points in the *form* of the Merino which the breeder is called upon particularly to eschew, are—a long, thin head, narrow between the eyes—a thin, long neck, arching downward before the shoulders—bad crops—back falling behind the shoulders—narrow loin—flat ribs—steep, narrow. hind quarters—long legs—thighs scarcely meeting at all—legs drawn far under the body at the least approach of cold. All these points were separately or conjointly illustrated in many of the Saxon flocks which have been recently swept from the country. The points to be avoided in the fleece have been sufficiently adverted to.

Having thus attempted to establish a standard for the Merino-breeder, it remains that we examine some of the most important principles, in breeding, by which that standard is to be reached or maintained.

The first great starting-point, among pure-blood animals, is that "like will beget like." If the sire and dam are perfect in any given point, the offspring will generally be; if either is defective, the offspring will (subject to a law presently to be adverted to) be half way between the two; if both are defective in the same point, the progeny will be more so than either of its parents—it will inherit the amount of the defect in both parents *added together.* There are exceedingly few perfect animals. Breeding, then, is a system of counterbalancing—breeding out—in the offspring, the defects of one parent, by the marked excellence of the other parent. *in the same points.* The highest blood confers on the parent possessing it the greatest power of stamping its own characteristics on its progeny; but blood being the same, the male sheep possesses this power in a greater degree than the female. We may, therefore, in the beginning, breed from ewes possessing any defects short of cardinal ones, without impropriety, provided we possess the proper ram for that purpose; but the flockmaster, aiming at a high standard of quality, should *gradually* throw out from breeding all ewes possessing even considerable defects. Every year should make him more rigorous in his selection. But from the beginning—and in the beginning more than at any other time—the greatest care should be evinced in the selection of the ram. If he has a defect, that defect is to be inherited by the whole future flock. If it is a material one, as, for example, a hollow back, bad crops, a thin fleece, or a highly uneven fleece, the flock will be one of low quality and little value. If, on the other hand, he is perfect, the defects in the females will be lessened, and gradually bred out. But it being difficult to find *perfect* rams, we are to take those which have the fewest and lightest defects, and none of

es, like those just enumerated. And these defects are t(
erbalanced by the decided excellence (sometimes running
ewe, in the *same points*. If the ram is a little too long
st-legged ewes should be selected for him ; if gummy
. ewes ; if his fleece is a trifle below the proper standard
he has been retained, as it often happens, for weight of
il excellence,) he is to be put to the finest and lightest
so on. Having a selection of rams, this system of coun
ld require little skill, if each parent possessed but *on(*
? was a trifle too thin fleeced, and good in all other par
require no nice judgment to decide that she needed to be
monly thick-fleeced ram. But most animals possess, te
degree, several defects. To select so that every one of
shall meet its opposite in the male, and *vice versa*, re
lentiful materials to select from, but the keenest dis
? time and the convenient method of selecting the ewes
ns, and the subsequent management, will be hereafter

uppose that the breeder has established his flock—that
ccessfully, and given them an excellent character. He
. serious evil. He must "breed in-and-in," as it is called
ed between animals more or less nearly related in blood
t rams from other flocks, to the risk of losing or changing
aracter of his flock, hitherto sought so sedulously, and
iuch care. It is contended by the opponents of in-and-in
inders diseases and all other defects hereditary, and that
se of size, to debility, and a general breaking up of the
apologists, on the other hand, insist that, if the parents
thy, incestuous connexion does not, *per se*, tend to any
lthiness in the offspring ; and they also claim, what must
it enables the skillful breeder much more rapidly to
a particular standard or model—and much more easily
unless it be true that, in course of time, they will dwin-
le. So far as the effect on the constitution is concerned,
y be, to a certain extent, true. But it is, perhaps, diffi-
ide with certainty when an animal is not only free from
ill tendency or predisposition toward it. A brother and
arently healthy—may be actually so—but may possess
hich, under certain circumstances, will manifest itself.—
inces do not chance to occur, they may live, apparently
it constitution, until old age. If bred together, their off-
already laid down, will possess the idiosyncrasy in a
Suppose the ram be interbred with sisters, half-sisters,
laughters, &c., for several generations, the predisposition
ir disease—in the first place slight, now strong, and con-
onger—will pervade, and become radically incorporated
ion of the whole flock. The first time the requisite ex-
rought to bear, the disease breaks out, and, under such
:h peculiar severity and malignancy. If it be of a fatal
k is rapidly swept away ; if not, it becomes chronic, or
iently recurring intervals. The same remarks apply, in
cts of the outward form which do not at first, from their
the notice of the ordinary breeder. They are rapidly
m st before thought of by the owner, they destroy the
. That such are the common effects of in-and-in breed-
Y

Ing, with such skill as it is ordinarily conducted, all know who have given attention to the subject; and for these reasons the system is looked upon with decided disapprobation and repugnance, as among all kinds of domestic animals, by nine out of ten of the best practical farmers of the Northern States.

How, then, shall the sheep-breeder avoid the effects of in-and-in breeding, and at the same time preserve the character of his flock? He should do so by seeking rams of the *same breed*, and possessing, *as nearly as possible, the characteristics which he wishes to preserve in his own flock.* If the latter rule is neglected—if he draws indiscriminately from all the different families or varieties of a breed—some large and some small—some long and some short-wooled—some medium and some superfine in quality— some tall and some squabby—some crusted over with black gum, some entirely free from it, &c. &c.—breeding will become a mere hotch-potch, and no certain or uniform results can be looked for. So many varieties cannot be fused into one, for a number of generations;* and it not unfrequently happens, as between the different classes of Saxons alluded to by Mr. Spooner,† that certain families can never be successfully amalgamated.

But suppose the breeder has reached no satisfactory standard—that his sheep are deficient in the requisites he desires? If the *desired requisites are characteristic of the breed* he possesses, he is to *adhere to the breed*, and select better animals to improve his own inferior ones. If he has an inferior flock of South-Downs, and wishes to obtain the qualities of the best South Dams, he should seek for the best rams of that breed. But if he wishes to obtain qualities *not characteristic of the breed he possesses*, he must *cross with a breed which does possess them.* If the possessor of South Downs wishes to convert them into a fine-wooled sheep similar to the Merino, he should cross his flock steadily with Merino rams—constantly increasing the amount of Merino and diminishing the amount of South-Down blood. To effect the same result, he would take the same course with the common sheep of the country, or any other coarse race. There are those who, forgetful that some of the finest varieties now in existence, of several kinds of domestic animals, are the result of *crosses*, bitterly inveigh against the practice of crossing, under any and all circumstances. As frequently conducted, where objects incompatible with each other are sought to be attained—as, for example, an attempt to unite the fleece of a Merino and the carcass of a Leicester, by crosses between those breeds—it is an unqualified absurdity. But under the limitations already laid down, and with the objects specified as legitimate ones, objection to crossing savors, in my judgment, of prejudice the most profound, or quackery the most unvarnished. The cry, " buy full-bloods," with such men, generally means, " buy *our* full-bloods!" It is neither convenient, nor within the means of every man wishing to start a flock of sheep, to start exclusively with full-bloods. With a few full-bloods to breed rams from, and to *begin* a full-blood flock, the Southern breeder will find it his best policy to purchase the best common sheep of his country, and gradually grade them up with Merino rams. In selecting the ewes, fair size, good shape, and a robust constitution, are the main points—the little difference that exists between the quality of the common sheep's wool is of no consequence. For their wool they are to look to the Merino; but good form and constitution they can and ought to possess, so as not to entail deep-rooted and entirely *unnecessary* evils on their progeny.

* This occasions the want of uniformity in the Rambouillet flock in France, which was begun by a promiscuous admixture of all the Spanish families.
† Quoted in Letter X.

I have already spoken, in this Letter, incidentally, of the effect on the fleece of the common sheep, by crossing with the Merino and breeding steadily toward the latter; and also of the mutton of this cross, as well as that of the Merino and the English breeds. The result of the cross with the common sheep has been sufficiently described. I would add a few remarks in relation to that with the South-Down and Leicester —both of which I have tried until sufficiently satisfied with the result. Resolved on making an experiment with a Down and Merino cross, a few years since, and finding it difficult to obtain Down ewes* of the proper quality, I obtained a small, compact, exceedingly beautiful, fine and even-fleeced Down ram,† and crossed him with a few large-sized Merino ewes. The half-blood ewes were bred to a Merino ram, and also their female progeny, and so on. The South-Down form and disposition to take on fat manifested itself, to a perceptible extent, in every generation which I bred,‡ and the wool of many of the sheep in the third generation ($\frac{7}{8}$-blood Merino and $\frac{1}{8}$-blood Down) was very even, and equal to medium, and some of them to good medium Merino. Their fleeces were lighter than the full-blood Merino, but increased in weight with each succeeding cross back toward the latter. Their mutton of the first, and even the second cross, was of a beautiful flavor—and it retained some of the superiority of South-Down mutton to the last.

I at the same time purchased a few Leicester ewes,‖ and, as in the preceding case, taking one cross of the blood, I bred toward the Merino. The mongrels, to the second generation (beyond which I did not breed them) were about midway between the size of the two parent stocks— with wool shorter, but far finer and more compact than the Leicester— their fleeces about the same in weight as in the present stocks§—and alto-gether they were a showy and profitable sheep, and well calculated to please the mass of farmers. Their fleeces lacked *evenness*—their thighs remaining disproportionately coarse and hairy; and making up my mind that this would always be a tendency of the sheep of this cross, I aban-doned them without farther experiment.

In relation to the number of crosses necessary before it is proper to breed from a mongrel *ram*, there is a difference of opinion. Mr. Livings-ton says :¶

"It is now so well established as not even to admit of the smallest doubt that a Merino in the fourth generation, from even the worst-wooled ewes, is in every respect equal to the stock of the sire. No difference is now made in Europe in the choice of a ram, whether he is a full-blood or a fifteen-sixteenths." . . . . "The French agriculturists say that however coarse the fleece of the parent ewe may have been, the progeny in the fourth generation will not show it."

I am constrained to differ with even this high authority. I admit that the only value of blood or pedigree, in breeding, is to insure the hereditary transmission of the properties of the parent to the offspring. As soon as a mongrel reaches the point where he stamps his characteristics on his progeny, with the same certainty that a full-blood does, he is equally valuable, provided he is, individually, as perfect an animal. But I do not

---

* To carry out the commonly received principle in breeding, that in crossing between different races, the ram of the smaller should be put to ewe of the larger one.
† This ram, obtained from Francis Rotch, Esq., was got by a prize ram of Mr. Ellman's, and from one of his choicest breeding-ewes, and showed infinitely more style, as well as fineness and evenness of wool, than the common Downs of our country. He was not larger than a large-sized Merino ram.
‡ These I finally put off to save myself the trouble of breeding several kinds of sheep on the same farm.
‖ Descended from the flock of the late Robert Adcock, of Otsego County, N. Y.—considered at the time equal to any flock in the State.
§ That is, about 5 lbs. I have put down the Leicester fleece, in my description of the breed, at 6 lbs., as this is the amount generally claimed for them; but in the few cases brought within my direct knowledge, they have never averaged it. My ewes above alluded to did not, I think, average quite 5 lbs.
¶ Essay on Sheep, pp. 181, 183.

believe that this can be depended upon, with any certainty, in rams of the fourth Merino cross. My only experience in this particular is in the observation of other men's flocks who have bred with high-grade rams.* These have invariably lacked the style and perfection of thorough-bred flocks. The sixth, seventh, or eighth cross might be generally, and the last perhaps almost invariably, as good as pure-blood rams, but I confess I should still prefer to adhere to the latter. Pure blood is a fixed standard, and were every breeder to think himself at liberty to depart from it, in his rams, each one more or less, according to his own judgment or caprice, the whole blood of the country would become adulterated. No man would be authorized to sell a ram of any cross, be it the tenth, or even the twentieth, as a full-blood.

It is all-important for those *commencing* flocks either of full-bloods, or by crossing, to select the choicest rams. A grown ram may be made to serve|| from 100 to 150 ewes in a season. A good Merino ram will, speaking within bounds, add *more* than a pound of wool to the fleece of the dam, on every lamb got by it, from a common-wooled ewe.§ Here is one hundred or one hundred and fifty pounds of wool for the use of a ram for a single season! And every lamb subsequently got by him adds a pound to this amount. Many a ram gets, during his life, 800 or 1,000 lambs! Nor is the extra amount of wool all. He gets from 800 to 1,000 half-blooded sheep, worth double their dams, and ready to be made the basis of another and higher stride in improvement. A good ram, then, is as important, and, it seems to *me*, quite as valuable an animal as a good farm-horse stallion! When the number of a ram's progeny are taken into consideration, and when it is seen over what an immense extent, even in his own direct offspring, his good or bad qualities are to be perpetuated, the folly of that economy which would select an inferior one is sufficiently obvious.

Every one desirous of starting a flock will find it his best economy, where the proper flocks to draw rams from are not near him, to purchase several of the *same breed*, of course, but of *different strains of blood*. Thus, ram No. 2 can be put on the offspring of No. 1, and *vice versa;* No. 3 can be put upon the offspring of both, and both upon the offspring of No. 3. The changes which can be rung on three distinct strains of blood, without in-and-in breeding close enough to be attended with any considerable danger, are innumerable.¶ But if these rams of different strains are bought promiscuously, without reference to similarity of characteristics, there may, and probably will be differences between them, and it might require time and skill to give a flock descended from them, a proper uniformity of character. Those who breed rams for sale should be prepared to furnish different strains of blood with the necessary individual and family uniformity.

* I have never knowingly bred with any other ram than a pure-blood, of any stock, or for any purpose.
|| By methods hereafter to be described.
§ That is, if the ewe at 3 years old sheared 3 lbs. of wool, the lamb at the same age will shear 4 lbs. of wool
¶ The brother and sister are of the *same* blood; the father and daughter, half; the father and grand-daughter, one-fourth; the father and great grand-daughter, one-eighth, and so on. Breeding between animals possessing one-eighth of the same blood, would not be considered very close breeding; and it is not uncustomary, in rugged, well-formed families, to breed between those possessing one-fourth of the same blood.

LETTER XII.

SUMMER MANAGEMENT OF SHEEP.

Tagging—necessity of—method of doing it...Burs—how avoided...Lambing—time of—Inclosures for—Mechanical Assistance—when rendered—assisting the Lamb—Feeding—necessary care in—Warming—Foster Ewes...Pens..." Pinning"...Numbering and Registering—advantages of—Von Thaër's System of Numbering—manner of doing it conveniently—Mr. Grove's form of a Register...Castration and Docking—proper time and method...Washing—time—necessary apparatus—" wetting"—manner of washing—ordinary waste in subsequent cleansing...Cutting the Hoofs—best time—implements—method...Time between Washing and Shearing...Shearing—proper conveniences for—catcher's business—directions to shearer—general directions...Shearing Lambs—shearing Sheep semi-annually—objectionable practices...Doing up Wool—Wool Table and Trough—handling fleece—arrangement on table—folding—rolling—tying—proper twine...Storing Wool—Wool-Room...Sacking Wool—methods...Sorting the Flock at shearing—how done...Marking Sheep—the proper way...Cold Storms after Shearing...Sun-scald...Ticks—how destroyed...Maggots—preventives...Cutting the Horns...Division of Flocks for Summer...Hoppling—Clogging, &c...Dangerous Rams...Fences...Salt...Tar...Water...Shade...Weaning Lambs...Full Feeding...Shepherd's Crook.

Dear Sir : Agreeably to your request, and that of various other Southern friends, I proceed to give directions for the practical management of sheep " plain and minute enough for the guidance of those entirely unacquainted with the subject." I will begin with their Summer Management.*

TAGGING.—If sheep are kept on dry feed through the winter, they will usually purge more or less, when let out to green feed in the spring. The wool around and below the anus becomes saturated with dung, which forms into hard pellets, if the purging ceases. But whether this takes place or not, the adhering dung cannot be removed from the wool in the ordinary process of washing. It forms a great impediment in shearing, dulling and straining the shears to cut through it when in a dry state, and it is often impracticable so to do. It is difficult to force the shears between it and the skin, without frequently and severely wounding the latter. Occasionally, too, flies deposit their eggs under this mass of filth prior to shearing, and the ensuing swarm of maggots, unless speedily discovered and removed, will lead the sheep to a miserable death.

Before sheep are let out to grass, each one should have the wool sheared from the roots of the tail down the inside of the thighs, over the surface included between the dotted lines in the cut. The wool should be sheared from off the entire bag of the ewe, that the newly dropped lamb may more readily find the teat, and from the scrotum, and so much space round the point of the sheath of the ram, as is usually kept wet. If the latter place is neglected, soreness and ulceration sometimes ensue from the constant maceration of the urine.

Fig. 16.

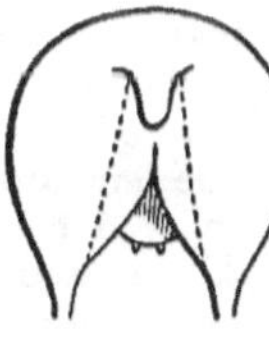

Sometimes each tagger catches and holds his own sheep, but it is, on the whole, better, I think, to have an assistant catch the sheep and hold them while they are tagged. The latter process requires a good shearer, as the wool must be cut off closely and smoothly, or the object is but half accomplished, and the sheep will have an unsightly and ridiculous appearance, when the remainder of their fleeces is taken off;

* I have not thought it necessary to mark with quotation points, various extracts in this Letter, from a series of Letters written by me a number of years since, and published in the " Valley Farmer "

and, on the other hand, it is not only improper to cut the skin of a sheep at any time, but it is peculiarly so to cut that on the bag of an ewe near lambing. The wool saved by tagging will far more than pay the expenses of the operation. It answers well for stockings and other ordinary domestic purposes, or it will sell for something like half the price of fleece wool.

Humanity and economy both dictate that care should be taken in handling sheep at all times, and it is especially important with ewes heavy with lamb. It is highly injurious and unsafe to chase them about and handle them roughly, for even if abortion, the worst consequence of such treatment, is avoided, they become timid and shy of being touched, rendering it difficult to catch or render them assistance at the lambing period—and even a matter of difficulty to enter the cotes where it is sometimes necessary to confine th*m at that time, without having them driving about pell-mell, running over their lambs, &c. It may not be known to every one, that if a sheep is suddenly caught by the wool when running, or is lifted by its wool, the skin is to a certain extent loosened from the body at the points where it is thus seized, and if killed a day or two afterward, blood will be found settled about those parts. A man knowing this, and subsequently guilty of such gratuitous brutality, richly deserves to be *kicked* out of the sheep-yard. When sheep are to be handled, they should be inclosed in a yard just large enough to hold them without their being crowded—so they shall have no chance to run and dash about. The catcher should stop them by seizing them by the hind leg close above the hock, or by clapping one hand before the neck and the other behind the buttocks. Then, not waiting for the sheep to make a violent struggle, he should throw his right arm over and about it immediately back of the shoulders, place his hand under the brisket, and lift the animal on his hip. If the sheep is very heavy, he can throw both arms around it, clasp his fingers under the brisket, and lift it up against the front part of his body. He then should set it carefully on its rump on the tagging-table, (which should be 18 or 20 inches high,) support its back with his legs, and hold it gently and conveniently until the tagger has performed his duty. Two men should not be permitted to lift the same sheep together, as it will be pretty sure to receive some strain between them. A good shearer and assistant will tag 200 sheep per day.

Where sheep receive green feed all the year round, as they will do in many parts of the South, and no purging ensues from eating the newly-starting grasses in the spring, tagging will not be necessary.

Burs, &c.—If sheep are let out in the spring into pastures where the dry stalks of the Burdock *(Arctium lappa)*, or the Hound's Tongue, or Tory-weed *(Cynoglossum officinale)*, have remained standing over the winter, the burs are caught in their now long wool, and, if numerous, the wool is rendered entirely unmarketable, and almost valueless. Even the dry prickles of the common and Canada thistles, where they are very numerous, get into the neck-wool of sheep, as they thrust their heads under and among them to crop the first scarce feed of the Northern spring; and, independently of injuring the wool, they make it difficult to wash and otherwise handle the sheep. The Burdock being a large and not very frequent plant, there is no excuse for its being found on the farm. The Hound's Tongue is very prevalent in forests and partly wooded pastures in the North, and it is not conspicuous enough to be easily eradicated, though careful sheep-farmers often do so. If sheep are let into pastures containing it, it must be only in the summer and fall, after shearing. The burs not sunk so deeply in the short wool, will wear cut during our winters—but no man thinks of letting his sheep into pastures containing it, before shearing in the spring. Indeed, sheep should be kept on the cleanest pastures—those free from these and all similar plants—during this period; and, in a region where they are pastured the year round, if such pests are not eradicated—which I should consider indispensable—the sheep should oe kept from contact with them for some months prior to shearing.

LAMBING.—Lambs are usually dropped, in the North, from the first to the fifteenth of May. In the South, they might safely come earlier. It is not expedient to have them dropped when the weather is cold and boisterous, as they require too much care; but the sooner the better, after the weather has become mild, and the herbage has started sufficiently to give the ewes that green food which is required to produce a plentiful secretion of milk. It is customary in the North to have fields of clover, or the earliest grasses, reserved for the early spring feed of the breeding ewes; and, if these can be contiguous to their shelters, it is a great convenience—for the ewes should be confined in the latter, on cold and stormy nights, during the lambing season.

If warm and pleasant, and the nights are warmish, I prefer to have the lambing take place in the pastures. I think sheep are more disposed to own and take kindly to their lambs thus, than in the confusion of a small inclosure. Unless particularly docile, sheep in a small inclosure crowd from one side to another when any one enters, running over young lambs, pressing them severely, &c. Ewes get separated from their lambs, and then run violently round from one to another, jostling and knocking them about. Young and timid ewes get separated from their lambs, and fre quently will neglect them for an hour or more before they will again approach them. If the weather is severely cold, the lamb, if it has nevei sucked, stands a chance to perish. Lambs, too, when just dropped, in a *dirty* inclosure, in their first efforts to rise, tumble about, and the membrane which adheres to them becomes smeared with dirt and dung—and the ewe refuses to lick them dry, which much increases the hazard of freezing.

Nevertheless, all this must be incurred in cold storms, and in sudden and severe weather; and, therefore, it should be the effort of every shepherd to teach his sheep docility. I have seen the late Mr. Grove walk about a barn filled with his Saxons, not only without their crowding from side to side, but many of them absolutely lying still while he stepped over them! I say it "must be incurred." I mean by this that it is the safest course with all breeds, and a matter of *necessity* with others. It takes but a very moderately cold night to destroy the new-born Saxon lamb, which (the pure blood) is yeaned nearly as *naked* as a child! During a severely cold period, of several days' continuance, it is almost impossible to rear them, even in the best shelter. The Merino, South-Down, and some other breeds, will endure a greater degree of cold with impunity.

Inclosures, when used for yeaning, should be kept clean by frequent litterings of straw—not enough, however, thrown on at one time, to embarrass the lamb about rising.

The ewe does not often require mechanical assistance in parturition.—Her labors will sometimes be prolonged for three or four hours, and her loud moanings will evince the extent of her pain. Sometimes she will go about several hours, and even resume her grazing, with the fore-feet and nose of the lamb showing at the mouth of the vagina. But, if let alone, Nature will generally finally relieve her. This might not do with the heavy English breeds. I should infer not, from the elaborate directions in the premises, by Youatt, Blacklock, and other English writers on Sheep; though with the comparatively small number of these varieties which I have bred, I have had no difficulty in this particular. Among the thousands and thousands of fine-wooled sheep which I have bred, I never have known a single instance of a false presentation of the fœtus, and never have had mechanical assistance rendered in to exceed half a dozen instances. The objection to interfering, except as a last resort, is that the ewe is frightened when caught, and her efforts to expel the lamb cease.— When aided, the gentlest force should be applied, and only in conjunction with the efforts of the ewe.

While the lamb is tumbling about and attempting to rise, and the ewe is licking it dry, it is better to be in no haste to interfere. A lamb that gets at the teat without help, and gets even a small quantity of milk, knows how to help itself afterward, and rarely perishes. If helped, it sometimes continues to expect it, and will do little for itself for two or three days.— The same is true when lambs are fed from a spoon or bottle.

But if the lamb ceases to make efforts to rise, particularly if the ewe has left off licking it while it is wet and chilly, it is time for the shepherd to render his assistance. It is better not to throw the ewe down, as is frequently practiced, to suckle the lamb, because instinct teaches the latter to point its nose *upward* in search of the teat. It is doubly difficult, therefore, to induce it to suck from the bag of the prostrate ewe; and when taught to do this, by being suckled so several times, I have invariably noticed that it renders it awkward about finding the teat *in the natural position*, when it begins to stand and help itself. Nothing is stupider than a weakly lamb! Carefully disengaging the ewe from her companions, with his *crook*, the assistant should place one hand before the neck and the other behind the buttocks of the ewe, and, then pressing her against his knees, he should hold her firmly and stilly, so that she shall not be constantly crowding away from the shepherd. The shepherd should set the lamb on its feet, inducing it to stand, if possible; if not, supporting it *on its feet* by placing one hand under its body—place its mouth to the teat, and encourage it to suck by tickling it about the roots of the tail, flanks, &c., with a finger. The lamb, mistaking this last for the caresses of its dam, will redouble its efforts to suck. Sometimes it will evince great dullness, and even apparent obstinacy, in refusing for a long time to attempt to assist itself, crowding backward, &c.; but the kind and gentle shepherd, who will *not sink himself to the level of a brute by resenting the stupidity of a brute*, will generally carry the point by perseverance. Sometimes milking a little into the lamb's mouth, holding the latter close to the teat, will induce it to take hold.

If the ewe has no milk, the lamb should be fed until the natural supply commences, with small quantities of the milk of a *new-milch* cow. This should be mixed, say half and half, with water—with enough molasses to give it the purgative effect of biestings, or the first milk—gently warmed to the natural heat (not scalded and suffered to cool), and then fed through a bottle with a sponge in the opening of it, which the lamb should *suck*, if it can be induced so to do. If the milk is *poured* in its mouth from a spoon or bottle, as already remarked, it is frequently difficult afterward to induce it to suck. And, moreover, unless milk is poured in the mouth slowly and with care—no faster than the lamb can swallow—a speedy wheezing, the infallible precursor of death, will show that a portion of the fluid has been forced into the lungs. I have known lambs frequently killed in this way.

If a lamb becomes chilled, it should be wrapped up in a woolen blanket, and placed in a warm room—giving a little milk as soon as it will swallow. A trifle of pepper is sometimes placed in the milk, and I think with good effect, to rouse the cold and torpid stomach into action. Some of the *Yankee* old ladies, under such circumstances, "*bake*" the lamb, as it is called—*i. e.*, put it in a blanket in a moderately heated oven, until warmth and animation are restored. Others immerse it in tepid water, and subsequently rub it dry. This is said to be an excellent method where the lamb is nearly frozen. I never have tried it. A good blanket, a warm room, and sometimes, perhaps, a little gentle friction, have always sufficed.

If a strong ewe, with a good bag of milk, chances to lose her lamb, she should be required to bring up one of some other ewe's pair of twins—or the lamb of some feeble or young ewe, having an inadequate supply of milk. Her own lamb should be skinned, as soon as possible after death, and the skin sowed over the lamb which she is required to foster. She will sometimes be a little suspicious for a day or two, and if so, she should be kept in a small pen with the lamb, being occasionally looked to. After taking well to it, the false skin may be removed in three or four days. If no lamb is placed on a ewe which has lost her lamb, and which has a full bag of milk, the milk should be drawn from the bag once or twice, or garget may ensue. If it does not, permanent indurations, or other results of inflammatory action will often take place, injuring the subsequent nursing properties of the animal. When milked, it is well to wash the bag for some time in cold water. It checks the subsequent secretions of milk, as well as abates inflammation. Garget will be treated under the head of Diseases of Sheep.

Sometimes a young ewe, though exhibiting sufficient fondness for her lamb, will not stand for it to suck; and in this case, if the lamb is not very strong and persevering, and especially if the weather is cold, it soon grows weak and perishes. The conduct of the dam in such cases is occasioned by inflammatory action about the bag or teats—and, perhaps somewhat by the *novelty* of her position! In this case the sheep should be caught and held until the lamb has exhausted the bag, and there will not often be any trouble afterward, though it may be well enough to keep them in a pen together until the fact is determined.

I have several times spoken of *pens*. They are necessary in the cases I have mentioned, and in a variety of others. It is therefore well for the flock-master to be always provided with a few of them for emergencies. They need not be to exceed eight or ten feet square, and should be built of light materials, and fastened together at the corners, so they can be readily moved by one, or, at the most, two men, from place to place, where they are wanted. Their position should be daily shifted when sheep are in them, for cleanliness and fresh feed. Light pine poles, laid up fence fashion, and each nailed or pegged to the lower ones, at the corners, as laid on, would make excellent ones. Two or three sides of a few of them should be wattled with twigs, and the tops partly covered to shelter feeble lambs from cold rains, piercing winds, &c.

Young lambs are subject to what is technically called "pinning,"—that is, their first excrements are so adhesive and tenacious that the orifice of the anus is closed, and subsequent evacuations prevented. The adhering matter should be entirely removed, and the part rubbed with a little dry clay to prevent subsequent adhesion. Lambs will frequently perish from this cause if not looked to for the first few days.

Z

NUMBERING AND REGISTERING.—This is not absolutely necessary for the *wool-grower*, though it is, in many points of view, a vast convenience to him, and leads to a degree of system in his efforts after improvement, and gives a definiteness and precision to the execution of his plans, otherwise unattainable. But the *breeder*—he who makes it his business more particularly to raise choice animals to sell for breeding purposes—is unworthy of the name, if he does not regularly number and register his sheep, so that he can trace the descent of any ram or ewe, through any number of generations. This is not merely to gratify an idle curiosity, or to furnish a purchaser with a sounding pedigree. Every breeder is under the necessity of directly breeding in-and-in, or of occasionally employing new strains of blood. If the latter step is often resorted to, the hazard is increased of changing the character of the flock.* If he numbers and registers his sheep, he can breed "closer,"† and consequently longer, without a change, without the hazard of confusion or mistake. Where half a dozen, or even three or four rams are used in the flock the same year, it would be beyond the power of any breeder, relying on his memory alone, to decide, six or eight or ten years subsequently, which were the daughters, grand-daughters, and great-grand-daughters of each. If the rams A and B be unrelated, A may be put to the daughters of B, and then B be put to the produce, (*i. e.*, his own grand-daughter, got by A,) without "close" breeding—because they possess but *one-quarter* of the same blood. Then the great-grand-daughter may be again put to A, because she possesses but one-quarter of *his* blood. As I remarked in my last Letter, with *three* strains of blood to start with, the breeder may ring innumerable changes, without ever trenching on that line which marks the boundaries of close breeding. He who pretends that he can preserve such multiplied classifications in his memory alone, is unworthy of the least confidence.

There is another very important consideration. Numbering and registering enables the breeder to trace *breeding effects* definitely to their *causes* Suppose that he finds that an unusual number of his young ewes are poor nurses—or exhibit some imperfection of form or wool. He can remove the *present effect* by throwing out the defective ones. But the undiscovered *cause* may still remain in operation. It may be a particular ram or the result of interbreeding between such ram, and ewes of a certain strain of blood. If this ram, or *perhaps others got by him*, be permitted to breed, or breed with a particular class of ewes, the evil creeps along in the flock, its cause remaining undiscovered. But if the breeder could fix the precise pedigree of every sheep, from an accurately kept register, he would soon ascertain what strains of blood, or the conjunction of what strains, produced the evil. By the same means, he could as readily trace the sources of particular excellence.

The system of numbering invented by the celebrated Von Thaër is far preferable to any other which I have seen.‡ It is as follows: ‖

---

* A ram of a new strain of blood, though of prime quality, and apparently possessing the same characteristics with the flock, does not always interbreed well with the flock in all those minute particulars which the *breeder* is bound to notice, though they might escape the eye of the ordinary flock-master. Every breeder, therefore, who has a flock that suits him, is exceedingly averse to an infusion of new blood, and resorts to it only as a matter of necessity.

† That is, he can breed in-and-in somewhat. "*Close*" breeding is breeding between *near* affinities, such as between brother and sister, which are of the *same* blood, or between a father and a grand-daughter begotten on a daughter, which would be three-fourths of the same blood, &c.

‡ It will not cause half the mutilation of the system given in the American Shepherd—is simple, and gives the *age*, which the former does not. Neither can this system of giving the *age* be ingrafted on that system of numbering.

‖ As furnished me by Mr. Grove, a number of years since, with this exception, that the point of the right ear cut square off, he made to stand for 700 instead of 500, as I have placed it. I made this change, as the notch and clip standing for 100 and 400, coming on the point of the same ear, there was no combination to express 500.

One notch over the left ear, (that which is on your left when the face of the sheep is *from* you,) stands for 1; two notches over the same, for 2. One notch under the left ear stands for 3. Three such notches carry up the number to 9. One notch over the right ear stands for 10; two such for 20. One notch under the same stands for 30; and three such for 90. Combinations of the above (three notches under each ear) would carry up the number to 99. These four classes of notches which express all parts of a hundred, are shown in the first of the annexed cuts. A sheep marked like fig. 17 would be No. 44.

Fig. 17.

Right ear.   Left ear.

No. 44—1841.

A notch in the end of loft ear, as in fig. 18, stands for 100; in right do. 200. In addition to these there are on the same cut two 1 notches, one 3 notch, one 10 do., and two 30 do. Adding the whole together, the sheep would therefore be No. 375.

Fig. 18.

No 375—1843.

As the 100 and 200 notches, together, make 300, no separate notch is required for the latter number. The point of the left ear cut square off, as in fig. 19, cut, stands for 400; the point of the right cut square off, for 500. The latter and the 100 notch would make 600, and so on.

Fig. 19.

No. 909—1848.

The lambs of each year and *each sex* are numbered from 1.

The age is expressed by round *holes* through the ears, standing for the year in which the sheep is born. As there is no possibility of making a mistake of *ten* years in the age of a sheep, these marks are the same between each tenth year of the century. Between 1840 and 1850, *no* hole would express 1840; one hole in the left ear, 1841; two holes in the left ear, 1842; one hole in the right ear, 1843; one hole in the right and one in the left, 1844; one hole in the right and two in the left, 1845; two in the right, 1846; two in the right and one in the left, 1847; two in each, 1848; three in the right, 1849; none in either, 1850—and the same for the next ten years. Examples are given in the preceding cuts. In other words, one hole in the left ear signifies 1, and one in the right 3, as applied to the years between each tenth of a century—and the combinations of these holes are made to express all the intermediate years, with the exception of the tenth.

Every ewe, when turned in with the ram, should be given a mark (entirely distinct from the mark of ownership) which will continue visible until the next shearing. Nothing is better for this purpose than Venetian red and hog's lard, well incorporated, and marked on with a cob. The ewes for each ram require a differently shaped mark, and the mark should also be made on the ram, or a minute of it in the sheep-book. Thus it can be determined at a glance by what ram the ewe was tupped, any time before the next shearing.

The holes in the ears, indicating the year, being the same on the whole annual crop of lambs, may be made at any convenient time. The holes re most conveniently made by a saddler's spring-punch, the cutting cylinder of which is about $\frac{3}{16}$ of an inch in diameter. If too small, the holes rill grow up in healing.

In numbering, it is difficult to prevent mistakes, if it is deferred until the lamb attains much size. If penned with the dams when a month or two old, hours will sometimes elapse before each lamb will suck—the only *certain* indication to which ewe it belongs. It being perfectly safe to perform this process when the lamb is only about a day old (or as soon as the lamb can walk, if it is a strong one), the shepherd carries the *notcher* in his pocket, and a little book, each page being ruled into six columns, and headed as in the register presently given. This constitutes the *day-book* which is subsequently drawn off on the Register.

The *notcher* which I use is of my own invention, and I have found it far preferable to any I have seen elsewhere. It consists of a saddler's spring-punch—the cutting cylinder being taken out, and a little sharp chisel of the same length being screwed in its place. The edge of the chisel describes a semi-ellipsis, cutting a notch out of the ear $\frac{1}{4}$ of an inch deep, and a little over $\frac{3}{16}$ wide at the base. A triangular cut in the ear, with so narrow a base, will grow together for some distance from the apex. This instrument is far more convenient than a chisel and block.

The shepherd, on finding a lamb of the right age to mark, goes quietly up to it, stopping it by the *neck* with his *crook* if it attempts to run away. The ewe will come near enough, in a moment or two, to be secured by the *crook*, and then the shepherd notes her number and age, and enters it in his pocket-book, and also by what ram tupped. The lamb then is numbered with the *notcher*, and this and its general appearance is noted down in the appropriate columns. If the ewe is too wild to be caught, the lamb may be notched—the number of the sire, &c., entered—and the number of the ewe subsequently ascertained in the pen.

I have two forms of Breeding Registers, originally furnished me by my lamented friend, the late Mr. Grove. One contains ten columns, the other eight. I have adopted the simplest one, omitting two of the columns, which leaves the Register in the following form :

BREEDING REGISTER—1845.

No. of Dam.	Tupp'd by Ram No.	Date of Lambing.	No. of Lamb. Rams.	No. of Lamb. Ewes.	Classification and Remarks.
22—40	16—39	May 4.	1		Coarsish—wrinkly—thick, short-legged, and stout—bad crops—ewe plenty of milk, and kind.
50—41	25—42	May 4.		1	Fine—thin—long-legged—will lack constitution—ewe kind—little milk.
6—42	7—43	May 5.	2 & 3		Small, but of good shape and fine wool—No. 3 wrinkly and like sire—No. 2 more like dam.—Ewe plenty of milk, but careless.
11—41	7—43	May 5.			The lamb was born dead, very small. Same last year. This ewe had better be thrown out of breeding.

The first entry above records the following facts : " The ewe No. 22, born in 1840, tupped by the ram No. 16 of 1839, dropped on the 4th of May a ram lamb, which was marked No. 1, its character being as described under the head of 'Classification and Remarks.' "

The column of " Remarks " is a very important one, if the minutes are made with accuracy and judgment. It should include an enumeration of all the prominent characteristics of the lamb, and of the appearances of the ewe as a breeder and nurse. These records will, in a single season, decide the character of a ram as a stock-getter, and that of the ewe, in a year or two, as a breeder and nurse.

EMASCULATION AND DOCKING.—These should usually precede washing, as at that period the oldest lambs will be about a month old, and it is safer to perform the operations when they are a couple of weeks younger.—Dry, pleasant weather should be selected. Castration is a simple and safe

SHEEP HUSBANDRY IN THE SOUTH.

process. Let a man hold the lamb with its back pressed firmly against his breast and stomach, and all four legs gathered in front in his hands.— Cut off the bottom of the pouch, free the testicle from the inclosing membrane, and then draw it steadily out, or clip the cord with a knife, if it does not snap off at a proper distance from the testicle. Some shepherds draw both testicles at once with their *teeth*. It is common to drop a little salt into the pouch. Where the weather is very warm, some touch the end of the pouch (and that of the tail, after that, is cut off) with an ointment, consisting of tar, lard, and turpentine. In ninety-nine cases out of a hundred, however, they will do just as well, here, without any application.

The tail should be cut off, say one and a half inches from the body, with a chisel on the head of a block, the skin being slid up toward the body with a finger and thumb, so that it will afterward cover the end of the stump. Severed with a knife, the end of the tail being grasped with one of the hands in the ordinary way, a naked stump is left which it takes · some time to heal.

It may occur to some unused to keeping sheep, that it is unnecessary to cut off the tail. If left on, it is apt to collect filth, and, if the sheep purges, it becomes an intolerable nuisance.

WASHING.—This is usually done here about the first of June. The climate of the Southern States would admit of its being done earlier. The rule should be to wait until the water has acquired sufficient warmth for bathing, and until cold rains and storms, and cold nights, are no longer to be expected.

Sheep are usually washed by our best flock-masters in vats. A small stream is dammed up, and the water taken from it in an aqueduct (formed by nailing boards together), and carried until sufficient fall is obtained to have it pour down a couple of feet or more into the vat. The body of water, to do the work fast and well, should be considerable—say 24 inches wide, and five or six deep—and the swifter the current the better. The vat should be say 3½ feet deep, and large enough for four sheep to swim in it. A yard is built near the vat, and a platform from the gate of the yard extends to and encircles the vat on three sides. This keeps the washer

Fig. 20.

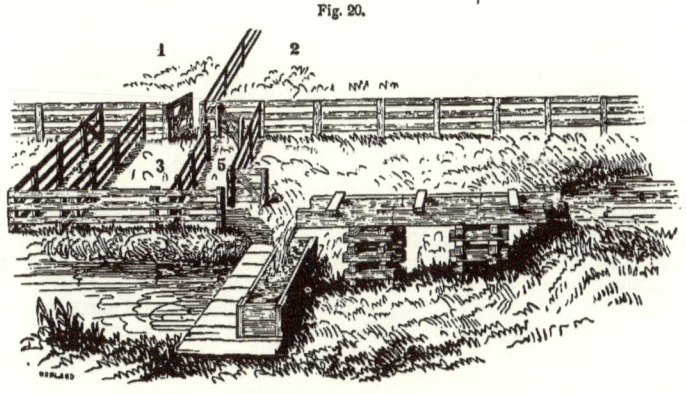

WASHING APPARATUS.

from standing in the water, and makes it much easier to lift the sheep in and out. The cut here given exhibits all the necessary appendages. The yard is built opposite the corners of two fields (1 and 2), to take advantage of the angle of one of them (1), to drive the sheep more readily into the yard (3). This yard should be large enough to hold the whole flock, if it does not exceed 200 ; and the bottom of it, as well as of the smaller yard (5), unless well sodded over, should be covered with coarse gravel, to avoid becoming muddy. If the same establishment is used by a number of flockmasters, graveling will be always necessary. As soon as the flock are confined in yard 3, the lambs are all immediately caught out from among them, and set over the fence into yard 4. This is to prevent their being trampled down, as it often happens, by the old sheep, or straying off if let loose. As many sheep are then driven out of yard 3 into the smaller yard 5 as it will conveniently hold. A boy stands by the gate next to the vat, to open and shut it (or the gate is drawn shut with a chain and weight), and two men, catching the sheep as directed under the head of tagging, commence placing them in the water for the preparatory process of "wetting." As soon as the water strikes through the wool, which occupies but an instant, the sheep is lifted out and let loose.* The vat should, of course, be in an inclosed field, to prevent their escape. The whole flock should thus be passed over, and again driven round through field 1 into yard 3, where they should stand, say, an hour, before washing commences. There is a large per centage of potash† in the wool oil, which acts upon the dirt, independently of the favorable effect which would result from thus soaking it for some time with water alone. If washed soon after a good shower, previous wetting might be dispensed with ; and it is not *absolutely necessary*, perhaps, in any case. If the water is warm enough to keep the sheep in it for the requisite period, they may be got clean by washing without any previous wetting—though the snowy whiteness of fleece which *tells* so on the *purchaser*, is not so often nor so perfectly attained in the latter way. Little time is saved by omitting "wetting," as it takes proportionably longer to wash, and it is not so well for the sheep to be kept such a length of time in the water at once.

When the washing commences, two and sometimes four sheep are plunged into the vat. When four are put in, two soak while two are washed. But this should not be done, unless the water is very warm, and the washers are uncommonly quick and expert. On the whole, it is rather an objectionable practice, for few animals suffer as much from the effects of a chill as sheep. If they have been previously wetted, it is wholly unnecessary. When the sheep are in the water, the two washers commence kneading the wool with their hands about the breech, belly, &c., (the dirtier parts,) and they then continue to turn the sheep so that the descending current of water can strike into all parts of the fleece. As soon as the sheep are clean, which may be known by the water running entirely clear, each washer seizes his own by the fore parts, plunges it deep in the vat, and taking advantage of the rebound, lifts it out, setting it gently down on its breech on the platform. He then, if the sheep is old or weak, (and it is well in all cases,) presses out some of the water from the wool, and after submitting the sheep to a process presently to be adverted to, lets it go. There should be no mud about the vat, the earth not covered with sod, being graveled. Sheep should be kept on clean pastures from washing to shearing—not where they can come in contact with

---

* Where there are conveniences for so doing, this process may be more easily performed by driving the sheep through a stream deep enough to compel them to swim. But *swimming* the compact-fleeced, fine wooled sheep for any length of time, as is practiced with the Long-Wools in England, will not properly cleanse the wool for shearing.

† Vauquelin, quoted by Youatt, says that it consists mostly of soapy matter with a basis of potash ; 2 Carb. of potash ; 3. Acetate of potash ; 4. Lime ; 5. Muriate of potash

the ground, burnt logs, &c.—and they should not be driven over dusty roads.

The washers should be *strong and careful* men, and protected as they are from anything but the water running over the sides of the vat, they can labor several hours without inconvenience, and without drinking whisky until they cease to know whether a sheep is well washed or well treated, as was the bad old fashion. Two hundred sheep will employ two expert men not over half a day, and I have known this rate much exceeded.

It is a great object, not only as a matter of propriety and honesty, but even as a matter of profit, to get the wool clean and of a snowy whiteness. It will always sell for more than enough extra, in this condition, to offset against the increased labor and the diminution in weight.

Mr. Lawrence wrote me, a few years since, that the *average* loss in American Saxon wool, in scouring, (after being washed on the back,) was 36 per cent., and in American Merino $42\frac{1}{2}$ per cent.!

CUTTING THE HOOFS.—The hoofs of fine-wooled sheep grow rapidly, turn up in front and under at the sides, and must be clipped as often as once a year, or they become unsightly, give an awkward, hobbling gait to the sheep, and the part of the horn which turns under at the sides holds dirt or dung in constant contact with the soles, and even prevents it from being readily shaken or washed out of the cleft of the foot in the natural movements of the sheep about the pastures, as would take place were the hoof in its proper shape. This greatly aggravates the hoof-ail, and the difficulty of curing it—and in England it is thought to *originate* the disease.

It is customary to clip the hoofs at tagging, or at or soon after the time of shearing. Some employ a chisel and mallet to shorten the hoofs, but then the sheep must be subsequently turned on its back to pare off the projecting and curling-under side crust. If the weather be dry, or the sheep have stood for some time on dry straw, (as at shearing,) the hoofs are as tough as horn, and are cut with great difficulty—and this is increased by the grit and dirt which adheres to the sole, and immediately takes the edge off from the knife.

The above periods are ill chosen, and the methods slow and bungling. It is particularly improper to submit heavily pregnant ewes to all this unnecessary handling at the time of tagging.

When the sheep is washed and lifted out of the vat, and placed on its rump on the platform, the gate-keeper advances with a pair of *toe-nippers*, and the washer presents each foot separately, pressing the toes together so they can be severed at a single clip. The nippers shown in the cut, can be made by any blacksmith who can temper an ax or chisel. They must be made strong, with handles a little

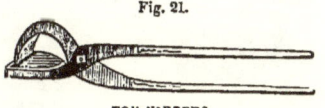

Fig. 21.

TOE-NIPPERS.

more than a foot long, the rivet being of half-inch iron and confined with a nut, so that they may be taken apart for sharpening. The cutting edge should descend upon a strip of copper inserted in the iron, to prevent it from being dulled. With this powerful instrument, the largest hoofs are severed with a moderate compression of the hand. Two well-sharpened knives, which should be kept in a stand or box within reach, are then grasped by the washer and assistant, and with two dexterous strokes to each foot, the side crust (being free from dirt, and soaked almost as soft as

a cucumber,) is reduced to the level of the soles.   Two expert men will
go through these processes in less time than it will take to read this description of them!

The closer the paring and clipping, the better, if blood is not drawn
An occasional sheep may require clipping again in the fall.

TIME BETWEEN WASHING AND SHEARING.—This depends altogether on
circumstances.   From four to six days of bright warm weather is suffi
cient.   If cold and rainy, or cloudy, more time must elapse.   I have known
the wool to remain in an unfit condition to shear a fortnight after washing
The rule is, the water should be thoroughly dried out, and the natural oil
of the wool should so far exude as to give the wool an unctuous feel and
a lively, glittering look.   If you shear it when dry, like cotton, before the
oil has exuded, you cheat yourself, and the wool will not keep so well for
long periods.*   If you leave it until it gets too oily, you cheat the manufacturer, or what more often happens, you lose on the price.

SHEARING—Is always done, in this country, on the threshing-floors of
our barns, sometimes on low platforms, but more commonly on the floor
itself.   The following cut represents a common Northern barn properly
arranged for this purpose.

Fig. 22.

SHEARING ARRANGEMENTS.

On the threshing-floor, three men are seen shearing—two of them using
a low table or platform, say 18 or 20 inches high.   The "bay"† (1, 2)
nearest the eye is divided by a temporary fence, one part (1) being used
for the yarding of the sheep, and the other (2) for doing up the wool, &c.
The inclosure 1 should communicate by a door with another and larger
yard outside of the barn.   Both of these should be well littered down with

* It is also very difficult to thrust the shears through this dry wool in shearing.
† The room for storing hay, grain, &c., which is always found on one, and sometimes on each side of the
threshing-floor in a Northern barn, is provincially termed a "bay"—and the low division between this and
the threshing floor a "breastwork."

straw, and fresh straw thrown on occasionally, to keep the sheep clean while shearing. No chaff, or other substances which will stick in the wool, should be used for this purpose. When the dew has dried off from the sheep, on the morning chosen for shearing, a portion of the flock sufficient to last the shearers half a day, is driven into the outside yard, and a convenient number into the bay (1). An assistant catches the sheep, lifts them off from the floor as already directed, and delivers them at the door through the "breastwork" (3) to each shearer. The shearer before taking the sheep, picks off any loose straws sticking to its wool, and if dung adheres to any of the feet, brushes it off with a little besom formed of twigs, hung up near the door for that purpose. The shearer then takes the sheep to his stand, and commences shearing.

The floor or tables used for shearing should be planed or worn perfectly smooth, so that they will not hold dirt or catch the wool. They all should be thoroughly cleaned, and, if necessary, washed, preparatory to shearing. It is the catcher's business to keep the floor constantly swept, dung removed, &c. Having a *new* stand or place swept for the shearer who has just finished his sheep, he catches him another, and then clears up the stand previously occupied. He first lifts the fleece, gathers it up so that it shall not be torn or drawn asunder, and turning his arms so as to invert it, (*i. e.*, bring the roots of the wool downward,) deposits it on the *folding-table* (4). He then picks up the "fribs" (small loose locks) left on the floor, which are deposited in a basket or on a corner of the table. Lastly, he sweeps the spot clean, to be again occupied by the shearer. An active fellow will tend four shearers, and do up the fleeces. But he should not be hurried too much, or he cannot give sufficient time to doing up. A small boy or two are handy to pick up fribs, sweep, &c.

If there are any sheep in the pen dirty from purging or other causes, they should first be caught out, to prevent them from dirtying the others.

It is difficult, if not impossible, to give intelligible practical instructions which would guide an entire novice in skillfully shearing a sheep. Practice is requisite. The following directions from the American Shepherd,* are correct, and are as plain, perhaps, as they can be made:

"The shearer may place the sheep on that part of the floor assigned to him, resting on its rump, and himself in a posture with one (his right) knee on a cushion, and the back of the animal resting against his left thigh. He grasps the shears about half-way from the point to the bow, resting his thumb along the blade, which affords him better command of the points. He may then commence cutting the wool at the brisket, and proceeding downward, all upon the sides of the belly to the extremity of the ribs, the external sides of both thighs to the edges of the flanks; then back to the brisket, and thence upward, shearing the wool from the breast, front, and both sides of the neck—but not yet the back of it—and also the poll or fore part, and top of the head. Now the 'jacket is opened' of the sheep, and its position and that of the shearer is changed, by being turned flat upon its side, one knee of the shearer resting on the cushion, and the other gently pressing the fore quarter of the animal, to prevent any struggling. He then resumes cutting upon the flank and rump, and thence onward to the head. Thus one side is complete. The sheep is then turned on to the other side, in doing which great care is requisite to prevent the fleece from being torn, and the shearer acts as upon the other, which finishes. He must then take his sheep near to the door through which it is to pass out, and neatly trim the legs, and leave not a solitary lock anywhere as a harbor for ticks. It is absolutely necessary for him to remove from his stand to trim, otherwise the useless stuff from the legs becomes intermingled with the fleece-wool. In the use of the shears, let the blades be laid as flat to the skin as possible, not lower the points too much, nor cut more than from one to two inches at a clip, frequently not so much, depending on the part and compactness of the wool."

In addition to the above, I would remark that the wool should be cut off as close as conveniently practicable, and even. It *may* be cut *too close*, so that the sheep can scarcely avoid "sun-scald," but this is very unusual.

* Pages 179, 180.

2 A

If the wool is left ridgy and uneven, it betrays that want of workmanship which is so distasteful to every good farmer.* Great care should be taken not to cut the wool twice in two, as inexperienced shearers are apt to do. It is a great damage to the wool. It is done by cutting too far from the point of the shears, and suffering the points to get too elevated. Every time the shears are pushed forward, the wool before cut off by the points, say a quarter or three-eighths of an inch from the hide, is again severed. To keep the fleece entire, so important to its good appearance when done up, (and therefore to its salableness,) it is very essential that the sheep be held easily *for itself*, so that it will not struggle violently. To hold it still by main strength, no man can do, and shear it well. The posture of the shearer should be such that the sheep is actually confined to its position, so that it is unable to start up suddenly and tear its fleece, but it should not be confined there by severe pressure or force, or it will be constantly kicking and struggling. Heavy-handed, careless men, therefore, always complain of getting the most troublesome sheep. The neck, for example, may be confined to the floor by placing it between the toe and knee of the leg on which the shearer kneels, but the lazy or brutal shearer who lets his leg rest directly on the neck, soon provokes that struggle which the animal is obliged to make to free itself from severe pain, and even perhaps to draw its breath!

Good shearers will shear, on the average, twenty-five Merinos per day, and a new beginner should not attempt to exceed from one-third to one-half that number. It is the last process in the world which should be hurried, as the shearer will soon leave more than enough wool on his sheep to pay for his day's wages.

It has been mentioned that but enough sheep should be yarded at once for half a day's shearing. The reason for this is that they shear much more easily, and there is less liability of cutting the skin, when they are distended with food, than when their bellies become flabby and collapsed for the want of it. This precaution, however, is often necessarily omitted in showery weather. It is very convenient to have the outside pen which communicates with the "bay," *covered*. On my farm, it is one of the regular sheep-houses. If it is showery over night, or showers come up on the day of shearing, a couple of hundred sheep may be run in and kept dry. And they can be let out to feed occasionally during the day on short grass. If let out in long wet grass, their bellies will become wetted. Wool *ought* not to be sheared, and *must* not be done up, with any water in it. —

SHEARING LAMBS, AND SHEARING SHEEP SEMI-ANNUALLY.—Shearing lambs is, in my judgment, every way an abominable and unprofitable practice—in this climate, at least. The lamb will give you the same wool at a year old, and you strip it of its natural protection from cold when it is young and tender, for the paltry gain of the *interest* on a pound or a pound and a half of wool for six months—not more than two or three cents—and this all covered by the expense of shearing.

I am aware that it is customary, in many parts of the South, to shear grown sheep twice a year; and there may be a reason for it where they receive so little care that a portion are expected to disappear every half-year, and the wool to be torn from the backs of the remainder by bushes, thorns, &c., if left for a longer period. But when sheep are inclosed, and

---

* I hold that man is not *half* a farmer who has not a dash of the æsthetic mixed up with his utilitarianism. Profit should not often be sacrificed to appearances, but where they are strictly compatible, he who disregards the latter betrays a sordid and uncultivated mind.

treated as domestic animals, there may be less barbarity in fall-shearing them than in the case of tender lambs, but I cannot conceive of any better reason for it than in the former case, on the score of utility. Any gain resulting from it cannot pay the additional expense it occasions.

DOING-UP WOOL.—The fleece has been deposited on the "folding table," and he whose business it is to do it up, first proceeds to spread it out, the *outer ends upward*, bringing every part to its natural relative position.— The table, with a fleece spread out on it, is represented in fig. 23. The table should be large— say five feet wide and eight long—that, if necessary, several *unspread* fleeces may be put upon it at the same time, and still give room for spreading one. It should be about three feet high. After the fleece

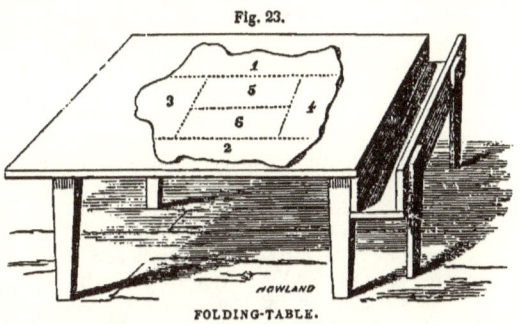

Fig. 23.

FOLDING-TABLE.

is spread, dung, burs, and all other extraneous substances are carefully removed from it with a pair of shears. It is then pressed together with the hands, so that it will cover but little if any more space than it would occupy on the skin of the animal, if that was placed *unstretched* on the table. About a quarter of the fleece, lengthwise, or from head to tail, (represented by 1 in the above cut,) is then turned or folded in *(inverting* it,) toward the middle. The opposite side (2) is next folded inward in the same way, leaving the fleece in a long strip, say 18 inches wide. The forward end (3) is then folded toward the breech, to a point (represented by dotted line) corresponding with the point of the shoulder. The breech (4) is next folded toward the head. The fleece now presents an oblong square represented by 5 and 6. On the breech, in a small, compact bunch—so they can be, subsequently, readily separated from the fleece—the clean fribs are placed. They do not include "trimmings," (the wool from the shanks,) which should not be done up in the fleeces. The fribs may be laid in at some earlier stage of the folding —but if thrown on top of the fleece, as is very customary, before it is folded at all, they *show through*, if the latter gets strained apart, as it frequently happens in the process of rolling—and being coarser and perhaps less white than the fine shoulder wool, they injure the appearance of the fleece. The fleece is now folded together by turning 5 over on to 6, and

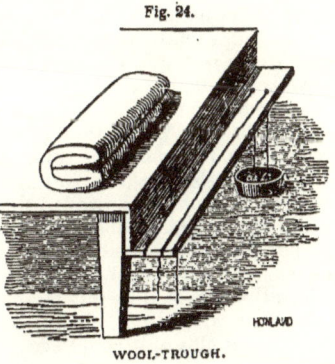

Fig. 24.

WOOL-TROUGH.

the tyer carefully sliding it around on the table with his arms, so that the shoulder shall be toward him, it appears as in fig. 24, ready to go into the wool-trough. The wool-trough, which is above represented with one of its sides off, to exhibit the interior arrangement, should form a part of the table, and should be about 9¼ inches wide and 9 deep, and its length corresponding with the width of the table, would be five feet. Near its back end, and about one-third of its width from each side, gimlet holes are bored just large enough for the passage of ordinary wool-twine. Two balls of twine are placed in a vessel beneath, the ends passed through the holes, and the whole length of the trough, and are fastened in front by being drawn into two slits formed by sawing a couple of inches into the bottom of the trough. The holes and slits should be small enough, so that the twine will be kept drawn straight between them.

The tyer placing his hands and arms (to the elbow) on each side of the fleece folded as above, now slides it into the trough. There are two methods of having it lie in the trough, represented by the following cuts. That on the left is the more ordinary, but not the best method. It will bring to the two *ends* of the done-up fleece (the parts most seen in the wool-room) the ridge of the back and two lines half way down each side of the sheep. The former is sometimes a little weather-beaten, and if any hay-seeds have fastened in the fleece, they show most on the back.* And the two lower lines are a little below the choicest wool.—

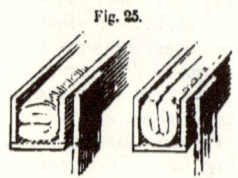

Fig. 25.

Placing it in the trough as in the right-hand figure, rolling would bring both ends of the fleece from the wool between four and five inches from the ridge of the back, the choicest part of the fleece. Besides, the edges of the breech fold, which is not so fine as the shoulder, which sometimes show by the first method of rolling, are always concealed by the last.

The wool being in the trough, the tyer steps round to the back end of it, and commences rolling the fleece from the breech to the shoulder. He rolls it as tightly as possible, pressing it down and exerting all the strength of his hands—minding, however, not to tear the outside fold—or strain it so apart as to exhibit the *outer* ends of the next inside layer or fold. When the rolling is completed, he keeps it tight by resting the lower part of his left arm across it, reaches over with the right, and withdrawing one of the ends of the twine from the slit, places it in the left hand. Then seizing the twine on the other side of the fleece with his right hand, he draws the twine once about the fleece with his *whole strength*, and ties it in a hard or square knot. The fleece will then keep its position, and the other twine is tied in the same way. The twines should be drawn with a force that would cut through the skin of a tender hand in a few moments.†

The twines are then cut within an inch of the knots, with a pair of shears. The fleece is *slid* out of the *end* of the trough, when it will be a solid, glittering mass of snowy wool, in the shape shown in the cut on the right. If well and tightly done up, however, the divisions given on the end of the fleece, in the cut, to exhibit the foldings, will not be perceptible—and nothing but an unbroken mass of the choicest wool of the fleece.

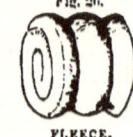

Fig. 26.

FLEECE.

The twine should be of flax or hemp, and of the diameter of ordinary sized hardware twine. Cotton might do, if smooth and hard enough so that no particles of it could become incorporated with the wool—in which event it does not separate from the wool in any of the subsequent processes, and receiving a different color from the dyes, spots the surface of the cloth.

* Hay-seed, or rather its chaff, will not wash entirely out of wool.

† It is customary with some tyers to wear a glove on the right hand—or cots on the two fore-fingers.

It is scarcely necessary to remark that it is considered perfectly fair by the purchaser, to take all the pains above recommended, to "put the best side out" in doing up wool, provided every fleece is done up by itself. He expects it, and graduates his prices accordingly. He who neglects it, therefore, cheats himself. But to do up coarser fleeces, or any parts of them, in finer ones—put in "trimmings"—leave in dung—or use unnecessary twine—are all base frauds. Sometimes the careless sheep-owner will have his wool filled with burs, which he cannot or will not remove. In that case he is bound to unequivocally apprise the buyer of the fact, and allow him to open fleeces until satisfied of the precise extent of the evil.

STORING WOOL.—Wool should be stored in a clean, tight, dry room. It is better that it should be an *upper* room, for reasons presently to be given, and it should be plastered, to exclude dust, vermin, insects, &c. Rats and mice love to build their nests in it, to which they will carry grain chaff and other substances, injuring much wool—and it is singular that if accessible to the common bumble-bee, numbers of their nests will be found in it. A north and pretty strong light is preferable for a wool-room.

When the wool-tyer removes each fleece from the trough, he places it in a long, high basket, capable of holding a dozen fleeces, and it is immediately carried to the wool-room—or he piles it on the clean floor in the inclosure in which his table stands, to be subsequently carried away. In either case, the fleeces are not thrown down promiscuously, which injures their shape, but are laid regularly one above another, on their sides. In the wool-room it is laid in the same way in smooth, straight north and south rows (supposing the light to be let in from the north) with alleys between, in which a man can pass to inspect the wool. The rows ought not, perhaps, to be more than two deep, so that the end of *every* fleece can be examined, but as it cannot be piled up more than about four fleeces high in this way, without liability of falling, it is customary to make the rows three or four fleeces deep—laying the lower ones a little wide, so that the pile may slightly recede as it goes up. In this way they may be piled six fleeces high. Where the character of the flock is known, or that of the seller relied on, it makes little difference. It is considered fairest to pile the fleeces without any discrimination as to quality, in the wool-room.

SACKING WOOL.—When the wool is sold, or when it must be sent away to find a market, it is put up in bales nine feet long, formed of 40-inch "burlaps." The mouth of the sack is sowed, with twine, round a strong hoop (riveted together with iron, and kept for the purpose,) and the body of it is let down through a circular aperture in the floor of the wool-room.* The hoop rests on the edge of the aperture, and the sack swings clear of the floor beneath. A man enters the sack, and another passes the fleeces down to him. After covering the bottom with a layer, he places a fleece in the center and forces down others around it, and so on to the top, which is then sowed up. Each fleece should be placed regularly with *the hands*, and then stamped down as compactly as possible, so that the bale when completed shall be hard and well filled in every part. The bulk of a given weight of wool will be greatly affected by the care with which this process is performed.

Those who do not expect buyers to come and look at their wool, sack it immediately after shearing. A temporary scaffolding is erected near

---

* It is to secure this convenience that the wool-room is best placed on the second floor.

the wool as deposited by the tyer, and one man tosses up fleeces to a sec
ond, who catches them and passes them down to the man in the sack. A
light frame, to suspend the sack, and *part way up it* a standing-place fo
the catcher, would be a convenient appendage to the establishment of
wool-grower who does not store his wool in a wool-room. With a set c
stairs up to his midway standing-place, an active fellow would keep th
treader supplied, without any assistance.

In the absence of any agreement, the price of wool, delivered at th
residence of the purchaser, does not include the cost of sacks and sacking
It is customary, however, for growers of small parcels, and those wh
keep no conveniences for sacking, to carry their wool tied up in sheets
&c., and deliver it to the purchaser at the nearest village or other point
where he has made arrangements for sacking.   .

SELECTION.—The necessity of annually weeding the flock, by excluding
all its members falling below a certain standard of quality, and what th
points are to which reference should be had in establishing that standard
have already been sufficiently adverted to in discussing the principles o
breeding. The time of shearing is by far the most favorable one for th
flockmaster to make his selection. He should be present on the shearing
floor, and inspect the fleece of every sheep as it is gradually taken off. I
there is a fault about it, he will then discover it better than at any othe
time. A glance, too, reveals to him every fault of form, previously con
cealed wholly or in part, by the wool, as soon as the newly shorn sheep
is permitted to stand on its feet. He takes down the number and age o
the sheep on his tablet, and if not sufficiently defective in form or quality
of fleece to call for its condemnation, in a pair of scales suspended near th
wool-tyer's table, he determines the weight of the fleece. If this, too, i
satisfactory, he marks "*retained*" opposite the sheep's number on his tab
let. If more or less defective in any point, he weighs this against th
other points—taking also into consideration the age of the sheep, its char
acter as a breeder, its nursing properties, quietness of disposition, &c.—
and then, in view of *all* these points, the question of retention or exclusio
is settled. A remarkably choice ewe is frequently kept until she dies o
old age. A poorish nurse or breeder would be excluded for the lightes
fault, and so on. I have been in the habit, for a number of years, of usin
a book kept for this purpose, each page being ruled and headed thus :

Number.	Qual. of Fleece.	Form.	Wt. of Fleece.	Conclusion.
27, '42	p.	f.	4½	r.
30, '44	o.	b.	4	e.

The figures in the first column signify No. 27 of the year 1842, and No
30 of the year 1844. The letters in the succeeding columns stand for the
words " prime," " fair," " ordinary," and " bad "—marking the gradation
of quality. The letters in the last column signify " retained," or " ex
cluded." Such a record will lead to far greater accuracy than by any
other method, and it is extremely valuable for purposes hereafter to b
stated.

If the sheep are not numbered, the flock-master should note each appear
ance, as above directed, have the sheep held by the neck by an assistant
or discharged by the shearer into a small pen at the door for that purpose
until the fleece is weighed, and then if he decides to exclude it, he give
it a small mark on the shoulder, consisting of Venetian Red and hog's lard
(conveniently applied with a brush or cob.)

MARKING SHEEP.—The sheep should be marked soon after shearing, or mistakes may occur. Every owner of sheep should be provided with a marking instrument, which will stamp his initials, or some other distinctive mark, such as a small circle, oval, triangle, square, &c., at a single stroke, and with *uniformity*, on the sheep. It has been customary here, to have the mark cut out of a plate of thin iron, with an iron handle terminated by wood. But one made by cutting a type or raised letter (or character) on the end of a stick of light wood, such as pine or basswood, is found to be better. If the pigment used be thin, and the marker be thrust into it a little too deeply, as often happens, the surplus will not run off from the wood, as from a thin sheet of iron, to daub the sides of the sheep, and spoil the appearance of the mark; and if the pigment be applied *hot*, the former will not, like the latter, get heated, and increase the danger of burning the hide. Various pigments are used. Many boil tar until it will assume a glazed, hard consistency, when cold, and give it a brilliant black color by stirring in a little lamp-black when boiling. It is applied when just cold enough not to burn the sheep's hide, and it forms a bright, conspicuous mark the year round. I have always used this, though the manufacturer would prefer the substitution of oil and turpentine for tar, as the latter is cleansed out of the wool with some difficulty. I boil it in a high-sided iron vessel (to prevent it from taking fire) on a small furnace or chafing-dish near where it is to be used. When cool enough, forty or fifty sheep can be marked before it gets too stiff. It is then warmed from time to time, as necessary, on the chafing-dish. The rump is a better place to mark than the side. The mark is about as conspicuous on the former, under any circumstances, and it is more so when the sheep are huddled in a pen, or when they are running away from you. And should any wool be injured by the mark, that on the rump is less valuable than that on the side. It is customary to distinguish ewes from wethers by marking them on different sides of the rump.

Many mark each sheep as it is discharged from the barn by the shearer. It consumes much less time to do it at one job, after the shearing is completed ; and it is necessary to take the latter course, if a hot pigment is used.

COLD STORMS AFTER SHEARING.—These sometimes destroy sheep, in this latitude, soon after shearing—particularly the delicate Saxons. I have known forty or fifty perish out of a single flock, from one night's exposure. The remedy, or rather the preventive, is to house them, or in default of the necessary fixtures to effect this, to drive them into dense forests. I presume, however, this would be a calamity of rare occurrence in the " sunny South."

SUN-SCALD—Might be more common. When sheep are sheared close in very hot weather—have no shade in their pastures—and particularly where they are driven immediately considerable distances, or rapidly, over burning and dusty roads, their backs are so scorched by the sun that the wool comes off. It is not common, however, here. You may see one such in a flock of a hundred. Let alone, the matter is not a serious one but the application of refuse lard to the back will accelerate the cure, and the starting of the wool.

TICKS.—These, when very numerous, greatly annoy and enfeeble sheep in the winter, and should be kept entirely out of the flock. After shearing, the heat and cold, the rubbing and biting of the sheep soon drive off the tick, and it takes refuge in the long wool of the lamb. Wait a fortnight after shearing, to allow all to make this transfer of residence. The boil refuse tobacco leaves until the decoction is strong enough to kill tick beyond a peradventure. This may be readily tested by experiment. Five or six pounds of cheap plug tobacco, or an equivalent in stems, &c. may be made to answer for 100 lambs. The decoction is poured into a deep, narrow box, kept for this purpose, and which has an inclined shelf one one side, covered with a wooden grate, as shown in the cut. One man holds the lamb by the hind legs, another clasps the fore-legs in one hand, and shuts the other about the nostrils to prevent the liquid entering them, and then the lamb is entirely immersed. It is immediately lifted out, laid on one side on the grate, and the water squeezed out of its wool. It is then turned over and squeezed on the other side. The grate conducts the fluid back into the box. If the lambs are regularly dipped every year, ticks will never trouble a flock.

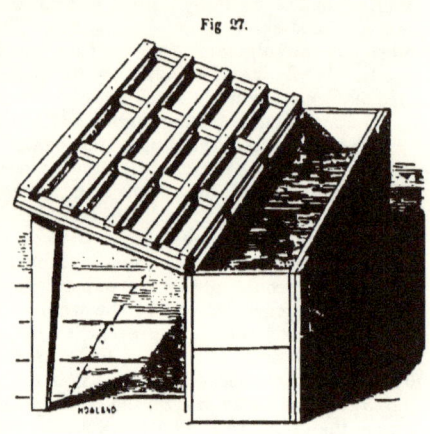

Fig 27.

DIPPING-BOX.

The effect of tobacco water in scab, will be hereafter adverted to.

MAGGOTS.—Rams with horns growing closely to their heads, are very liable to have maggots generated under them, particularly if the skin on the surrounding parts gets broken in fighting, and these, if not removed, soon destroy the sheep. Both remedy and preventive is boiled tar—or the marking substance heretofore described. Put it under the horns, at the time of marking, and no trouble will ever arise from this cause. Sometimes when a sheep scours in warm weather, and clotted dung adheres about the anus, maggots are generated under it, and the sheep perishes miserably. Preventive: remove the dung. Remedy: remove the dung and maggots, the latter by touching them with a little turpentine, and then apply sulphur and grease to the excoriated surface.

Maggot flies, says Blacklock, sometimes deposit their eggs on the *backs* of the long, open-wooled English sheep, and the maggots during the few days before they assume the pupa state, so tease and irritate the animal, that fever and death are the consequence. Tar and turpentine, or butter and sulphur, smeared over the parts are given as the preventives. The Merino and Saxon are exempt from these attacks.

SHORTENING THE HORNS.—A convolution of the horn of a ram sometimes so presses in upon the side of the head or neck, that it is necessary to shave or rasp it away on the under side, to prevent ultimately fatal effects. The *point* of the horn of the ram and ewe both not unfrequently turn in so that they will grow into the flesh and sometimes into the eye, unless shortened. The toe-nippers will often suffice on the thin extremity of a horn, but if not, a fine saw must be used. The marking time is the best one to attend to this.

DIVISION OF FLOCKS.—It is customary at, or soon after shearing, to make those divisions in large flocks, which utility demands. It is better to have not to exceed two hundred sheep run together in the pastures, though the number might perhaps be safely increased to three hundred, if the range is extensive. Wethers and dry ewes to be turned off, should be kept separate from the nursing-ewes, and if the flock is sufficiently numerous to require a third division, it is customary to put the yearling and two-year-old ewes and wethers and the old, feeble sheep together. It is better in all cases to separate the rams from all the other sheep, at the time of shearing, and to inclose them in a particularly well-fenced field. If put even with wethers, they are more quarrelsome, and when cool nights arrive, will worry themselves and waste their flesh in constant efforts to ride the wethers. The Merino ram is a quiet animal compared with the common-wooled one, but poor fences, or fences half the time down, will tempt him to jump, and if once taught this trick, he becomes very troublesome as the rutting period approaches, unless hoppling, yoking, clogging, or "poking" is resorted to—either of which causes the animal to waste his flesh and strength, and are the causes of frequent accidents.

—

HOPPLING, CLOGGING, &c.—Hoppling is done by sowing the ends of a .eathern strap (broad at the extremities so that it will not cut into the flesh) to a fore and hind leg, just above the pastern joints—leaving the legs at about the natural distance apart. Clogging is fastening a billet of wood to the fore leg by a leather strap. Yoking is fastening two rams two or three feet apart, by bows around their necks, inserted in a light piece of timber, say two by three inches in size. Poking is done by inserting a bow in a short bit of light timber, into which bit (worn on the under side of the neck) a rod is inserted which projects a couple of feet in front of the sheep. These, and similar devices, to prevent rams scaling fences, may be employed as a last resort, by those improvident farmers who prefer by such troublesome, injurious, and at best, insecure means, to guard against that viciousness which they might, so much more easily, have prevented from being acquired. —

DANGEROUS RAMS.—From being teased and annoyed by boys, or petted and played with when young—and sometimes without any other stimulant than a naturally vicious temper—rams occasionally become very troublesome by their propensity to attack men or cattle. I know of one for which his owner has refused $250, which will permit no man to enter the field with him without making an immediate onset on him. I have known several that would knock down the ox or horse which presumed to dispute the possession of a lock of hay with them. A ram which is known to have acquired this propensity should at once be *hooded*, and, if not valuable, at the proper season converted into a wether by "cording." But the courage thus manifested, is usually the concomitant of great strength and vigor of constitution—and of a powerfully developed frame. If good in other particulars, it is a pity to lose the services of such an animal. I have in several such instances hooded them, by covering their faces with leather in such a manner that they could only see a little backward and downward. They must then, however, be kept apart from the flock of rams, or they will soon be killed or injured by blows, which they cannot see to escape.

It sometimes happens that a usually quiet tempered ram will suddenly exhibit some pugnacity when you are salting or feeding the flock. If you turn to run, you are immediately knocked down, and the ram learns, at that single lesson, the secret of his mastery, and the propensity to exercise it. The ram giving his blow from the summit of the parietal and the posterior portion of the frontal bones on the *top* of the head, (and not from the forehead,) couches his head so low when he makes his onset, that he does not see forward well enough to swerve suddenly from his right line, and a few quick motions to the right and left enable you to escape him. Run in upon him, as he dashes by you, with pitchfork, club, or *boot-heel*—punishing him severely by blows, (about the head if the club is used,) and giving him no time to rally until he is thoroughly cowed.*

FENCES.—Poor fences will teach ewes and wethers to jump, as well as rams, and for a jumping *flock* there is no remedy but immoderately high fences, or extirpation. One jumper will soon teach the trick to a whole flock, and if one by chance is bought in, it should be immediately hoppled or killed. The last is by far the surest and safest remedy.

SALT.—Salt, in my judgment, is indispensable to the health of sheep, particularly in the summer—and I know not a flock-master among the hundreds, nay, thousands with whom I am acquainted, who differs with me in this opinion. It is common to give it once a week while the sheep are at grass.

It is still better to give them free access to salt at all times, by keeping it in a covered box, open on one side, like the following:

A large hollow log, with holes cut along the side, for the insertion of the heads of the sheep will make a respectable substitute. A sheep having free access to salt at all times, will never eat too much, and it will take its supply when and in what quantities Nature demands, instead of eating voraciously at stated periods, as intermediate abstinence will stimulate it do. When fed but once a week, it is better to have a stated day, so that it will not be forgotten, and it is well to lay the salt on flat stones, though if laid in little handsfull on the grass, very little will be lost.

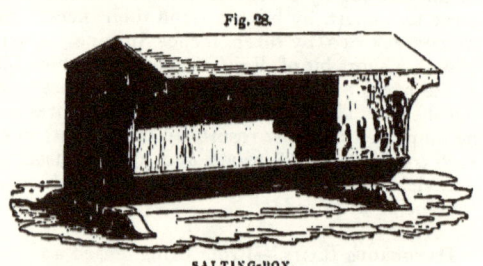

Fig. 28.

SALTING-BOX.

TAR.—This is supposed by many to form a very healthful condiment for sheep. The nose of the sheep is smeared with it, and it is licked and swallowed as the natural heat of the flesh, or that of the weather, causes it to trickle down over the nostrils and lips. Others, suffering the flock to get unusually salt hungry, place tar upon flat stones, or in troughs, and then scatter salt on it, so that both shall be consumed together. Applied to the nose, in the nature of a cataplasm, I have no doubt that it is advantageous in catarrhs—and put on the same place, at the proper periods, it may perhaps, by its odor, repel the visitations of the fly *(Œstris ovis)*, the eggs of which produce the "grub in the head." As a *medicine* it may be valuable, and even as a *detergent* in the case specified, but as a condiment

---

* This may be pronounced harsh "measure for measure," and some may think it would tend to increase the viciousness of the animal. Repeated instances have proved the contrary to me. And if their mastery is once acknowledged, it is never forgotten by them.

imply, for a perfectly healthy animal, I confess I have no confidence in its itility. —

WATER.—Water is not indispensable in the summer pastures, the dews nd the succulence of the feed answering as a substitute. But my impres- iot is decided that free access to water is advantageous to sheep, particu- arly to those having lambs ; and I should consider it a matter of import- nce on a sheep farm, to arrange the pastures, if practicable, so as to iring water into each of them. —

SHADE.—No one who has observed with what eagerness sheep seek liade in hot weather, and how they pant and apparently suffer when a hot un is pouring down on their nearly naked bodies, will doubt that, both as matter of humanity and utility, they should be provided, during the hot ummer months, with a better shelter than that afforded by a common rail ence. Forest-trees are the most natural and best shades, and it is as con- rary to utility as it is to good taste to strip them entirely from the sheep- ralks. A strip of stone-wall or close board fence on the south and west ides of the pasture, will form a passable substitute for trees. But in the bsence of all these, and of buildings of any kind, a shade can be cheaply onstructed of poles and brush, in the same manner as the sheds of the ime materials for winter shelter, which will be described in my next Letter.
—

WEANING LAMBS.—Lambs should be weaned at four months old. It is etter for them, and much better for their dams. The lambs when taken way should be put for several days in a field distant from the ewes, that ιey may not hear each other's bleatings. The lambs when in hearing of ιeir dams, continue restless much longer, and they make constant and equently successful efforts to crawl through the fences which separate ιem. One or two tame old ewes are turned into the field with them to ιacli them to come at the call, find salt when thrown to them, and eat ιain, &c., out of troughs when winter approaches.
The lambs when weaned should be put on the freshest and tenderest ιed. I have usually reserved for mine the grass and clover sown, the pre- ιding spring, on the grain fields which were seeded down.
The dams, on the contrary, should be put for a fortnight on short, dry ιed, to stop the flow of milk. They should be looked to, once or twice, ιd should the bags of any be found much distended, the milk should be- ιawn and the bag washed for a little time in cold water. But on short ιed, they rarely give much trouble in this particular. When properly ιied off, they should be put on good feed to recruit, and get in condition. ιr winter. —

FALL FEEDING.—In the North, the grass often gets very short by the ιth or 15th of November, and it has lost much of its nutritiousness from ιpeated freezing and thawing. At this time, though no snow has yeι llen, it is best to give the sheep a light daily foddering of bright hay— ι a few oats in the bundle. Given thus for the ten or twelve days which ιecede the covering of the ground by snow, fodder pays for itself as well ι at any other time during the year. I have usually fed oats in the bun- ιε, or threshed oats, (about a gill to the head,) in the feeding-troughs, ιrried to the fields for that purpose.

THE CROOK.—This implement has been several times alluded to as a ιnvenient one for catching sheep. It is made in the form exhibited in

the cut, of ¾-inch round iron, drawn smaller toward the point—and the
point made safe by a knob. The other end is furnished
with a socket, which receives a handle six or eight feet
long. The manner of using it is thus described in Mr. Stephens's admirable "Book of the Farm":

Fig. 29.

" The hind-leg is hooked in at *a*, from behind the sheep, and it fills up
the narrow part beyond *a*, while passing along it until it reaches the loop,
when the animal is caught by the hock, and when secured, its foot easily slips through the loop. Some caution is required in using the crook,
'or should the sheep give a sudden start forward to get away, the moment it feels the crook the leg will be drawn forcibly through the narrow
part, and strike the bone with such violence against the bend of the loop
as to cause the animal considerable pain, and even occasion lameness for
some days. On first embracing the leg, the crook should be drawn
quickly toward you, so as to bring the bend of the loop against the leg as
high up as the hock, before the sheep has time even to break off, and being secure, its struggles will cease the moment your hand seizes the leg."

No flock-master should be without this implement, as it
saves a vast deal of yarding, running, &c., and leads to
a prompt examination of every improper or suspicious appearance, and a timely application of remedy or preventive—which would often be deferred if the whole flock had
to be driven to a distant yard, to enable the shepherd to
catch a particular sheep.

SHEPHERD'S
CROOK.

Dexterity in the use of the crook is speedily acquired by any one; and
if a flock are properly tame, any one of its number can be readily caught
by it, at salting-time—or, generally, at other times, by a person with whom
the flock are familiar. But it is at the lambing-time, when sheep and lamb
require to be so repeatedly caught, that the crook is more particularly ser
viceable. For this purpose, at this time alone, it will pay for itself ten
times over in a single season, in saving *time*, to say nothing of the advan
tage of the sheep.

LETTER XIII.

WINTER MANAGEMENT OF SHEEP.

Use of Rams—proper age, number, &c.—selecting ewes for—different methods of coupling—way to treat rams...Division of Flocks for Winter...The Hospital...Yards—when necessary...Feeding-Racks—various plans of—the Box Rack—the Hole Rack—the Sparred Rack—the Hopper Rack—their respective advantages—improvements suggested...Troughs...Grain-Boxes...Barns and Sheds—necessity of shelter at the North—the common Northern Sheep-Barn...Stells—the Outside Stell—Ancient Stells—Inside Circular Stell—Circular Stell fitted up with racks...Tree-Coverts...Cheap Sheds—fitted between stacks, barracks, &c...Value of Barracks for the Preservation of Fodder...The Main Sheep-Barn of the Farm or Plantation, with Shearing-Floor, &c.—arrangements for breeding-ewes...Feeding sheep in yards with other stock—Improper—reasons...Hay-Holders...Winter Dry Feed for Sheep—Variations in Feed—German views on this subject—proper kinds of fodder—Boussingault's Table of the Nutritive Equivalents of different kinds of Fodders...Effect of Food in the Production of Wool—De Reaumur's Table showing the Effects of Food in this particular...Effect of Food in producing Fat and Muscle...Fattening Wethers in the North...Feeding Grain to Store-Sheep in Winter—when practiced at the North—economy of so doing—kinds of grain preferred—necessity of regularity in quantity—difficulty of raising the condition of poor sheep in the winter...Feeding Roots, Browse, &c.—Roots a Substitute for Grain—to what sheep they may be fed—Hemlock Browse—when and in what manner useful—substitutes for...Winter Feed of Breeding-Ewes...Necessity of regularity in the times of feeding sheep...Salt...Water.

Dear Sir: As the turning out of the rams usually takes place, here, on the first day of winter, I will describe the proper accompanying arrangements, as the first step in winter management.

USE OF RAMS.—The period of gestation in the ewe averages five months. Merino rams are frequently used from the first to the tenth year, and even longer. The lambs of very old rams are not supposed to be as vigorous as those of youngish or middle-aged ones, but where rams have not been overtasked, and have been properly fed, I confess I have been able to discover very little difference in their progeny on account of age. A ram lamb should not be used, as it retards his growth, injures his form, and I think, permanently impairs his vigor and courage. A yearling may run with 30 ewes, a two-year-old with from 40 to 50, and a three-year-old with from 50 to 60. Some very powerful, mature rams will serve 70 or 80 ewes; but 50 is enough, where they *run with* the ewes. I am satisfied that an impoverished and overtasked animal does not transmit his individual properties so decidedly to his offspring as one in full vigor.

Several rams running in the same flock excite each other to an unnatural and unnecessary activity, besides injuring each other by constant blows. It is, in every point of view, bad husbandry, where it can be avoided, and, as usually managed, is destructive to everything like careful and judicious breeding. The nice adaptation which the male should possess to the female, already discussed under the head of Principles of Breeding—counterbalancing her defects with his own marked excellence in the same points, and, in turn, having his defects counterbalanced by her excellencies—how shall this be accomplished, where half a dozen or more rams are running promiscuously with two or three hundred ewes ?

Before the rams are let out, the flock-master should have all the breeding-ewes brought together in one yard. He has carefully inspected his stock rams and noted every defect and peculiarity of their fleeces and forms. The breeding register is before him to settle every pedigree, provided his stock rams are nearly enough connected with some portions of the flock to render it necessary to guard against in-and-in breeding. The shepherd catches a ewe and places her before him. The pedigree being determined, he first notes her form, and then opening the wool on the shoulder, thigh and belly, notes the length, thickness, quality, and style of the staple. If he kept the minutes at shearing recommended by me (under the head of "Selection") it will save much time and lead to far more accurate classification. When every point in the ewe is determined, he decides which ram, on the whole, is best calculated to perpetuate her excellencies both of fleece and carcass, and best counterbalance her defects in their mutual offspring. With a pigment composed of Venitian red and nog's lard, he then, as has been already mentioned under the head of Registering, gives the ewe a mark which will last until the next shearing, which will show by what ram she was tupped. Those selected for each ram are placed in *different* inclosures, and the chosen ram placed with them. In four weeks' time, the rams are withdrawn, and the flocks doubled or otherwise rearranged for winter, as may be necessary. This looks like taking considerable trouble, but having practiced it for years on my farm, and having always made these selections myself, I know that in reality the trouble is very slight—*nothing*, when the beneficial re-sults are taken into consideration. With a couple assistants, to catch, a day would suffice for effecting the proper classification and division of several hundred ewes.

Where choice rams are scarce, so that it is an object to make the ser-vices of one go a great way—or where it is impossible to have separate inclosures, (as on farms where there there are a great number of breeding-ewes, or where the shep-herd system is adopted to the exclusion of fences,) the following method may be resorted to. Build a hut containing as many apartments as you wish to use rams, with an alley between them. That part of fig. 30 which is surrounded by black lines repre-sents the hut divided into four apartments, each fur-nished with a feeding-box and trough in one corner. Gates or bars open from each apartment into the alley, and at *each end* of the alley. The dotted lines inclose a yard just sufficient to hold the flock of breeding-ewes.

Fig. 30.

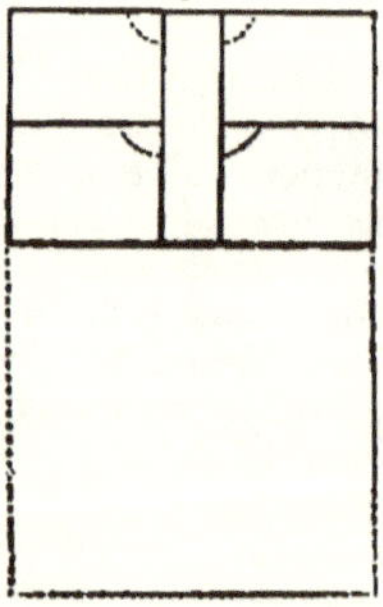

A couple of strong rams (of any quality) for about every hundred ewes, are then aproned, their brisk-ets rubbed with Venitian red and hog's lard, and they are let loose among the ewes. Aproning is performed by sewing a belt of coarse sacking broad enough to extend from the fore to the hind legs, loosely but strongly round the body. To prevent its slipping forward or back, straps are car-ried round the breast and back of the breech. It is indispensable that it be made *perfectly secure*, or all the labor of this method of coupling will be far worse than thrown away. The pigment on the brisket should be renewed every two or three days—and it will be necessary, usually, to change the "teasers," as these aproned rams are called, about once a week, as they do not long retain their courage under such unnatural cir-cumstances. Twice a day the ewes are brought into the yard in front of the hut. Those marked on their rumps by the teasers are taken into the alley. Each is admitted to the ram for which she is marked *once*, and then goes out at *the opposite end of the alley from which she entered*, into a separate field from that containing the flock from which she was taken. A powerful and vigorous ram from three to seven years old, and properly fed, can thus be made to serve from 150 to even 200 ewes, with no greater injury than from running loose with 50 or 60.

This, too, looks like a great amount of labor to attain the result sought, but having had it formerly practiced for two years on my farm, I know that when conducted with system, and by a prompt and handy shepherd, it consumes no great amount of time.

Rams will do better, accomplish more, and last two or three years longer, if daily fed with grain, when on service, and it is well to continue it, gradually decreasing the quantity, for a few days after they are withdrawn from the flock of ewes. A ram should receive the equivalent of from half a pint to a pint of oats, daily, when worked hard. They are much more conveniently fed when kept in huts. If suffered to run at large, they should be so thoroughly tamed that they will eat from a measure held by the shepherd. Careful breeders thus train their stock-rams from the time they are lambs. It is very convenient, also, to have them halter-broke, so that they can be led about without dragging or lifting them. An iron ring attached to one of the horns, near the point, to which a cord can be attached for leading, confining, &c., is very useful and handy. If rams are wild, it is a matter of considerable difficulty to feed them separately, and it can only be effected by yarding the flock and catching them out. Some breeders, in addition to extra feeding, take the rams out of the flocks nights, shutting them up in a barn or stable by themselves. There is no objection to this practice, and it is a great saving of their strength.

Rams should not be suffered to run with the ewes over a month, at least in the North. It is much better that a ewe go dry than that she have a lamb later than the first of June. And after the rutting season is over, the rams grow cross, frequently striking the pregnant ewes dangerous blows with their heavy horns, at the racks and troughs.

DIVISION OF FLOCKS.—If flocks are shut up in small inclosures during winter, according to the Northern custom, it is necessary to divide them into flocks of about 100 each, to consist of sheep of about the same size and strength. Otherwise the stronger rob the weaker, and the latter rapidly decline. This would not be so important where the sheep roam at large, but even in that case some division and classification are necessary, —or, at all events *best*. It is best, indeed, as already stated, even in summer. The poorer and feebler can by this means receive better pasture, or a little more grain and better shelter in winter.

By those who grow wool to any extent, breeding ewes, lambs, and wethers are invariably kept in separate flocks in winter; and it is best to keep yearling sheep by themselves with a few of the smallest two-year-olds, and any old crones which are kept for their excellence as breeders, but which cannot maintain themselves in the flock of breeding-ewes.

THE HOSPITAL.—Old and feeble, or wounded sheep, late-born lambs, etc., should be placed by themselves, if the number does not even exceed a score. They require better feed, warmer shelter, and more attention. But after all, unless the sheep are of a peculiarly valuable variety, it is better to sell them off in the fall at any price,—or to give them to some poor neighbor who has time to nurse them, and who may thus commence a flock.

YARDS.—Experience has amply demonstrated, that in the climate of the Northern and Eastern States—where no grass *grows* from four to four and a half months in the winter—and where, therefore, all that can be obtained from the ground is the repeatedly frozen, innutritious herbage left in the fall—it is better to keep sheep confined in yards, excepting where the ground is covered with snow. If suffered to roam over the fields at other

times, they get enough grass to take away their appetite for dry hay, but
not enough to sustain them; they fall away, and towards spring they become weak, and a large proportion of them frequently perish. I speak,
of course, of flocks of some size, and on properly stocked farms. A few
sheep, with a boundless range, would do better.

Some of our flock-masters let out their sheep occasionally for a single
day, during a thaw; others keep them entirely from the ground until let
out to grass in the spring. I prefer the former course, where the sheep
ordinarily get nothing but *dry* fodder. It affords a healthy laxative, and
a single day's grazing will not take off their appetite from more than one
succeeding dry feed. It is necessary, here, to keep the sheep in the yards
until the feed has got a good start in the spring, or they, particularly
breeding-ewes, will get off from their feed, and get weak at the most crit-
ical time for them in the year.

Yards should be firm-bottomed, dry,—and they should, (in *this* climate,)
be kept well littered with straw.

My impression is that the yarding system will never be practiced to any
extent in the South. It certainly should not be, where sheep can get their
living from the fields. How far, and under what circumstances, they will
do this, has already been sufficiently discussed in my preceding Letters.

FEEDING-RACKS.—When the ground is frozen, and especially when
covered with snow, the sheep eats hay better on the ground than anywhere
else. When the land is soft, muddy, or foul with manure, they will scarcely touch hay placed on it. It should then be fed in racks.

These are of various forms.
Figure 31 gives the common box
rack, in the most general use in
the North. It is ten feet long,
two and a half wide, the lower
boards a foot wide, the upper
ones about ten inches, the two
about nine inches apart, and the
corner posts three by three, or

Fig. 31.

BOX RACK.

three and a half by two and a half inches. The boards are spiked on these
posts by large flat headed nails wrought for the purpose, and the lower
edges of the upper boards and the upper edges of the lower ones are
rounded so they shall not wear the wool off from the sheep's necks. The
lower boards and the opening for the heads, should be two or three inches
narrower for lambs. If made of light wood, as they should be, a man
standing in the inside and middle of one of these racks, can easily carry it
about—an important desideratum. Unless overfed, sheep waste very little hay in them.

A capital shed or barn rack is represented in the following cut. The

Fig. 32.

HOLE RACK.

holes are eight inches wide, nine inches high, and eighteen inches from
center to center. Sheep do not crowd and take advantage of each other

ac much with these as with box racks. But they would be too heavy and unnecessarily expensive for a common out-door rack. Fig. 32 represents a *box*, the front formed of a board nailed on horizontally, but they are usually formed by nailing the boards perpendicularly, the bottoms on the sill of a barn, and the tops to horizontal pieces of timber.

In the South, as in England, racks will not be so necessary for that constant use to which they are put in colder countries, as for *depositories* of dry food, for the *occasional* visitation of the sheep. In soft warm weather, when the ground is unfrozen, and any kind of green herbage is to be obtained, sheep will scarcely touch dry fodder—though *the little they will then eat will be highly serviceable to them*. But in a sudden freeze, or on the occurrence of cold storms, they will resort to the racks, and fill themselves with dry food. By an instinct beautifully illustrative of the providence of the Creator, sheep anticipate the coming storm, and eat an extra quantity of food to sustain the animal heat, during the succeeding depression of temperature. They should always have racks of dry fodder to resort to in such emergencies.

These occasionally used racks should have covers or roofs to protect their contents from rain, as otherwise the feed would be often spoiled before but a small portion of it was consumed. Hay or straw saturated with water, or soaked and dried, is only eaten by the sheep as a matter of absolute necessity. The common box rack (fig. 31) would answer the purpose very well by placing on the top a triangular cover or roof formed of a couple of boards, (one hung at the upper edge with iron or leather hinges so that it could be lifted up like a lid;) making the ends tight; drawing in the lower edges of the sides so that it shall not be more than a foot wide on the bottom; inserting a floor; and then mounting it on and making it fast to two cross sills four or five inches square to keep the floor off from the ground, and long enough to prevent it from being easily overturned. The lower side board should be narrower than in fig. 31, on account of the increased hight given its upper edge by the sills.

Still better, but somewhat more expensive, would be a rack of the same construction, with the sides like those of fig. 32.

Or, the sides might consist of rundles as in fig. 33. In either of the preceding, the top might be nailed down, and the fodder inserted by little doors in the ends.

The following form and description of an English rack is from the "Book of the Farm."*

Fig. 33.

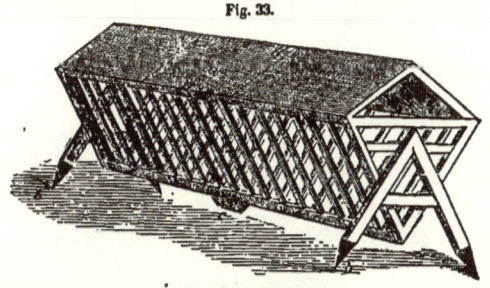

SPARRED RACK.

"I have found," says Mr. Stephens, "this form convenient, containing as much straw at time as should be given, admitting the straw easily into it, being easily moved about and

---

* It will be found in the reprint of this splendid work, in The Farmers' Library, vol. ii, p. 449.

easy access to the sheep, and being so near the ground as to form an excellent shelter. 1 is made of wood, is 9 feet in length, 4½ feet in hight, and 3 feet in width, having a sparred rack with a double face below, which is covered with an angled roof of boards to throw off the rain. The rack is supported on two triangular-shaped tressels *b*, shod with iron at the points, which are pushed into the ground, and act as stays against the effects of the wind from either side. The billet *c*, fixed on the under or acute edge of the rack, rests upon the ground, and in common with the feet, supports it from bending down in the middle. The lid *a* is opened on hinges when the fodder is put into the rack. . . . . Such a rack is easily moved about by two persons, and their position should be changed according to a change of wind indicative of a storm.'

I used racks formed of rounds (or "sparred") for several years, and found them decidedly *objectionable*. The sheep grasping a lock of hay in its mouth, brings the head to its *natural position*, and then draws in the *adhering* fibres in the process of mastication. But when eating from a rack, *it will not pick up the hay which it drops under foot*. In the box or hole racks (figs. 31 and 32) most sheep will not withdraw their heads from the openings, as they can there hold them in the ordinary position for mastication, and as, if they *step back* to do so, they are very liable to be crowded out of their places. The hay, therefore, is not drawn out of the rack, and if any is dropped, it falls within it and is saved. At a sparred rack, the sheep will not keep its nose between the rundles (in a horizontal or upward position) until it *detaches* a mere mouthfull of hay. It will particularly when partly sated, *twitch out* its fodder prior to mastication and all which scatters off and drops to the ground, is trampled under foot and wasted, except for the mere purpose of manure. A considerable loss will always result from this cause.

And there is another objection to this form of rack, particularly where it runs down to an acute edge on the bottom, as in fig. 33. The sheep frequently drawing the hay from the lower part, will shake down from above hay-seeds and chaff into the wool on their head and necks; and the wind will sometimes carry these as far as their shoulders and even their backs. As heretofore remarked, these cannot be washed out, and they materially lower the market value of the wool.

The following rack has been used and is highly approved by my friend George Geddes, Esq., of Fairmount, N. Y., to whom I am indebted fo the drawing and description of the cut. It serves both for a rack and feeding-trough.

Fig. 34.

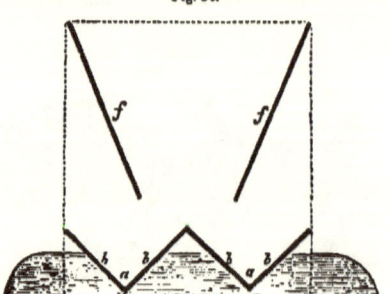

THE HOPPER-RACK.

" The above is intended to represent a section of what I think the best sheep-rack I have been.

"A piece of durable wood about 4½ feet long, 6 or 8 inches deep, and 4 inches thick has two notches, *a, a*, cut into it, and two troughs made of inch boards, *b, b, b, b*, placed in these notches, and nailed fast, constitutes the foundation. If the rack is to be 14 feet long, three sills will be required. The ends of the rack are made by nailing against the side of the sill-boards that reach up as high as it is desired to have the rack, and nails driven through these end-boards into the ends of the side-boards f, f, secure them. The sides may be farther strengthened by pieces of board on the outside of them, and fitted into the trough. A roof may be put over all if desired. With a roof, the fodder is kept entirely from the weather, and no seeds or chaff can get into the wool."

TROUGHS.—Threshed grain, chopped roots, &c., when fed to sheep should be laid in troughs. With any of the preceding forms of racks, ex cepting fig. 34, a separate trough would be required. For a number of years I have used those of the following form, and have found them every way satisfactory.

Fig. 35.

SHEEP-TROUGH.

One of the side-boards is usually about ten and the other eleven inches wide. The feet are commonly of two-inch plank, rising high enough on the sides to keep the sides of the trough firm in their places.

In our snowy climate they are turned over after feeding, and when falls of snow are anticipated, one end is laid on the yard fence.*

The following elaborately ingenious contrivance for keeping grain where sheep can feed on it *at will*, is from the "Book of the Farm," and I append the author's description of it.†

Fig. 36.   Fig. 37

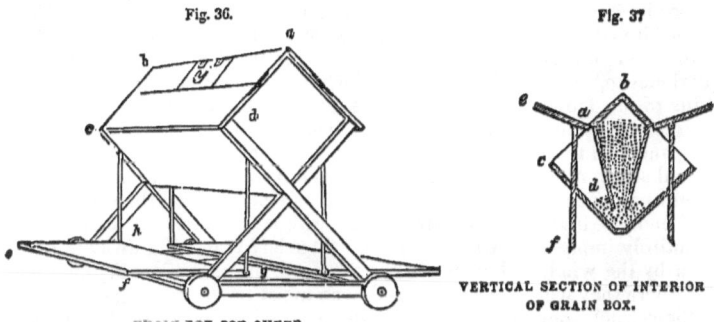

GRAIN BOX FOR SHEEP.

VERTICAL SECTION OF INTERIOR
OF GRAIN BOX.

"There is a mode of preserving corn (grain) for sheep on turnips which has been tried with success in Fife. It consists of a box like a hay-rack, in which the grain is at all times kept closely shut up, except when sheep wish to eat it, and then they get it by a simple contrivance. The box $a\,b$ contains the grain, into which it is poured through the small hinged lid y. The cover $c\,d$ concealing the grain, is also hinged, and when elevated the sheep have access to the grain. Its elevation is effected by the pressure of the sheep's fore-feet upon the platform $e\,f$, which, moving as a lever, acts upon the lower ends of the upright rods g and h, raises them up, and elevates the cover $c\,d$, under which their heads then find admittance into the box. A similar apparatus gives them access to the other side of the box. The whole machine can be moved about to convenient places by means of

* To you, Sir, living on the ocean shore of South Carolina, and who, I think, have not visited the North. in the depth of winter, the idea of a farmer's finding the racks used by him the day before, buried under from eighteen inches to three feet of snow, and having to *dig them out*, may be rather an odd one! But, nevertheless, it is a matter of no *very* rare occurrence, at least at the lowest depth mentioned
† See Farmers' Library, vol. ii., No. 10, p. 476.

four wheels. The construction of the interior of the box being somewhat peculiar, another fig. 37, is given as a vertical section of it, where *b* is the hinged lid by which the grain is put into the box, whence it is at once received into the hopper *d*, the bottom of which being open, and brought near that of the box, a small space only is left for the grain to pass into the box, the hopper forming the grain-store, *a* is the cover of the box raised on its hinges by the rod *f*, acted-upon by the platform *e f*, fig. 35; and, when in this position, the sheep put their heads below *a* at *e*, and eat the grain at *d*. Machines of similar construction to this have also been devised to serve poultry with grain at will."

I never have thought it best in feeding or fattening *any animals*, or, at all events, any quadrupeds, to allow them grain *at will*—preferring stated feeds; and the same remark is applicable to fodder. If this system is departed from in using depository racks, as heretofore recommended, it is because it is rendered necessary by the circumstances of the case. A Merino store-sheep, allowed grain *ad libitum*, would stand a chance to inflict an injury on itself, and I cannot but believe that grain so fed would generally be productive of more injury than benefit.

BARNS AND SHEDS, &c.—Sheep barns and sheds, at the North, are frequently made very elaborate contrivances—particularly on *paper*. But expensive barns, with feeding-cellars and other arrangements for keeping sheep *within doors* during a greater portion of the winter, would, it strikes me, be entirely out of place in the South. Even in our rigorous climate none but the breeders of Saxons pretend to make a regular practice or feeding under cover.

Humanity and economy both dictate, here, that sheep be provided with shelters to lie under nights, and to which they can resort *at will*. In our severe winter storms, it is sometimes necessary, or at least by far the best, to feed under shelter for a day or two. It is not an uncommon circumstance in New-York and New-England, for snow to fall to the depth of 20 or 30 inches within 24 or 48 hours, and then to be succeeded by a strong and intensely cold west or north-west wind of two or three days' continuance,* which lifts the snow, blocking up the roads, and piling huge drifts to the leeward of fences, barns, &c. A flock without shelter will huddle closely together, turning their backs to the storm, constantly stepping and thus treading down the snow as it rises about them. Strong, close-coated sheep do not seem to suffer as much from the cold, for a period, as would be expected. But it is next to impossible to feed them enough or half enough, under such circumstances, without an immense waste of hay —entirely impossible, without racks. The hay is whirled away in an instant by the wind, and even if racks are used, the sheep leaving their huddle where they were kept warm and even moist by the melting of the snow in their wool, soon get chilled and are disposed to return to their huddle. Imperfectly filled with food, the supply of animal heat is lowered, and at the end of the second or third day, the feeble ones have sunk down hopelessly, the yearlings and oldish ones have received a shock which nothing but careful nursing will recover them from, and even the strongest have suffered an injurious loss in condition.

Few holders of more than 40 or 50 sheep now attempt to get along here without some kind of shelters. The following (fig. 38) is a very common form of a Northern sheep-barn with sheds. The sheds front the south, or, what is a better arrangement, one fronts the east, and the other, being turned to a right angle to the direction of this, fronts the south. I have represented hole racks, as in fig. 32, running round the sheds, as although not yet in general use, they are undoubtedly the best in such sit

* These terrible wind-storms are of much longer continuance in many parts of New-England.

uations. The sheds are not usually framed or silled,—but are supported by posts of some durable timber set in the ground. The roofs are formed of boards "battened" with slabs. The barn has no partitions within, and is entirely filled with hay.

Fig. 38.

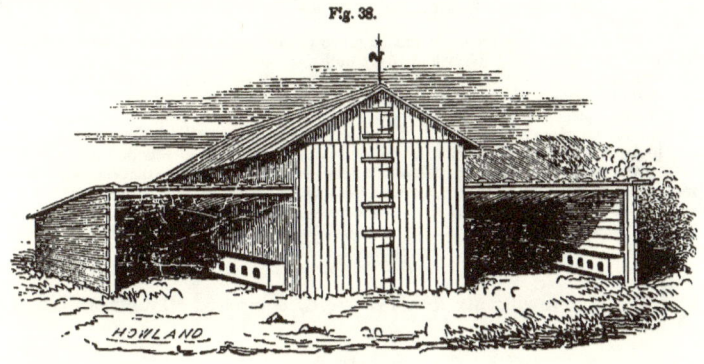

SHEEP-BARN.

There are many situations where these open sheds are very liable to have snow drifted under them by certain winds, and they are subject in all cases in severe gales, to have the snow carried *over* them to fall down in large drifts in front, which gradually encroach on the sheltered space, and are very inconvenient—particularly when they *thaw.* I therefore much prefer sheep-houses covered on all sides, with the exception of a wide door-way for ingress and egress, and one or two windows for ventilation when it is necessary. They are convenient for yarding sheep, for the various process· es where this is required, as for shearing, marking, sorting, "doctoring,"

Fig. 39.

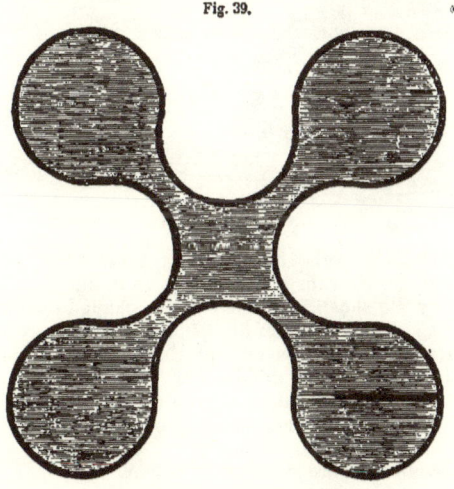

THE OUTSIDE STELL.

&c., and especially so, for lambing places or the confinement of newly shorn sheep in cold storms. They should be spacious enough, so that in addition to the outside racks, others can be placed temporarily throu the middle when required.

In many parts of Scotland, "Stells," as they are called, are made i of to shelter sheep. Fig. 39 on the preceding page is the form of one giv in " The Book of the Farm," and the author's description of it :

" In a storm, their provender cannot be given to the sheep upon snow, safely and conv iently, as ground-drift may blow and cover both ; and no place is so suitable for the purp as a *stell.* . . . . It may be formed of planting or high stone-wall. Either will afford sl ter ; but the former most, though most costly, as it should be fenced by a stone-wall. this class I conceive the form represented (fig. 39) a good one, and which may be ell acterized as an *outside* stell. . . . . The circumscribing strong black line is a stone-wall feet high ; the dark ground within is covered with trees. Its four rounded projections sll ter a corresponding number of recesses embraced between them, so that let the wind bli from what quarter it may, two of the recesses will be always sheltered from the storm. Tl size of this stell is regulated by the number of sheep kept ; but this rule may be reme bered in regard to its accommodation for stock, that each recess occupies about ¼ part of space comprehended between the extremities of the 4 projections ; so that in a stell cover 4 acres—which is perhaps the least size they should be, every recess will contain ½ an acr

The two following are forms of stells, composed of stone-wall, witho planting.

Fig. 40.

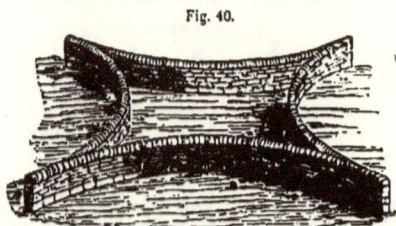

Fig 41.

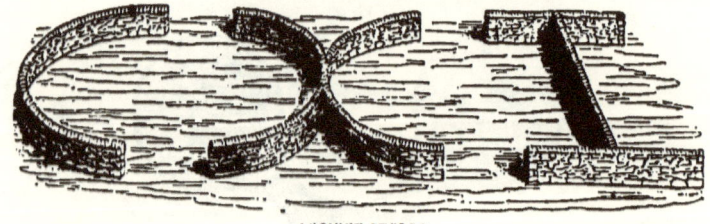

ANCIENT STELLS.

Figures 42 and 43, on the following page, are forms of circular stell the first made by stone-walls and planting, as in fig. 39. The open spac *a* is occupied by the sheep, and *b* is a funnel-shaped opening to it.

On the whole I should consider fig. 42 preferable to any of the precedir forms. Figure 43 represents one of the same form, but without tl planting, with a stack in the middle, &c. Either of the stells which a formed in part of trees, would be convenient in severe winds, would for excellent shades in summer, and would constitute highly ornamental o jects on the farm, and in the landscape. On the most northerly of tl Southern mountains, where considerable snow falls, they might even l good contrivances for winter shelter. They might also be convenient c the lowlands farther south, provided the shelter of evergreens could l made dense enough to protect the sheep from the winter *rains.* In th case, the stell or covert might be of any shape, and ought to have no ce al opening. It would be merely a dense clump of evergreen trees, for e sheep to take refuge under in storms of rain, and it might be surround-l on the outside with a tight board fence or stone-wall, if much exposed

Fig. 42.

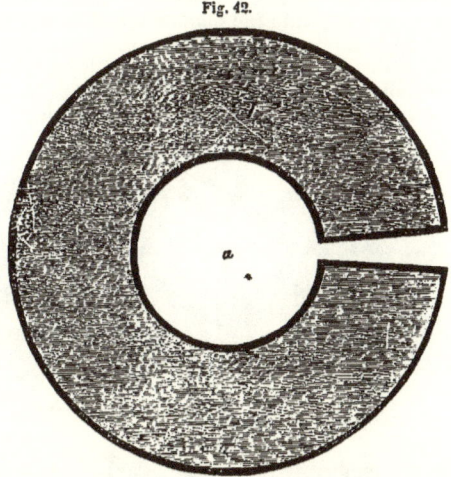

THE INSIDE CIRCULAR STELL.

the sweep of cold winds. As the sheep would lie among the trees, a imp 50 or 60 feet in diameter—though 100 feet would be better—would ffice for 100 sheep.

Fig. 43.

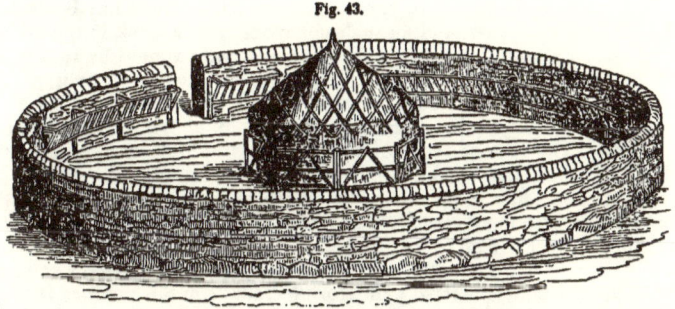

THE CIRCULAR STELL FITTED UP WITH HAY-RACKS.

But in determining upon the best winter shelters, for the various re ons in the South, the fact must not be lost sight of that *cold rains*, or ins of any temperature, *when immediately succeeded by cold or freezing ·ather, or cold, piercing winds*, are more hurtful to sheep than even snow-·rms—and that consequently sheep must be adequately guarded against em. There must also be suitable shelter from *any* storms to which the untry is subject, *in the lambing season.* Any person with the least ex-·rience can determine whether an inclosed clump of trees will answer ese purposes, in his own immediate region.

I think it very probable that in the Gulf States, and some of the lower tlantic ones—particularly in regions near the ocean—these tree coverts,

surrounded by fences to break the winds, would be found sufficient.
sections infested with wolves, they might also be made to answer for *fol*,
by carrying the fence to the requisite hight, to bar the ingress of the wo
But farther north, and on the high lands and mountains, better shelte
would, I am inclined to think, in the end, be found more economical.

The simplest and cheapest kind of shed is represented in the followin
cut (fig. 44). It is formed by poles or rails, the upper ends resting on
strong horizontal pole supported by crotched posts set in the ground.
may be rendered rain-proof by pea-haulm, straw, or pine boughs.

Fig. 44.

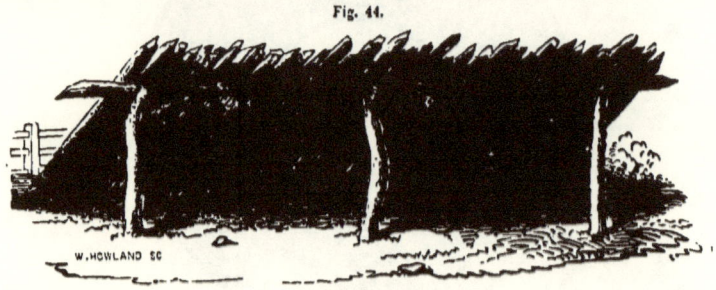

W.HOWLAND SC

SHED OF RAILS.

In a region where lumber is very cheap, planks or boards (of sufficien
thickness not to spring downward and thus open the roof) battened wit.
slabs, may take the place of the poles and boughs; and they would mak
a tighter and more durable roof. If the lower ends of the boards or pole
are raised a couple of feet from the ground, by placing a log under them
the shed will shelter more sheep.

These movable sheds may be connected with hay-barns, "hay-barracks,'
stacks, or they may surround an inclosed space with a stack in the middl
like fig. 43. In the latter case, however, the yard should be *square*, in
stead of *round*, on account of the divergence in the lower ends of th
boards or poles, which the round form would render necessary.

Sheds of this description are frequently made, in the North, betwee
two stacks. The end of the horizontal supporting pole is placed on th
stack-pens, when the stacks are built, and the middle is propped b
crotched posts. The supporting-pole may rest, in the same way, on th
upper girts of two hay-barracks; or two such sheds (at angles with eacl
other) might form wings to this structure. The "barrack," as it is pro
vincially termed in the North, would, it strikes me, afford a most econom
ical and a most convenient way of storing fodder in the South. It is ea
sily movable, so that it possesses the same advantage that stacks do, i
manuring different parts of the field or farm. On the other hand, the fod
der cannot be drenched by a winter rain, as in a partly fed out stack
Hay can be more rapidly stored in it than on a stack at any time, and yo
can pitch into it to the last moment, when threatened with rain, withou
stopping to round up the top as is necessary in a stack. The outside i
not weather-beaten and damaged, as is the case with the sides, and fre
quently with a considerable of the top of a stack. Fig. 45 (on the nex
page) represents the form of a barrack. It is 12 feet square on the bot
tom, and the frame is formed by girting together four strong poles, 16 fee
long, at the bottom, and 6 feet from the bottom. Boards 6 feet long ar
nailed perpendicularly on the girts. Two-inch holes are bored at con
venient distances through the corner poles, so that the roof, which rest

on pins thrust through these holes, can be raised or lowered, at pleasure. It is occasionally lowered as the fodder gets lower in the barrack, so that rain or snow shall not drive under it. It seems to me that this structure would be remarkably well adapted to the storing and feeding out of un-threshed peas, which, as has been remarked in a former Letter, are so advantageously raised at the South, and constitute so admirable a feed for sheep.

Fig. 45.

BARRACK.

On all large sheep-farms convenience requires that there be one barn of considerable size, to contain the shearing-floor, and the necessary conveniences about it for yarding the sheep, &c. This should also, for economy, be a hay-barn, (where hay is used,) and from its necessary size (for the shearing-floor), it should hold hay for 400 sheep. It may be constructed in the corner of four fields, so that four hundred sheep can be fed from it, without making improperly sized flocks. At this barn it would be expedient to make the best shelters, and to bring together all the breeding-ewes on the farm, if their number did not exceed 400. Thus the shepherd would be saved much travel at all times, and particularly at the lambing-time, and each flock would be under his almost constant supervision.

I offer the following ground-plan of a barn with fixtures, &c., as one which I think will be found well adapted to the purpose above specified. The upper is the north part of the plan.

Fig. 46.

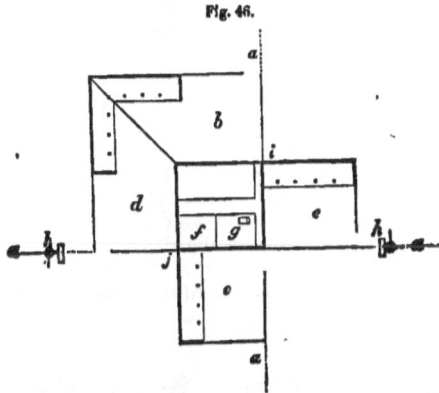

The dotted lines *a, a, a, a,* are the fences dividing four fields, which would corner at the south-east corner of the barn. The barn is surrounded by double lines, and the sheds by double lines on the backs and ends—the dots in front of them, representing the crotched posts supporting their front. The single black lines round the yards, represent tight board fences, which screen the four yards *b, c, d, e,* from every wind. There are two pumps and troughs at *h, h,* which accommodate the whole four fields, if a want of springs or streams in them render these necessary. The sheds are so ar-

2 D

ranged that even without the screens they entirely shut out the north and west winds—the prevailing and severe winter ones of *this* region—and if other ones are more prevalent in other regions, the sheds can be changed accordingly. Each of the sheds is 50 feet long and 12 feet wide—six square feet being the smallest *proper* allowance of sheltered area for *each* sheep. The barn is 48 feet square, a floor 13 feet wide running east and west through the center, for shearing and for the drawing in of hay. An alley 4 feet wide and 8 feet high (boarded up on the side toward the mow, and covered at the top) cuts off the lower part of each bay from the east wall of the barn. This is for carrying hay into the yards *b, c*. It is carried into the yards *d, e*, from the large doors at each end of the shearing-floor (or from smaller ones cut through them.) The south bay is represented as divided by a temporary fence, cutting it into two pens *f, g*. The outside inclosure *e*, for yarding the sheep, communicating by a door with *f*, and *g* being used as a room to tie up wool in, presents precisely the same arrangement which is exhibited in the cut of the shearing-barn (fig. 22) in Letter XII.

The barn here given (fig. 46) is probably larger than would be necessary for 400 sheep, in most parts of the South. Its necessary size is a question to be entirely determined by the climate. For large flocks of sheep, I should regard the storage of *some* hay or other fodder for winter as an indispensable *precautionary* measure, at least, in *any* part of the United States; and, other things being equal, the farther north, or the more elevated the land, the greater would be the necessary amount to be stored.

The shearing-floor shortened to 30 or 35 feet, would still, perhaps, be sufficiently commodious, and this would reduce the dimensions of the barn east and west 13 or 18 feet; and one of the bays might be dispensed with. But having constructed so large, so smooth, and so tight a barn-floor as the shearing one ought to be, it would be good economy to use it for the threshing of grain. One of the bays, therefore, might be used for the storage of grain in the sheaf. I have always considered this an excellent arrangement in a *Northern* barn of this description, as in our cold climate the sheep require much straw litter in their sheds, yards, &c. Thrown out to them daily, as threshed, much bright straw and chaff will be consumed by them—particularly of greenish cut oats.

The yards *c, e*, in fig. 46 are represented but the width of the barn, 48 feet. If these were reduced too much, by diminishing the size of the barn, the shed of *c* could be carried farther west at *j*, and that of *e* farther north at *i*, being connected with the barn by wind-breakers, composed of a tight board fence, as high as the summit of the sheds. Or, what would perhaps be better, the fences thrown forward in a straight line from the *ends* of these two sheds might be continued until they intersected each other, and a fence from their point of intersection to the south-east corner of the barn would divide the two yards.   —

FEEDING SHEEP WITH OTHER STOCK.—Sheep should not run or be fed, *in yards*, with any other stock. Cattle hook them, often mortally. Colts tease and frequently injure them. It is often said that "colts will pick up what sheep leave." Well-managed sheep rarely leave anything—and if they chance to. it is better to rake it up and throw it into the colts'yard, than to feed them together. If sheep are not required to eat their feeds pretty clean, they will soon learn to waste large quantities. But if sheep are overfed with either hay or grain, it is not proper to compel them by starvation to come back and eat it. They will not unless sorely pinched. Clean out the troughs,—or rake up the hay, and the next time feed less.

HAY-HOLDERS.—Where hay or other fodder is thrown out of the upper doors of a barn into the sheep-yard, as it always must necessarily be in a barn constructed like fig. 38, or any *mere hay*-barn, or where it is thrown from a barrack or stack, the sheep immediately rush on it, trampling it and soiling it, and the succeeding forkfulls fall on their backs, filling their wool with dust, seed and chaff. This is avoided by hay-holders—yards 10 feet square—either portable by being made of posts and boards, or simply a pen of rails, placed under the doors of the barns, and by the sides of each stack or barrack. The hay is pitched into the holder, in fair weather enough for a day's foddering at a time, and is taken from this by forkfulls and placed in the racks. I would here offer a necessary caution in relation to the use of *rails* or *poles*, for stack-pens or hay-holders. The poles should be so small as to entirely prevent the sheep from inserting their heads between them after hay. A sheep will often insert its head where the opening is wide enough for that purpose, shove it along or get crowded along, to where the opening is not wide enough to withdraw the head, and it will hang there until observed and extricated by the shepherd. If, as it often happens, it is thus caught when its fore parts are elevated by climbing up the side of the pen, it will continue to lose its fore footing in its struggles, and will soon choke to death.

WINTER DRY FEED FOR SHEEP.—The proper dry winter fodder for sheep has already been repeatedly alluded to, in general terms. Volumes have been expended on this subject, particularly in Germany—and curious and elaborate systems of feeding given. In Germany great stress is laid on *variety* in the winter fodder. In the German Farmer's Encyclopædia, the following table of the proper variations and amounts of feed is given by PETRI.

TABLE 15.

Day.	Lbs.	Loth, equal ½ oz.	Morning.	Lbs.	Loth, equal ½ oz.	Noon.	Lbs.	Loth, equal ½ oz.	Evening.
1		21	hay		21	hay		21	hay
2	1	1	rye straw	1	22	hay	1	1	rye straw
3		23	bean straw		26	vetch-hay		23	bean straw
4	1		wheat straw	1		sainfoin	1		wheat straw
5	1	6	oat straw		21	hay	1	6	oat straw
6	1	6	artichoke stalk	1	19	red clover	1	6	artichoke stalk
7	1	8	turkey wheat	1	12	lucern	1	8	turkey-wheat str'w
8	1	8	buckwheat straw	1	16	hay	1	8	buckwheat straw
9	1	6	oat straw		7	horse-beans	1	6	oat straw
10		19	red clover		19	red clover		19	red clover
11		18	sainfoin		18	sainfoin		18	sainfoin
12	1	6	millet straw	1	6	millet straw	1	6	millet straw
13		30	lentil straw		21	hay		30	lentil straw
14		30	pea straw		21	hay		30	pea straw
15		30	barley straw	1		artichoke stalk		30	barley straw
16	1	10	horse-bean straw	1	10	horse-bean straw	1	10	horse-bean straw
17	1	1	rye straw	1	11	oat straw	1	1	rye straw
18	1	3	wheat straw	1	9	oat straw	1	3	wheat straw
19	1	6	rye straw	1		turkey-wheat	1	3	wheat straw
20	1	6	oat straw	1		turkey-wheat	1	6	oat straw
21	1	3	wheat straw		22	artichoke stalk	1	6	oat straw
22		30	lentil straw	1	30	vetch straw		30	lentil straw
23	1	6	oat straw	1	6	wheat straw	1	6	oat straw

The same writer gives the following as the proper winter feed of a ewe, the month preceding lambing :

TABLE 16.

1st day..	In the morning,	¾ lb.	of good oat straw.
	.. noon	½ ..	of good hay of clover.
	.. evening..	¾ ..	of good barley straw.
2d day ..	.. morning .	¾ ..	of millet straw.
	.. noon	2 ..	of potatoes with 4 oz. of chopped straw, and 4 oz. of oats.
	.. evening..	¾ ..	of barley straw.
3d day ..	.. morning .	¾ ..	of hay.
	.. noon	½ ..	of hay,
	.. evening..	1 ..	of wheat, oat, barley or buckwheat straw.
4th day..	.. morning .	¾ ..	of summer straw.
	.. noon	½ ..	of chopped straw, with 3 oz. oats and 3 oz. bran, moistened with water.
	.. evening,.	¾ ..	of winter straw.
5th day..	.. morning .	¾ ..	of hay.
	.. noon	2 ..	of potatoes with ½ lb. of chopped straw.
	.. evening..	¾ ..	of winter straw.
6th day..	.. morning .	¾ ..	of hay.
	.. noon	as in 4th day.	
	.. evening..	1 lb.	of straw.

All this would be infinitely "more nice than wise," in any part of the United States. Variations of dry fodder are well enough, but hundreds and thousands of Northern flocks receive nothing but ordinary hay, consisting mainly of Timothy, *(Phleum pratense,)* some Red and White Clover, *(Trifolium pratense et repens,)* and frequently a sprinkling of June or Spear grass, *(Poa pratensis,)* during the entire winter. Others receive an occasional fodder of corn-stalks and straw—and some farmers give a daily feed of grain through the winter. Where hay is the principal feed, it may be well, where it is convenient, to give corn-stalks (or "blades") every fifth or sixth feed, or even once a day. Or the daily feed, *not of hay*, might alternate between blades, pea-straw, straw of the cereal grains, &c. Should any other fodder besides hay be the principal one, as, for example, corn blades or pea-haulm, each of the *other* fodders might be alternated in the same way. It is mainly, in my judgment, a question of convenience with the flock-master, provided a *proper supply of palatable nutriment within a proper compass*, is given. Hay, clover, properly cured pea-haulm, and corn-blades are palatable to the sheep, and each contain the necessary supply of nutriment in the quantity which the sheep can readily take into its stomach. Consequently, from either of these, the sheep can derive its entire subsistence. The same remarks *may*, possibly, apply to greenish cut oat and barley straw; but it would not, I apprehend, be economical or altogether safe to *confine* any kind of sheep to the *straw* of the cereal grains unless some of those little hardy varieties of sheep which would be of no value in this country. Experiment will readily show the flock-master what kinds of food are palatable and agree with the health of his flock. The following exceedingly valuable Table, prepared by Boussingault, will give the value of various kinds of feed in comparison with ordinary natural meadow hay, as ascertained by himself, Von Thaër, Block, and other distinguished Agricultural Chemists. The results are obtained by chemical analysis, and by actual experiments in feeding. The amount of nitrogen in 100 parts is made the chemical test of value, as it shows the quantity of fibrin, albumen, and casein, (by multiplying by 6.3.) The experimental result is obtained by weighing the animal and the feed, and giving him enough of each to maintain him in good condition.

TABLE 17.
FODDERS.
TABLE OF THE NUTRITIVE EQUIVALENTS OF DIFFERENT KINDS OF FODDERS.

Kinds of Food.	Standard water per cent.	Nitrogen per cent.	Nitrogen per cent. in the article not dried.	Theory.	Block.	Petri.	Mayer.	Thaer.	Pabst.	Remarks.
Ordinary natural meadow hay....	11.0	1.34	1.15	100	100	100	100	100	100	
Do. of fine quality..............	14.0	1.50	1.30	98						
Do. select....................	18.8	2.40	2.00	58						
Do. freed from woody stems.....	14.0	2.44	2.10	55						
Lucern hay....................	16.6	1.66	1.38	83		90		90	100	90, 90 Dombasle,
Red clover hay, 2d year's growth.	10.1	1.70	1.54	75	100	90		90	100	[Crud.
Red clover cut in flower, green, do.	76.0		0.64	311	430			450	425	
New wheat straw, crop 1841......	26.0	0.36	0.27	426	200	360	150	450	300	500 Rieder.
Old wheat straw................	8.5	0.53	0.49	235						
Do. do. lower parts of the stalk...	5.3	0.43	0.41	280						
Do. do. upper part of do. and ear..	9.4	1.42	1.33	86						
New rye-straw.................	18.7	0.30	0.24	479	200	500	150	666		
Old do.......................	12.6	0.50	0.42	450						
Oat-straw....................	21.0	0.36	0.30	383	200	200	150	190	200	400 Schwertz.
Barley do....................	11.0	0.30	0.25	460	193	180	150	150	200	400 do.
Pea do......................	8.5	1.95	1.79	64	165	200	150	130	150	90 Pohl.
Millet do....................	19.0	0.96	0.78	147		250				
Buckwheat do.................	11.6	0.54	0.48	240		200				
Lentil do....................	9.2	1.18	1.01	114	160	200		130	150	
Vetches cut in flower and dried into hay..................	11.0	1.16	1.14	101		125			100	
Potato tops..................	76.0	2.30	0 55	209		300				
Field-beet leaves.............	88.9	4.50	0.50	230	600			600		
Carrot do....................	70.9	2.94	0.85	135						
Jerusalem artichoke stems......	86.4	2.70	0.37	311				325		
Lime-trees, young shoots........	55.0	3.25	1.45	79	73					
Canada Poplar do.............	62.5	2.29	0.86	134	67					
Oak do......................	57.4	2.16	0.92	125	83					
Acacia do. (autumn)...........	53.6	1.56	0.72	160						
Drum cabbage................	92.3	3.70	0.28	411	556	500	250	429	600	
Swedish turnip...............	91.0	1.83	0.17	676		300		300	250	
Turnip......................	92.5	1.70	0.13	885	533	600	290	526	450	
Field-beet (1838).............	87.8	1.70	0.21	548	366	400	250	460	250	
Do. white Silesian............	85.6	1.43	0.18	669	366					
Carrots.....................	87.6	2.40	0.30	382	205	250	225	300	250	380 Boussingault.
Jerusalem artichokes (1839).....	79.2	1.60	0.33	348						280 do.
Do. (1836)...................	75.5	2.20	0.42	274						
Potatoes (1838)..............	65.9	1.50	0.36	319	216	200	150	200	200	280 Boussingault.
Do. (1836)...................	79.4	1.80	0.37	311						
Do. after keeping in the pit......	76.8	1.18	0.30	383	400					
Cider apple pulp dried in the air..	6.4	0.63	0.59	195						
Beet-root from the sugar mill....	70.0		0.38	303						
Vetches in seed...............	14.6	5.13	4.37	26	30	54		66	40	
Field-beans	7.9	5.50	5.11	23	30	54	50	73	40	
White peas (dry)..............	8.6	4.20	3.84	27	30	54	48	66	40	
White haricots...............	5.0	4.30	4.58	25		39				
Lentils......................	9.0	4.40	4.00	29						
New Indian Corn..............	18.0	2.00	1.64	70		52				59 Boussingault.
Buckwheat	12.5	2.40	2.10	55		64				
Barley (1836)................	13.9	2.02	1.76	65	33	61	53	76	50	
Barley-meal.................	13.0	2.46	2.14	54						
Oats (1838).................	20.8	2.20	1.74	68		71		86	60	
Do. (1836)..................	12.4	2.22	1.92	60						
Rye (1838)..................	11.5	2.27	2.00	58						
Wheat (1836, Alsace)..........	10.5	2.33	2.09	55	27	52	46	64	40	
Do. from highly manured soil....	16.6	3.18	2.65	43						
Recent Bran.................	37.1	2.18	1.36	85	105					{ Some specimens { are twice as rich.
Wheat husks or chaff..........	7.6	0.94	0.85	135	160					
Rice (Piedmont)..............	13.4	1.39	1.20	96						
Gold of Pleasure seed (Madia)....	8.0	4.00	3.67	31						
Do. cake....................	11.2	5.70	5.06	23						
Linseed cake................	13.4	6.00	5.20	22	42	180				
Colza do....................	10.5	5.50	4.92	23						
Madia do....................	6.5	5.93	5.51	21						
Hemp do....................	5.0	4.78	4.21	27						
Poppy do...................	6.8	5.70	5.36	21						
Nut do......................	6.0	5.59	5.24	22						
Beech-mast do...............	6.2	3.53	3.31	35						
Arachis (Pindars) do...........	6.6	8.89	8.33	14						
Dry acorns..................			0.80	143						
Refuse of the wine-press, air-dried	48.2	3.31	1.71	68		62			75	

The great value of pea-haulm, as shown in the above Table, is worthy of the particular notice of the Southern flock-master. Also that of millet straw, another crop peculiarly congenial to the Southern States, provided it can be cured so that sheep will eat it. Corn-stalks are not, unfortunately, included in the Table. According to Petri, 100 *pounds* of corn "straw," (including *stalks* and *leaves*, I suppose,) contains but ⅔ as much nutriment as the same weight of " aromatic meadow hay," and not so much by ¼ as an equal weight of oat or pea straw, which *he* makes *equivalent* to each other! My opinion is that this by no means indicates the comparative value of *well cured* corn-stalks. No analysis of them now occurs to me, in any authority which I have on hand. Mr. Ellsworth, of the Patent Office, stated in the Cultivator in 1842, that the *juice* of corn-*stalks*, on Beaumé's Saccharometer, is equal in saccharine matter with that of the cane in this country, five times greater than that of the Northern sugar-maple, *(Acer saccharinum,)* and three times that of beet! The daily experiments of our farmers demonstrate the absurdity of placing corn-stalks below the value of the cereal straws. Cured green and bright they are a highly valuable fodder, and are relished by all herbivorous animals. My friend, James M. Ellis, Esq. of Onondaga, N. Y., one of the best managing flock-masters of this State, has fed corn-stalks largely to his sheep for several years and with decided succes.

——

EFFECT OF FOOD IN THE PRODUCTION OF WOOL.—The fact has been before alluded to that well fed sheep produce more wool than poorly fed ones. The question now arises—if the effect on the condition (flesh) of the sheep is the same, will one kind of food produce more wool than another? No doctrine is more clearly recognized in Agricultural Chemistry, than that animal tissues derive their chemical components from the same components existing in their food.[*] The analyses of Liebig, Johnston, Scherer, Playfair, Boeckmann, Mulder, &c., show that the chemical composition of wool, hair, hoofs, nails, horns, feathers, lean meat, blood, cellular tissue, nerves, &c. are nearly identical. The organic part of wool, according to Johnston,[†] consists of carbon 50.65, hydrogen 7.03, nitrogen 17.71, oxygen and sulphur 24.61. The inorganic constituents are small. When burned, it leaves but 2·0 per cent of ash. The large quantity of nitrogen (17.71) contained in wool, shows that its production is increased by highly azotized food. This is fully verified by the experiments made on Saxon sheep, in Silesia, by Reaumur, whose Table I append. A striking correspondence will be found to exist between the amount of wool and the amount of nitrogen in the food.

TABLE 18.

Kinds of food.	Increase of weight in live animal.	Produced wool.		Produced tallow.		Nitrogen per cent. in food.
		lbs.	ns.	lbs.	oz.	
1000 pounds of raw potatoes, with salt.......	46½	6	8¾	12	5½	0.36
1000 " " " without salt......	44	6	8	10	14¾	0.36
1000 .. raw mangel-wurzel	38	5	3½	6	5¾	0.21
1000 .. peaso	131	14	11	41	6	3.63
1000 .. wheat	155	13	13¾	59	9	2.09
1000 .. rye, with salt.................	90	13	14½	35	11½	2.00
1000 .. rye, without salt.............	83	12	10½	33	8½	2.00
1000 .. oats.........................	146	9	12	40	8	1.70
1000 .. barley.......................	136	11	6½	60	1	1.00
1000 .. buckwheat....................	120	10	4½	33	8	2.10
1000 .. good hay.....................	58	7	10½	12	14	1.15
1000 .. hay, with straw, without other fodder	31	15	8	6	11	
1600 .. whisky, still-grains or wash....	35	6	1	4	0	

[*] For full information on this whole subject, see Liebig's Animal Chemistry, Part I and II.
[†] See Johnston's Agricultural Chemistry—Lecture XVIII. Analyses of the horny tissues, by Scherer, will be found in the Appendix to Liebig's Animal Chemistry.

The singular difference stated in the Table, between the amount of wool produced by "good hay," and "hay with straw without other fodder," I confess is scarcely credible to me. It may be a misprint in the Table from which I copy.

The peculiar value of pease not only in increasing the wool, where they rank *first*,[*] but in the average comparative increase which they produce in *all* the tissues, is again worthy of notice.

EFFECT OF FOOD IN PRODUCING FAT AND MUSCLE.—The increase of fat and muscle, as of wool, depends upon the nature of the food. It would be foreign from my purpose to enter into an elaborate theoretical examination of this subject. Liebig, in Parts I. and II. of his Animal Chemistry, has covered the whole ground, and to him I take the liberty to refer you. Mr. Spooner, writing for England, where the production of flesh and fat is the primary object of Sheep Husbandry, has given a synopsis of Liebig's positions, analyses, &c., in his chapter (XXI.) on Feeding and Fattening—and the substance of this is again repeated by Mr. Morrell in his chapters on the same subjects, in The American Shepherd. To either of the latter I would refer you for *sufficient* details for practical purposes, or for *full* information, to Liebig.

The Tables of Boussingault and Reaumur, already given, (Tables 17 and 18,) sufficiently indicate the value of the various grains, straw, roots, &c., in fattening.

It is not very common, in the North, for *wool-growers* to fatten their wethers, for market, by extra winter feeding. Some give them a little more generous keep the winter before they are to be turned off, and then sell them when they have attained their maximum fatness the succeeding fall. When winter fattening is attempted, sheep require warm, dry shelters, and should receive, in addition to all the hay they will eat, meal twice a day in troughs—or meal once and chopped roots once. The equivalent of from half a pint to a pint of (yellow) corn meal per head per diem is about as much as ordinary flocks of Merino wethers will profitably consume, though in selected flocks consisting of large animals, this amount is frequently exceeded. —

FEEDING GRAIN TO STORE-SHEEP IN WINTER.—The expediency of feeding grain to store-sheep in winter depends much upon circumstances. If in a climate where they can obtain a proper supply of grass or other green esculents, it would, of course, be unnecessary. Neither is it a matter of *necessity* where the ground is frozen or covered with snow for weeks or months, provided the sheep be supplied plentifully with *good* dry fodder. Near markets where the coarse grains find a good and ready sale, it is not usual in the North, to feed grain. Remote from markets, it is generally fed by the holders of large flocks. Oats are commonly preferred, and they are fed at the rate of a gill a head per day. Some feed half the same amount of (yellow) corn. Fewer sheep—particularly lambs, yearlings, and crones—get thin and perish, where they receive a daily feed of grain; they consume less hay; and their fleeces *are increased in weight*. On the whole, therefore, it is considered good economy. Where no grain is fed, three daily feeds of hay are given. It is a common and very good practice to feed greenish cut oats *in the bundle*, at noon, and give but two feeds of hay—one at morning and one at night. A few feed greenish cut peas in the same way. In warm, thawing weather when sheep get

---

[*] With the exception of "hay and straw"—the given product of which, in wool, I have already stated must undoubtedly be misprinted.

to the ground, and refuse dry hay, a little grain assists materially in keeping up their strength and condition. This may furnish a useful hint for many parts of the South. When the feed is shortest in winter, in the South, there are many localities where sheep would get enough grass to take off their appetite for dry hay, but not *quite* enough to keep them in prime condition. A moderate daily feed of oats or pease placed in the depository racks, would keep them strong, in good plight for the lambing season, and increase their weight of wool.

Few Northern farmers feed Indian corn to store-sheep. It is considered " too hot and stimulating," and sheep are thought to be more liable to become " cloyed " on it than on oats, pease, &c. I never have fed it to sheep sufficiently to speak advisedly on this point. A neighboring flock-master whose admirable arrangements for keeping sheep are only equaled by his usual success, lost most of a large flock of lambs a few winters since. They received all they would eat of the best hay, and, as the owner *supposed*, a half gill of corn a head per day. They were in fine order in the beginning, and for some time into the winter. During a thaw, when they got a little off from their feed, and looked " hollow," the shepherd, without the knowledge of the owner, increased the feed of corn. This caused them to eat still less hay, and the shepherd not only continued but increased the allowance of the corn as their appetite for hay diminished. In a short time they ate scarcely any hay, and soon after began to eat their corn very irregularly. Their stomachs were now so completely deranged, that they would not eat *anything*, in quantities sufficient for their subsistence, and they perished rapidly and miserably. The same consequences might doubtless have ensued from feeding *other grains*, in the same improper manner. But I am inclined to think that the evil would have been less rapid and remediless with some other grains. I do *not* consider yellow corn a very safe feed, at least for lambs and yearlings. From the obviously different character of the larger Southern varieties, I presume they would be less, and very probably not at all, objectionable for sheep feed. Half a gill of yellow corn, or a gill of oats per head, is a sufficient daily allowance of grain. While there can be nothing more absurd than the German *starving* system to increase the fineness of the wool, excessive fatness is not to be aimed at, especially in breeding-ewes. Store sheep should be kept in good, fair, *plump* condition. Lambs and yearlings may be as fat as they will become on *proper feeding*.

It will not do to suffer sheep to get thin in the winter, with the idea that their condition can at any time be readily raised by better feed, as with the horse or ox. It is always difficult, and unless properly managed, expensive and hazardous, to attempt to raise the condition of a poor flock in the winter—especially if they have reached that point where they manifest *weakness*. If the feeding of a liberal allowance of grain be suddenly commenced, fatal diarrhea will frequently supervene. All extra feeding, therefore, must be begun very gradually, and it does not seem, in any case, to produce proportionable results.

I have seen it stated that sheep will eat cotton-seed and thrive on it. If this be true, this must, of course, be a far more remunerating application of that product, than as a mere *manure* to soils.

FEEDING ROOTS, BROWSE, &c., IN WINTER.—Ruta-bagas, Irish potatoes, &c., make a good substitute for grain, as an extra feed for grown sheep. I prefer the ruta-baga to the potato in equivalents of nutriment. I do not consider either of them, or any other root, as good for lambs and yearlings as an equivalent in grain. Sheep may be *taught* to eat nearly all the cul tivated roots. This is done by withholding salt from them, and then feed-ing the chopped root a few times rubbed with just sufficient salt to induce them to eat the root to obtain it; but not enough to satisfy their appetite for salt before they have acquired a taste for the roots.

It is customary with some of our flock-masters to cut down from time to time, in the winter, and draw into the sheep-yards, young trees of the hemlock *(Abies canadensis)*. The foliage is greedily eaten by sheep, af-ter being confined for some time to dry feed. I have known sheep, un-doubtedly, I think, killed by overeating it. This browse is commonly used for some supposed medicinal virtues. It is pronounced "healthy for sheep." The popular supposition is that it is a tonic and stimulant. If this be true, which I will not pause to inquire, of what *good* use are tonics and stimulants to *healthy* animals? With sheep, as with horses, and even with men, *preventive* medicines are productive of injury in a thousand cases, where they are of benefit in one. There could be no objection, cer-tainly, to sheep's eating the foliage of the hemlock, if it was constantly accessible to them. Their instincts, in that case, would teach them whether, and in what quantities, to devour it. But when entirely confined to dry feed for a protracted period, sheep will consume hurtful and even poisonous succulents—and of the most wholesome ones, hurtful *quantities*. As a mere *laxative*, an occasional feed of hemlock may be beneficial; but in this point of view, a day's run at grass in a thaw, or a feed of roots, would produce the same result. In a climate where grass is obtained most of the time, I should consider browse for medicinal purposes entirely unnecessary. —

WINTER FEED OF BREEDING-EWES.—Until two or three weeks pre-ceding lambing, it is only necessary that breeding-ewes, like other store-sheep, be kept in good plump ordinary condition. Nor are any separate arrangements necessary for them, after that period, in a climate where they obtain sufficient succulent food to provide for a proper secretion of milk. In backward seasons in the North, where the grass does not start prior to the lambing time, careful flock-masters feed their ewes chopped roots, or roots mixed with oat or pea meal. This is, in my judgment, excellent economy.* —

REGULARITY IN FEEDING.—If there is one rule which may be consider-ed more imperative than any other in Sheep Husbandry, it is that the ut-most regularity be preserved in feeding. First, there should be regularity as to the *times* of feeding. However abundantly provided for, when a flock are foddered sometimes at one hour and sometimes at another—sometimes three times a day and sometimes twice—some days grain and some days none—*they cannot be made to thrive.* They will do far better on *inferior keep*, if fed with strict regularity. In a climate where they re-quire hay three times a day, the best times for feeding are about sunrise in the morning, at noon, and an hour *before dark* at night. Unlike cattle and horses, sheep do not eat well *in the dark*, and therefore they should have time to consume their feed before night sets in. Noon is the common time for feeding grain or roots, and is the best time if but two fodderings of hay are given. If the sheep receive hay three times, it is not a matter of much consequence with which feeding the grain is given, only that the practice be uniform.

It is also highly essential that there be regularity preserved in the *amount* fed. The consumption of hay will, it is true, depend much upon the

* For the effect of the various esculents on the quantity and quality of the milk, see Liebig's Animal Chem.

2 E

weather. The keener the cold, the more sheep will eat. In the South much would also depend upon the amount of grass obtained. In many places a light daily foddering would suffice—in others, a light foddering placed in the depository racks once in two days would answer the purpose. In the steady cold weather of the North, the shepherd readily learns to determine about how much hay will be consumed before the next foddering time. And this is the amount which should, as near as may be, be *regularly* fed. In feeding grain or roots there is no difficulty in preserving *entire regularity*, and it is vastly more important than in feeding hay. Of the latter a sheep will not overeat and surfeit itself. Of the former it will. And if not fed grain to the point of surfeiting, but still over-plenteously, it will expect a like amount at the next feeding, and failing to receive it, will pine for it and manifest uneasiness. The effect of such irregularity on the stomach and system of *any* animal is bad—and the sheep suffers more from it than any other animal. I would much rather that my flock receive no grain at all, than that they receive it without regard to regularity in the amount. The shepherd should be required to *measure* out the grain to sheep in all instances—instead of *guessing* it out—and to measure it to each separate flock.

———

SALT.—Sheep undoubtedly require salt in winter. Some salt their hay when it is stored in the barn or stack. This is objectionable, as you thus constitute yourself the judge, or controller in a matter, where the appetite of the sheep is a much safer guide. It may be left accessible to them in the salt-box (fig. 28) as in summer, or it is an excellent plan to give them an occasional feed of brined hay or straw. This last is done in warm thawing weather, when their appetite is poor, and thus serves a double purpose. With a wisp of straw sprinkle a thin layer of straw with brine—then another layer of straw and another sprinkling, and so on. Let this lie until the next day, for the brine to be absorbed by the straw, and then feed it to all the grazing animals on the farm which need salting.

———

WATER.—Unless sheep have access to succulent food or clean snow, water is indispensable. *Constant* access to a brook or spring is best, but in default of this, they should be watered, at least *once a day*, in some other way.

———————

THE BEST THINGS TO IMPORT ARE BIPEDS,

On two-legged animals, for they not only *consume* largely of the products of the cotton-grower, the wool-grower, and the iron-master, (and always in proportion to their wages,) but by their labour, they produce and add largely to the elements of the best sort of commerce and free-trade—commerce and free-trade among ourselves. We therefore rejoice much more when we see amounts of the importation of *men and women*, than of cattle and sheep. There is no better sign of the prosperity of a country than when you see men flocking into it from all parts of the world; and 't the labour of the country had been *steadily protected*, as it was some years since, we should by this time have imported annually, more by a million, than we now do. Talk of military glory!—the glory of success in the work of " blood and slaughter!" —there is no glory to be compared with that administration of the affairs of a country, which wins for it the regard and admiration of the world, and makes of it a great magnet, attracting the talents, the capital, and the labour of men of all nations. One year of such bloodless and beneficent glory is worth an eternity of fame, won by arms and by conquest.

Immigration, says a New York paper, for the last month has been quite large, yet we see that, as compared with the same seven months of last year, instead of going on rapidly augmenting, as it would do under a common-sense (not party) system, it had actually fallen off. The Journal of Commerce gives it as 34,810 souls. Of the whole number 16,169 were from Ireland, 8449 from Germany, 4788 from England, and 1386 from Scotland. The following table will show the immigration at New York for the years 1849 and 1850, up to the 1st of August:—

	Year 1849.	Year 1850.			Year 1849.	Year 1850.
January	3,258	15,154	May		37,406	44,146
February	8,819	3,206	June		28,985	11,763
March	9,630	5,569	July		31,934	34,910
April	19,934	14,627				
			Total		144,856	185,075

Immigration less th s year

LETTER XIV.

ANATOMY AND DISEASES OF SHEEP.

Character of American ovine veterinary works—of the English...Anatomical details of the latter valuable—necessity of cutting clear from their systems of pathology and therapeutics—reasons...Exciting causes of disease even in adjacent localities in England not the same—popular superstitions on the subject...Necessarily greater differences as between remote countries possessing different climates, etc....Ravages of rot in Europe—scarcely known in most parts of America...Exciting causes apparently the same in both...Hoof-ail, though retained here by contagion, not primarily produced by the same causes as in England...Various European diseases not known here...Difference in the pathology of the *same* diseases in this country and in England...The English ones accompanied with more inflammatory action—the American of an asthenic or sinking character...Pathological differences require a corresponding difference in therapeutics...English system of therapeutics objectionable for th ' above reason—on account of its expensiveness—and, for popular purposes, by the extent of its pharmacopœias...The proper ovine veterinary system to be adopted—manner of classifying diseases...Anatomy of the sheep—how far to be studied—directions to beginners...The Omentum...The Rumen...The Reticulum...The Maniplus...The Abomasum...The functions of the different Stomachs...The Duodenum...The Jejunum...The Ileum...The Cœcum...The Colon...The Rectum...The Mesentary...The process of digestion...The Spleen...The Pancreas...The Liver...The Kidneys...The Bladder...The Uterus and Vagina.

Dear Sir : Most of the veterinary works which have appeared in this country in relation to the Sheep, Horse, and other domestic animals, have been made up simply of medical recipes ; or, if they have given systems of veterinary nosology and pathology, these systems have been mere transcripts of those of European, and particularly of English writers.

I have examined all, I believe, of the most celebrated late English authors, scientific and empirical,* on the diseases of the Sheep and their cures. For anatomical and general pathological details, the works of some of the former possess great value, and compare favorably with the treatises on the same topics by the most eminent physicians and surgeons. This is particularly true of the work on Sheep by the late Mr. Youatt—the fountain-head from which most of the later English writers on the same subject have so liberally drawn, and will probably continue so to do for a century to come. For minute accuracy of description, particularly in the department of pathology—for elaborate research into both facts and authorities—for clearness and sparkling vivacity of style, this gentleman, it seems to me, is entirely without a competitor among the English veterinarians, and his works will bear reading alongside those of a Cooper, a Louis, and a Chapman.

I have hesitated whether to transcribe entire Mr. Youatt's treatise on the Anatomy of the Sheep. It would be the sheerest affectation—not to say plagiarism—to publish a mere abridgment of his remarks, or their substance dressed up in other words, as some late English writers have done, for the purpose of setting up pretensions to that originality which Mr. Youatt has left so little room for in this department. But as these Letters, Sir are *published* for the benefit of the *many*, rather than to instruct those already versed to any considerable extent in Veterinary Science, I have been led to doubt whether *any* systematic treatise on Anatomy is necessary. On the whole, I have come to the conclusion that farther than to exhibit the

---

* I do not use the word "empirical" here in its invidious sense. I mean to describe by it a class of writers *versed in experiments merely*, as contradistinguished from those who possess a scientific knowledge of physiology, pathology, therapeutics &c

localities of disease, explain certain operations in the animal economy, an
render terms intelligible, it would be time thrown away.

In pathology somewhat, and to a much greater extent in the systems o
therapeutics adopted, I have found it necessary to cut clear from all Eng
lish ovine veterinarians. If this is regarded as presumptuous, I have onl[y]
to say that the testimony or opinions of that man are worth little who s[o]
far pins his faith on another's views, as to disregard the plain evidence o
his own senses. The salutary rule of the law is, each witness testifies t[o]
what *he* has *seen*, and to what, crediting the assertions of his own senses,
he knows. It is for the investigating tribunal to decide what weight shal
be attached to the testimony. That tribunal, in the present case, is th[e]
public.

But in reality, a discrepancy of views on the above subjects, does not *ne[-]*
cessarily imply an error on either side. The pathology of diseases fre[-]
quently does not coincide, as between different climates and countries, an[d]
sometimes, singularly enough, between contiguous localities in the same
country. This is especially true as regards the origin or exciting caus[e]
of disease. Where the atmospheric, alimentary, and all other *observabl[e]*
conditions are nearly identical, occult causes which baffle the closest an[d]
most scientific scrutiny, not unfrequently either periodically or regularl[y]
scourge man or beast with disease in one locality, while another one is al[-]
most uniformly exempt from these attacks. What English pathologist, fo[r]
example, has ever assigned a physical cause which would answer, quanti-
tatively, as a criterion to decide on the proportionable prevalence of the
same malady in other regions—or the existence of which would even prove
that the disease existed at all—for the frequent appearance of goitre (*bronchocele*) among the inhabitants of Derbyshire, and the comparative exemption from it of the inhabitants of contiguous counties ?* The theatres of
its especial visitation, in other parts of the world, seem to be equally determined by chance—though undoubtedly dependent upon physical causes
which have as yet eluded observation.

It is not astonishing, therefore, that the ignorant down to our own times,
and even the enlightened, until a period comparatively recent, should have
sought the incomprehensible causes of many diseases, in the regions of the
preternatural. Among brutes especially, which were supposed to be more
given up to such influences, these phenomena were conveniently assigned,
by our English and Scotch ancestors, to

————"some dev'lish cantrip slight"

of "warlocks and witches"—the malevolence of an offended fairy or spite-
ful gnome.†

* I understand that the inhabitants of the *adjoining* counties of Stafford, Nottingham and Leicester are comparatively exempt from the attack of goitre.

† In Burns's inimitable Tam O'Shanter, some of the singular powers once exercised
————"by withered beldams auld and droll
          *   *   *   *   *   *
          Lowping and flinging on a crummock"—
and sometimes, though far more rarely, by "ae winsome wench and walie," to turn aside the established laws of Nature and God's providence, are thus enumerated in describing one of the diabolical sisterhood :
          "Mony a beast to dead she shot,
          And perished mony a bonny beast,
          And shook baith meikle corn and bear,
          And kept the country-side in fear."

No one will understand that the witch, in full league with the Devil, had any occasion for mortal fire-arms, in "shooting" the beasts of her victims. Murrain, and in some cases death, followed a glance of her "evil eye." And even the witches of Burns are tame every-day bodies, compared with those which swell the infernal *dramatis personæ* of Faust, or mingle in the gloomy horrors of Macbeth.

Two centuries ago, and even less, there was not a parish in England, a hill or dell in Scotland, or even a colonized nook in the wild woods of America, where witchcraft was not rife ; and *multitudes* in every rank in life were consigned to the gallows, the faggot, strangling, &c., for this crime, by the *highest judicial* [au]

Equally unphilosophical, and not less mischievous in its effects on the progress of medical science, are those religious views, widely prevalent even at the present day, which in every epizo'tic as well as epidemic scourge, recognize only a direct Theocratic infliction, operating without the intervention of physical causes. If these doctrines do not, as when carried to their full extent among the Mussulmans—who yield a passive non-resistance to plague and conflagration as the direct expression of God's will—lead to an entire abandonment of remedial measures, they at least deter scrutiny into the inducing *natural causes*, and thus occasion a neglect of all *preventive*, and a much less perfect understanding of appropriate *remedial* action.

Between countries widely separated—where their climates and other circumstances exhibit considerable differences—it would naturally be expected that still greater discrepancies would appear in their local nosology. England and the United States are subject to several corresponding ovine diseases, yet it is notorious that some of the most destructive ones of the former are unknown, or next to unknown, in the latter. The rot, according to Mr. Youatt, destroys a million of sheep annually in the British Isl-

ards—and in 1830-1, the number swept off much exceeded two million
Its ravages are equally fatal in Germany, and more so in Egypt. It
also common in France, Spain, Australia, &c. There is nothing sufficie-
ly marked in its diagnosis to effectually distinguish it from some other d
eases, to a person possessing no previous practical acquaintance with .
or no more veterinary knowledge than is common among farmers; a
when a slow train of wasting symptoms have occurred, and the structu-
of the liver is found disorganized, after death, it is not uncommon in tl
country to pronounce it a case of the rot. The same mistake, accordi
to Dr. Coventry (late Professor of Agriculture in the University of Ed-
burgh), is often made by even the shepherds and flock-masters of Europ
There are other diseases besides the rot which specifically attack the i
tegrity of the liver. Even *fasciola* or flukes in the liver, the most infal-
ble diagnostic, to the common eye, of the rot, also, according to Dr. C-
entry, accompany *hepatitis chronica*. I will not take upon me to deny th-
the rot ever exists in the Northern States, but I have yet to see, or he-
of, adequately authenticated, the first undoubted instance ; and this wou
go to show that if isolated cases of it do sometimes occur, it has dwindl-
from the wholesale destroyer of Europe to an obscure and occasional d-
ease. The same remarks apply to existence of the disease in the Southe-
Atlantic and Gulf States, judging from the statements of my correspon
ents, and from the agricultural newspapers. I cannot learn from eith-
of these sources that anything analogous to this malady is common in the
States. According to Mr. Cockerel, of Tennessee, and Mr. Flower, of Il-
nois, the rot *does* prevail in our Western States; and the latter gentlema
who has, I presume, seen the disease in Europe, and who ought therefo
to be familiar with its *prae-mortem* and *post-mortem* appearances, stat
that it occurs in Southern Illinois " from suffering sheep to pasture on la-
that is overflowed with water ;" and he adds, " even a crop of green oa-
early in the fall before a frost comes, has been known to rot young sheep

It is worthy of remark that Mr. Livingston—equally distinguished f-
research and observation—does not include the rot in his list of Americ-
ovine diseases. This affords a strong corroboration of the position I ha-
assumed in relation to the existence of this disease in the North-easte-
States, and those of the Southern ones lying east of the Apalachians.‡

The Hoof-ail, though introduced here by contagion, and kept in consta
existence by the same means, does not appear, in the common phrase,
originate spontaneously, as in Europe ; or, in other words, to be excited l
any other causes than contagion. I have never known an instance goin
even colorably, to prove the contrary of this proposition.

Acute dropsy or Red-water, I judge to be an exceedingly rare disease
the Northern States, though the author of the American Shepherd thin-
differently.‖

Enteretis, or inflammation of the coats of the intestines; blain, or i
flammation of the cellular tissue of the tongue ; and a whole train of oth
diseases—including most of the frightful list of infectious or contagio-
European epizoötics—seem to be unknown in this country.

Why there should be so wide a difference between the ovine nosolo-
of Europe and the United States, is a matter of curious and interesti-
speculation. Whether it will always remain so, or whether the advent

* Youatt on Sheep. p. 445.
† See remarks of Dr. Coventry, quoted at some length in Mountain Shepherd's Manual, p. 20.
‡ I limit the remark to the States lying (mostly) east of these mountains, because they would proba
ne the only ones, at the time at which Mr. Livingston wrote, with the Sheep Husbandry of which he wo
ce supposed to be familiar.
‖ American Shepherd, p. 359.

t) Europeah diseases is only delayed here for more artificial systems of f;ding, breeding, or perhaps more artificial systems of Agriculture af· f;ting the aliment of the sheep, or other and unexplainable causes, time ajne must determine.

If we look for these differences in the observable differences of climate, v find no satisfactory solution of the problem. The climate of England i essentially different from our own—but that it is a favorable one for the ralthy development of all the animal tissues, her large, strong, long-fed population, as well as her well-developed animal kingdom, abundntly attest. The atmosphere of England is a moist and humid one, and r,isture is thought to be one of the necessary predisposing causes of both r; and hoof-ail. Of the origin of the former disease, Mr. Youatt rnarks : *

The rot in sheep is evidently connected with the soil or state of the pasture. It is confd to wet seasons, or to the feeding on ground moist and marshy at all seasons. It has reence to the evaporation of water, and to the presence and decomposition of moist veget-ae matter. It is rarely, or almost never, on dry and sandy soils and in dry seasons; it is rely wanting on boggy or poachy ground, except when that ground is dried by the heat of tl summer sun, or completely covered by the winter rain. In the same farm there are cer-ti fields on which no sheep can be turned with impunity. There are others that seldom o never give the rot."

Mr. Youatt continues his descriptions of these predisposing conditions a great length, and his final conclusion is, in substance, that the miasmata, o gases exhaling from the decomposition of vegetable substances, are the cases of the rot. Mr. Spooner adopts the same views; indeed, they are uiversally received among scientific veterinarians.

If these views are correct, the evil lies not in a generally humid *atmo-s*ere, but in a generally or temporarily humid *soil;* and that they are t'e *quo ad hoc,* is proved by the fearful ravages of the disease in the Jest atmosphere of Germany, in the clear, dry atmosphere of the South) France, and under the torrid skies of southern Spain, where rain does n; fall for months.

Boggy or fenny soils, where decaying vegetable substances are con s'ntly exhaling their gases, are to be found in all parts of the United ites—more or less, in every township, and almost every school district o New-York and New-England. Sheep pasture on such lands, promis-cusly with other stock, in every county—and, in the latter States, at lst, with entire impunity from the rot.

Humidity of soil is also supposed to be the most prominent cause in oginating hoof-ail, or producing it otherwise than by contagion. Mr. Youatt and Professor Dick attribute the disease most often to the effect o sand and dirt forced into the pores of the hoof, when macerated by nisture. The following is the language of Professor Dick :

The finest and richest old pastures and lawns are particularly liable to give this disease, so are soft, marshy and luxuriant meadows. It exists to a greater or less extent in every ation that has a tendency to increase the growth of the hoofs without wearing them ay. . . . . The different parts of the hoof, deprived of their natural wear, grow out of tir proper proportions. The crust, especially, grows too long: and the overgrown parts er break off in irregular rents, or by overshooting the sole allow small particles of sand . dirt to enter into the pores of the hoof. These particles soon reach the quick, and set the inflammation already described and followed by all its destructive effects."†

The same writer assigns another cause for it—inflammation induced by a improper bearing of the foot, caused by the unnatural growth of the horn c wet pastures.

Mr. Spooner attributes the disease to decaying vegetables—" roots and

Youatt on Sheep. p. 451.   † See Dick, quoted by Youatt, p. 527, 528.

leaves of the grasses in a state of rottenness "—brought in contact with th
sheep's foot when "blanched and weakened by continual moisture ! " *
There is another point of difference in the pathology of ovine disease
in this and the old world, judging from the details furnished by the Eng
lish veterinarians. Most of the pyrexial diseases, in England, are accom
panied, at least in their initiatory stages, with active inflammatory symp
toms. Fever runs high, and decidedly antiphlogistic treatment is called
for. On the other hand, so far as my observation and inquiries have ex
tended, the ovine diseases of the United States are usually of an asthenic
nature—characterized by debility from the outset. The difference in the
physical character, feeding, and ordinary state of fatness of the sheep of
the two countries, offers, perhaps, a sufficient explanation of these facts·
The gross, high-fed English sheep, forced forward by bountiful feeding to
an unnaturally precocious maturity, is always in a high state of plethora
and predisposed, therefore, to inflammatory action. A slight derangement
of any function, produced by a cold, by an error in feeding, or by any other
causes, is sufficient to make the organs exercising those functions the seat of
such action. On the other hand, the sheep of the United States, kept mainly
for wool-growing purposes, is rarely raised above a moderately fleshy or
medium condition. And, unexcited by an unnaturally plethoric habit, the
weak vascular and muscular system of the animal little predisposes it to
inflammatory disease.
A difference in the pathological character of disease requires a corresponding difference in the system of therapeutics adopted. The English
system of therapeutics is decidedly objectionable, here, *first*, on the account just named ; *secondly*, from its expensiveness ; and, *thirdly*, (for
popular purposes,) by the extent and complexity of its pharmacology.
1. As has been already remarked, most of the English ovine diseases
commence with pyrexiæ—and the fever is synochal or inflammatory in its
type. The subject is strong, plethoric, and full of blood. Antiphlogistic
treatment is clearly called for. Accordingly, depletion, by bleeding or
purgatives, or both, is first and promptly resorted to by the English veteri
narian. In the United States, also, most important constitutional diseases
commence with pyrexiæ, but the fever in its first *discovered* stage is almost
uniformly of a low, sinking, typhoid type, accompanied with great pros
tration of muscular energy. The animal is in a leanish or only moder
ately fleshy condition. It has been confined to dry, and perhaps rather
unnutritious food—for most of the list of constitutional maladies, here,
make their attacks in the winter, and old, lean, and feeble sheep are usually the first victims. A sheep is observed drooping, and indifferent to
food. It is caught and examined. Whatever organ or portion of the system is laboring under attack, bleed so as to produce a constitutional impression, (which the English veterinarians almost invariably recommend,
where they recommend bleeding at all,) and follow this with an active
purgative, and in four cases out of. five the sheep will, in the expressive
phrase of the English shepherds, "take the ground " ; it will never rise
from the ground more without assistance, and will soon become unable to
stand when set upon its feet. Growing weaker and weaker, it soon refuses to eat, and death supervenes. These remarks are not designed to
apply to stall-fed wethers, or other very high-conditioned sheep.
2. The English, and indeed the European method of treating diseases
is too *expensive* for this country. In curing hoof-ail, *e. g.*, Mr. Youatt, after
recommending washing in chloride of lime, and cauterizing, says :

* This seems to me a most unphilosophical cause to be assigned by a veterinarian of the standing of Mr
Spooner.

"If the foot has been in a manner stripped of its horn, and especially if a considerable por\
tion of the sole has been removed, it may be expedient to wrap a little clean tow round\
the foot, and to bind it tightly down with a tape, the sheep being removed to a straw-yard,\
or some inclosed space, or to a drier pasture. . . . . The foot *should be dressed every day,*\
each new separation of horn removed, and every portion of the fungus submitted to the\
caustic."[*]

Mr. Spooner recommends daily, and not less troublesome treatment.[†]\
The Mountain Shepherd's Manual recommends daily treatment,[‡] and this\
s the case, I believe, with nearly all, if not all, of the foreign veterinarians.\
Professor Pictet, of Switzerland, in addition to daily applications, fumiga-\
ions, etc., innumerable, goes a step beyond "tow pledgets and tape band-\
ges." He says:

"In order to prevent any dirt, &c., from getting into the wound, the diseased foot should\
be placed in a little boot, the sole of which is of leather or felt, and the upper part of cloth,\
in order to fasten it round the leg of the sheep."

This disease rages most when haying and harvesting are at their hight,\
in the Northern States—in July and August—and when the labor of day\
ands costs from seventy-five cents to a dollar per head per diem. Half\
he flocks in the country can then be bought for $1 25 per head. How\
upon *daily* parings, cauterizings, embrocations, fumigations, etc., including\
he expense of drugs and Professor Pictet's *gaiter-boots*, would reach an\
expense equivalent to the price of a *sound sheep*, it requires not the exer-\
ise of much arithmetic to determine! It would certainly be more eco-\
nomical to *kill* sheep of any ordinary grade in the first instance!

The same remark will apply to the English system of treating nearly\
all important diseases. The labor bestowed on it would be worth more,\
ere, than the value of the sheep.

3. The English ovine veterinary pharmacopæia is too extensive and\
complex for popular use. The prescribed formulæ are so compound in\
their character—so minute oftentimes in their quantitative proportions—re-\
quire so much skill for their chemical and mechanical admixture—and,\
lastly, and more important than all the rest, they demand so much med-\
ical knowledge for their proper and timely administration—that they can\
be generally used with safety and advantage only by professional veteri-\
narians, a class entirely wanting, unless occasionally in cities, in the United\
States. Besides, our ordinary country drug-stores are usually lacking in\
many of the articles included in the European prescriptions||—and no one,\
without possessing considerable medical knowledge, could decide what\
effect it would have on the prescription to subtract this or that ingredient.\
might neutralize its effects, or even render it pernicious.

A veterinary system for anything like *popular use*, in this country, must\
be exceedingly simple in its remedies, and in its rules for their administra-\
on. As it is impossible to describe the various symptoms which may\
exhibit themselves in a disease, so as to be understood by all, it is unsafe\
prescribe a constant change of medicines, applicable to the several\
ates which have caused those symptoms to appear. Indeed, changes in\
medicine should only be made consequent on those distinct crises of dis-\
ase which can be detected and understood by the most ordinary observer.\
rescriptions, therefore, inapplicable, or at least unsafe, in any stage *from*\
e *distinct crisis* of disease *to another*, should, as far as practicable, be\
oided. True, such a system of therapeutics will be very imperfect, par-\
ularly in the treatment of serious constitutional maladies. But it will go

Youatt, p. 529.   † Spooner, (endorsing the views of Mr. Read,) p. 438 to 442.\
Quem vide, p. 27.\
Not unfrequently the *most important* ones, as I know from repeated experience.

as far as the knowledge of the uninstructed practitioner will safely admit of—and if, even in cases of constitutional disease, it should simply cause him to *do no hurt by his interference*, and prevent him from resorting to some miserably ignorant empiric*—*the most important object*, perhaps, *would be attained*. It is infinitely safer in such diseases to rely on unaided Nature to effect the cure, than to submit a sheep, or any other animal, to the drugging and dosing of a person ignorant of the true nature of the disease, and of the remedies which he employs. It is better to *do too lit tle* than to *do too much;* and in all cases where *it is not known what to do*, it is better to *do nothing.*

Lord Western, in a letter to Mr. Bischoff, says :†

"I have little to say on the medical treatment of sheep; my study is prevention by suffi: cient wholesome food, with a constant and abundant supply of salt in every yard and every field. . . . . When sheep are taken ill, there is little hope for them, and rarely any use in administering medicines."

If the latter portion of this remark is true among the educated, intelli gent and experienced veterinarians of England, how much more must it be so among those destitute of even the first rudiments of veterinary sci ence! In relation to some of the more serious constitutional maladies, af ter considerable experience and observation, I feel constrained to express the opinion that the remark *is*, to a considerable extent, true. The sheep is almost as unsatisfactory a patient to deal with, in some such cases, as the hog, of which it is frequently said, with no great exaggeration, "that if he is seriously sick he is sure to die, and the more you do for him the sooner he will die!" "Then why give a therapeutic system at all in a class of diseases where it will do so little good?" In the first place, the cases are perhaps few where judicious prescriptions will not *somewhat* diminish the tendency to a fatal result; but the great reason, after all, is, that every man having a sick animal *will* dose and physic it, or *will* permit some officious neighbor to do so, or *will* call in that most dangerous of all *epizoötics*, the cattle-doctor. It is therefore better in the most hopeless cases, to give a few simple directions, based on sound med ical principles, which will not, at all events, *aggravate the disease*, and which will *tend* to alleviate or suppress it, rather than to surrender the helpless animal over to the additional tortures inflicted by ignorance and quackery. Fortunate it is that well-managed sheep, in this country, are so little subject to such diseases !

In classifying diseases, I shall depart from the system adopted by You att, Spooner, etc., who arrange them with reference to the parts of the sys tem they more especially attack, as, for example, "diseases of the brain,"

---

* The self-matriculated "cattle doctor" is a decidedly interesting personage. His qualifications are nu merous, and it is somewhat difficult to find them all brilliantly combined in the same person. He should be the most ignorant man in the town, particularly in everything relating to the anatomy and physiology of man or beast. He should be equally ignorant of the chemical and medicinal properties of nearly all the drugs used by him. His prescriptions, to give them due potency, should consist of a great number of in gredients—a large portion of them bearing very "*hard names*." He should flank and fortify these, at least in all difficult cases, with substances possessing rare occult virtues, entirely unknown to "human phys? cians," such as the "blood of black cats," the "entrails of fowls," "human faeces," simples culled under pe culiar circumstances—

"Root of hemlock, digged i' the dark,
*   *   *   *   *
*   *   *   slips of yew,
Slivered in the moon's eclipse."

He should decidedly affect the mysterious, and should always repel the attempted intrusions of ordinary humanity—the profane vulgar--into the arcana of his high art. He should have half a dozen maladies, such as "baked in the manyfolds," "overflow of the gall." "kidney disease." "rising of the lights," "strained across the loin," etc., to which he can promptly assign all the ills which beasts are heir to. He should never mistake a disease or a remedy. If the patient dies, it should invariably be in consequence of a deviation from his directions!

† Blacht & vol. II.

"diseases of the digestive organs," &c. This method of classification, though not without its advantages, and though it would seem, at first view, to present an arrangement most convenient for reference, examination and comparison, in the end, leads, I think, to confusion and misunderstanding

ANATOMY OF THE SHEEP.

He who breeds sheep to any considerable extent, should make himself familiar with the anatomical structure of some of the parts of the animal— particularly with the arrangement, size, natural appearance, consistency and contents of the several viscera; to some extent with the circulatory system; with the alimentary and respiratory organs; with the brain, and the whole osseous structure of the head. He should be in the constant habit of making more or less extended examinations of all these structures, as opportunity occurs by the slaughter of sheep for economic purposes; and when the animal dies from disease, such examination should be in no ordinary case omitted by the flock-master who is *desirous of making him self thoroughly acquainted with his business.* He will require some instruc tion, in the outset, to enable him to make such dissections understandingly and properly; but he can readily obtain this from any educated physician or surgeon. There are no sufficiently wide differences in the anatomical structure of the sheep and of the human being, to give the surgeon the least difficulty in pointing out the arrangement, uses, &c., of the several parts of the former, unless it be in the conformation of the stomachs.— Here, the structure of the sheep, like that of other ruminating animals, dif- fers widely from man, but that physician or surgeon must have been singu- larly limited in his physiological investigations, who has not made himself acquainted with it. At all events, a glance at a veterinary work, while conducting a dissection, will enable him to understand, and explain it to the learner. The learner while making his examinations in company with,. and under the direction of the surgeon, should *perform every manipulation :* his own hand should handle, remove, test the consistency, &c. of the parts: —alone wield the saw and guide the scalpel. This is an important rule if he would *understand* and *remember.*

The subjects of a portion of the examinations should be sheep killed in: full health. It is necessary to be familiar with the healthy appearance of all the parts, so as to distinctly recognize all departures from it—the effect: of any diseased or abnormal action.

The sides of a lean sheep are more translucent, after being skinned, than- those of a fat one, and therefore the former makes a better subject, if the circulatory system is to be examined. On the sides of the thorax and ab-- domen, at a little distance from the spine, the veins and arteries of those parts can often be traced with beautiful distinctness, without any dissection of the intercostal muscles.

Subjects should be examined which have had their blood drawn (by hav- ing their throats cut), and also those which have died with all their blood in them. Some of the viscera—*e. g.* the lungs, veins and arteries—will present very different appearances under these different circumstances; and this fact not understood might frequently lead to very erroneous con-- clusions in *post mortem* examinations.

I will give a very general description of the parts I have mentioned as necessary to be studied—designed merely for those who have no previous knowledge of the subject.

After the animal has been neatly skinned, place it on a low table, an as- sistant grasping its fore-legs, and holding it firmly on its back. Then slit open the belly from the middle of the *sternum*, or cartilaginous connection between the ribs, to the anus. In making this and all similar incisions hold the edge of the knife upward, guarding its point with the fore-finger so that the viscera shall not be wounded. The abdomen—the whole cavity of the trunk back of the diaphragm or "midriff"—is now laid open. It is usually necessary for a better examination of the parts to make cross incisions part way between the diaphragm and anus, extending down on each side several inches toward the backbone.

I shall describe the viscera in the order in which I have usually examined them.

On opening the abdomen the *omentum* or caul is found covering the intestines. It is a thin, and, in a normal state, colorless and transparent structure, formed of two membranes, between which extend streaks of fat in the form of a net.

The external appearance of the stomachs is given in the following cut of those of a young sheep which died of disease. Their arrangement is slightly different in the animal.

Fig. 47.

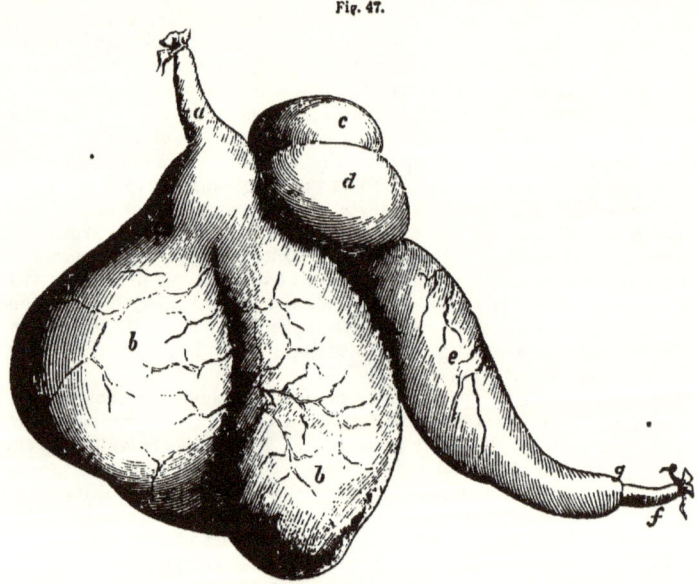

THE STOMACHS.

a. The œsophagus or gullet, entering the rumen or paunch.
b. b. The rumen, or paunch, occupying three-fourths of the abdomen.
c. The reticulum, or honey-comb—the 2d stomach.
d. The maniplus, or many folds—the 3d stomach.
e. The abomasum, or 4th stomach.
f. The commencement of the duodenum or first intestine.
g. The place of the pylorus, a valve which separates the contents of the abomasum and duodenum.

The walls of the rumen or paunch consist of four coats or tunics—1st, the peritoneal or outer coat; 2d, the muscular; 3d, the mucous, covered with papillæ, or little protuberances, from which (or glands under which) is secreted a peculiar fluid to soften and prepare the food for re-mastication; and, 4th, the inner or cuticular coat, a thin, entirely insensible membrane, which defends the mucous coat from abrasion or erosion.

ır honey-comb is composed of the same number of coats ınctions. But the mucous coat, in addition to minute . with elevations arranged in pentagons and sexagons somewhat resembling a honey-comb, except that the ı shallower.

as the same four coats. Its *floor* is a continuation of ınal. From its *roof* depend many parallel folds of the re thicker and stronger than in the other stomachs— its floor. The cuticle is covered toward the edges of ı, bony processes, shaped like fangs, or cones bent in a ınd pointing toward the entrance of the stomach. The d or leaf contains muscles which impart to it the power orcible motion. There are forty-two of these folds in he sheep—occasionally forty-eight. They do not all roach the œsophagean canal, but are disposed in groups central ones of each nearly reaching the canal or floor he others on each side growing shorter and shorter, so of irregular reëntering angles.

is the digesting stomach, where the gastric juices are e the pultaceous food is converted into chyme. It is ı its lower extremity connects with the intestines, as The cuticular lining of the three preceding stomachs The mucous coat is disposed in the form of *rugœ* or ıged longitudinally with the direction of the stomach, brane the gastric juices are secreted.

ı size of the four stomachs will be sufficiently seen in

hagus enters the rumen, it terminates in what is called nal, a continuation of the former constituting the roof bottom or floor of this canal is formed of divided por- e upper parts of the rumen and reticulum—muscular ' as they are sometimes denominated—which may re- ı the food will pass over them into the third and fourth may open, permitting the food to fall between them, .oor, into the first and second stomachs. It is probable these lips, as food passes over them, depends some- ınical effect, and somewhat upon the will of the animal. ıceous food fit for immediate digestion glide over them. ıd of the sheep, like that of other ruminating animals, little preparatory mastication; and these untriturated through the first opening above described into the ru- however, that the animal can, at will, also cause water opening into the first stomach. This would be neces- economy, and the water is always found there. ıas entered the rumen, the muscular action of that vis- ıake the circuit of its different compartments, and, in r swallowed forces it on and up to near the opening ' entered. In its passage it is macerated by a solvent ted by the mucous coat. The papillæ of that coat are ce the mechanical action of the contents of the stomach, ıertain extent, to aid in triturating them. The food t of the stomach, and is ready for re-mastication, ac- ani, in from sixteen to eighteen hours. By a muscular b, a portion of it is then thrown over the membraneous guards the opening from this into the second stomach

The reticulum contracts upon it, forming it into a suitable pellet to be returned to the mouth, and also covers it with a mucus secreted in this stomach. By a spasmodic effort (always perceptible externally when the sheep or cow commences rumination) the pellet is forced through the roof of the reticulum, by the opening before described, and returned to the mouth by the contractions of the spiral muscle of the œsophagus or gullet, for mastication.

This explanation of the functions of the second stomach is not accepted by all the physiologists who have examined this subject. Some contend that all the solider portions of the food are returned directly from the rumen for re-mastication; that when raised to the floor of the œsophagean canal, the hard parts are carried up to the mouth—the more pultaceous ones (but still not sufficiently pultaceous for the fourth stomach) passing into the reticulum, where they are again macerated—the fluid squeezed out of them by a contraction of the stomach and allowed to pass on to the fourth stomach—and then the drier parts raised, like those from the paunch, for re-mastication. More solid and indigestible substances "may be submitted two or more times to the process of rumination." Such appear to be the views of Mr. Spooner.[*]

According to this theory, both stomachs are created substantially for one and the same purpose, and one would seem to be unnecessary. And where would be the use of the opening from one stomach into the other? And if the second stomach, like the first, is simply for the maceration and return of food, why the superior thickness and strength of the coatings of the former? Being of a volume greatly inferior to that of the latter, it certainly would require less strength, if the functions of both were the same.

The main support for this, as it seems to me, erroneous theory, is found in the fact that the contents of the reticulum, after death, are usually found considerably *more fluid* than those of the rumen. I conceive that but small portions of solid food are introduced at one time from the rumen into the reticulum—not enough to give to the liquid contents of the latter viscus the consistency of those of the former—proceeding on the supposition that the reticulum of the living animal is *filled* with fluid, as usually found after death. But why may not a portion of this fluid have escaped by the valve—been decanted, as it were, from the paunch to the reticulum, *after death?* I see no violence in this supposition. If this is not so, the *uniform* fluidity of the contents of the reticulum would be, it seems to me, fatal to the theory based on it—for, according to Spooner and others who adopt it, after the reticulum has "become moderately full," it contracts on its contents, expressing the liquid from the solid parts, which said liquid is forced into the œsophagean canal, and escapes into the fourth stomach. The solid parts would be thus left comparatively dry. Sheep penned up for butchery often do, as every one has observed, ruminate until within a few seconds of the time that all their natural visceral functions are suddenly suspended by death—and when, therefore, this suspension would, at times, as a matter of course, take place at all the different stages of rumination and preparation for rumination—how happens it that the reticulum is not often found with its liquid parts expressed—containing nothing but the solids, just prepared for re-mastication? Or if it be supposed that the act of forcing out the liquid, and forcing up the solids into the œsophagus, are coincident or simultaneous, why is not this stomach *sometimes* found *entirely empty?* Can it be supposed that this fluid (I have *uniformly* found the fluid mixed with considerable quantities of the solid food) is so instantaneously re-supplied?

* Spooner, p. 162-3.

If so, by what process? I think there are other reasons which support the view I have taken, but I will not push the discussion, there not being, so far as I am aware, any questions to be solved by it which directly and practically affect the interests or the practices of the sheep-breeder.

Let us now observe the course pursued by the food, and the process to which it is submitted, after rumination. It now glides *over* the trap-doors which open into the first and second stomachs. As it passes over the floor of the third, or the maniplus, the pendant leaves of this viscus, armed with their beak-like protuberances, seize the advancing mass, and squeezing out the fluid and the more finely comminuted portions of the food which escape with it, commence triturating the bulkier fibrous portions between their folds. Their bony papillæ give to these folds something of the mechanical action of rasps, in grinding down the vegetable fibre. The food being now reduced to an entirely pultaceous state, passes into the fourth stomach, or abomasum, where it is acted upon by the gastric juice, and converted into chyme. The amount of food found between the folds of the maniplus, after death, depends upon the time that has elapsed since rumination. It is dry and hard, compared with the contents of the other stomachs.

The entrance to the fourth stomach—the *cardiac* opening—is closed against regurgitation or vomiting, by a sort of valve, composed of a portion of one of the *rugæ*, before alluded to, which line the interior of this stomach. The *pylorus* is also closed by a valve, which prevents a premature passage of the contents of the stomach into the intestines.

The intestines are exhibited in fig. 48, copied from Mr. Youatt's work. Before the duodenum enters into (or changes its name to) the jejunum, and about 18 inches from the pylorus, it is perforated by the biliary duct—*ductus choledochus*—which brings the bile eliminated by the liver, from the gall-bladder, and also the fluid which is secreted by the *pancreas*, or sweetbread, which last is introduced into the biliary duct two inches from its entrance into the duodenum, by another duct or small tube. The compound fluid thus introduced into the duodenum exercises various important offices in the digestive and assimilating processes. The bile is supposed to aid in the separation of the chyme into chyle and fecal matter—or the nutritive parts of the food which are assimilated into blood, from the innutritious parts which are discharged as excrement. It also prevents a putrid decomposition of the vegetable contents of the intestines, and serves various other useful purposes.

The chyle—a white albuminous fluid, with a composition differing but little from that of blood—is taken from the intestines by a multitude of minute ducts called *lacteals*, which traverse the mesentary, constantly uniting as they advance, so as to form larger ducts. These enter the mesenteric glands—small glandular bodies attached to the mesentary—after the passage of which the chyle begins to change its color. The lacteals still continue to unite and enlarge, and finally terminate in the *thoracic duct*. In this the chyle is mingled with the *lymph* secreted from a portion of the *lymphatics*—another exceedingly minute system of absorbent ducts, which open on the internal and external surfaces of the whole system. From the thoracic duct, the chyle is conveyed to the heart, and enters into the circulation as blood.

—

THE SPLEEN.—With the appearance of the spleen or milt—in the sheep a dark, firm, spongy viscus, attached to the rumen, and lying on the left side of the belly—all are sufficiently familiar. Its uses and functions in the animal economy are not well understood, and 1. has in some instances been removed from the living animal without the apparent derangement

Fig. 48.

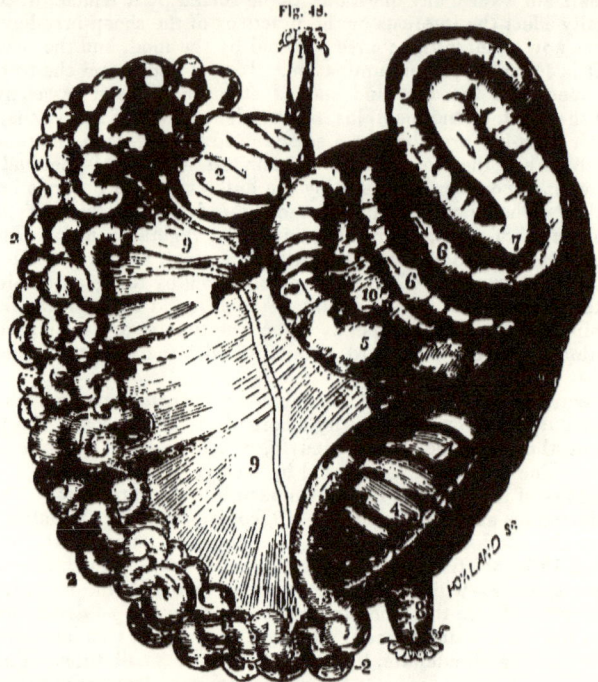

THE INTESTINES AND MESENTARY.

1 The duodenum.   2. The jejunum.   3. The ileum
4 The cœcum, being the anterior prolongation of the colon, or first large intestine. The ileum opens
   into this (on the back side as presented in the cut), about twelve inches from its extremity—the
   opening being defended by a valve.
5 The large anterior portion of the colon, retaining its size (about three times that of the smaller intestines) for about two feet.
6. 6. The colon tending toward the center.
7. 7. The returning convolutions of the colon.
8 The rectum or straight gut, communicating with the anus.
9. 9. The mesentary, or that portion of the peritoneum which retains the intestines in their places.
10. The portion of the mesentary supporting the colon, &c.
The united length of those intestines is upward of sixty feet !

of any function.   Mr. Youatt conjectures that its main office is to supply the coloring matter of the blood.

THE PANCREAS.—The pancreas or sweet-bread, with the appearance of which all who have noticed the entrails of a sheep, are also familiar, is a glandular body found on the left side of the belly.   It has a series of ducts which unite into a larger one, and, as before stated, this discharges a transparent, albuminous, and somewhat acid fluid into the biliary duct near the point where it enters the duodenum.   This fluid acts some un-known, but probably necessary part in preparing the chyme for the offices it is to perform in the animal economy.

THE LIVER.—The liver is much larger in proportion, in the sheep, than in the horse and ox, and it is twice the proportionate size of that of Mar

It is situated mostly on the right side, between the maniplus and diaphragm. It is supplied with arterial blood, and receives the venous blood which is conveyed from the intestines, from which it separates the bile, and conveys it to the *gall-bladder*. The bile having undergone certain changes in this bladder, is conveyed, as already stated, by the biliary duct, to the duodenum. The venous blood, after the gall is separated from it, is returned to the lungs, to undergo the process which fits it again to en ter into the circulation. _____

THE URINARY AND GENERATIVE ORGANS.

Though it might seem the more natural order to complete the examination of the circulatory and respiratory organs, before taking up those named at the head of this paragraph, I shall, adhering to my first arrangement to follow the order which I have uniformly pursued in making dissections, first complete the description of those of the abdominal cavity.

THE KIDNEYS.—The kidneys are two bean-shaped glands firmly attached to the roof of the abdomen, and usually imbedded in fat. They are supplied with blood by large arteries, and, having filtered out the urine from it, they discharge the latter through two ducts, termed *ureters*, into the bladder. The passage of these ducts through the walls of the latter is in an oblique course, so that it is closed by pressure from within, and thus the urine cannot return. _____

THE BLADDER.—The bladder joins the *urethra*, in the pelvis, and its posterior part is attached to the floor of that cavity. The anterior part, where the diameter is larger, floats free in the abdomen. A circular muscle or *sphincter* closes the entrance into the urethra, to prevent the continuous escape of the urine, and this relaxes when the muscular coat of the bladder contracts for the periodical expulsion of that fluid. The urethra is but a few inches long in the ewe, and opens into the vagina. It is much longer 'n the ram, as it extends the whole length of the penis. _____

THE UTERUS AND VAGINA.—The vagina is several inches in length and opens into the uterus or womb by a circular opening which becomes closed after impregnation. They are situated between the rectum above, and the bladder below. They are mostly within the pelvis in the unpregnant animal, but the womb rises into the abdomen when it encloses a fœtus. The womb is a cylindrical body with two "horns" or branches. At the interior extremity of each horn are protuberances, of a red color, called *ovaries* which are supposed to contain the germs of the offspring.

LETTER XV.

ANATOMY OF THE SHEEP (Continued)—DISEASES AND THEIR TREATMENT.

The Thoracic Viscera...The Diaphragm...The Thorax...The Heart, Arteries, Capillaries, and Veins...
The Lungs...The Windpipe, Larynx and Pharynx...The Thyroid and Parotid Glands...The Head and
its structures...The Brain...The Nerves...The Teeth...The Lower Extremities...The Biflex Canal...
Febrile diseases—those of Europe which are not common here...Ophthalmia—popular remedies—proper
treatment...Pneumonia—symptoms—Mr. Spooner's prescription for...Bronchitis—symptoms—treatment
...Catarrh—ordinarily not dangerous—preventives...Malignant Epizoötic Catarrh—prevalence in the
Northern States—character of the disease has not been understood—prevalence in author's flock—how
produced—symptoms—post-mortem appearances—character of the disease ascertained—Nosology—
treatment, &c...The Rot—its diagnosis—post-mortem appearances—description of the Fluke—causes of
the Rot—treatment...Diarrhea—cause—diagnosis—treatment...Dysentery—cause—difference between
it and diarrhea—treatment...Garget—seat and origin of the disease—treatment...Nervous Diseases...
Apoplexy—unrecognized cases of it—several cases detailed—symptoms—treatment...Phrenitis...Tet-
anus...Epilepsy...Rabies...Neither of them common in this country...Paralysis—symptoms—treatment. .Colic—symptoms—attributed to intussusception—true cause—treatment.

THE THORACIC VISCERA.

Among these, for convenience, I will include the diaphragm.

THE DIAPHRAGM.—The diaphragm or midriff is a muscle extending entirely across the inner cavity of the body, separating the abdomen from
the thorax or chest. Its structure is unique, and beautifully adapted to the
functions it has to perform. Its outer margin is muscular, giving it the necessary power of contraction, while toward the middle it changes into a
transparent tendonous substance. Through this tendonous substance pass
the œsophagus, the *aorta*, and the *vena cava*.

If the parts of the diaphragm which immediately surround these vessels
had been muscular, every contraction of the former in the act of respiration, would have compressed the latter, and therefore interfered with the
passage of the food to the stomach, and the circulation of the blood. In a
state of rest the diaphragm is convex toward the thorax. When contracted and flattened, therefore, it enlarges the cavity of the thorax, and air
rushes into the lungs. Its alternate contractions and relaxations mainly
produce the act of respiration or breathing.

THE THORAX.—Without injuring the diaphragm, divide the sternum and
brisket of the sheep longitudinally through the center, with a fine saw, and
on pulling the lower extremity of the ribs slightly apart, the thorax will be
disclosed in its natural arrangement. It consists of three cavities, formed by
the doublings of the *pleura*, a thin serous membrane, which lines the whole
interior of the chest. Two outer and larger cavities (the right one being
the largest), contain the lungs—a third and smaller one, lying between the
posterior portions of the former, contains the heart. The œsophagus passes through the upper portion of the thorax, over the lungs and heart, and
between them and the spine, to the lower portion of the neck.

THE HEART, ARTERIES, CAPILLARIES, AND VEINS.—With the size and
general appearance of the heart, all are familiar. Enclosed in a membraneous sac—the *pericardium*—it hangs suspended by its superior attachments to the roof of the thorax, its lower extremity nearly reaching to its
floor, and pointing toward the left side. The heart has two cavities on
each side, termed *auricles* and *ventricles*. The chyle and venous blood are

discharged into the right auricle, and thence into the right ventricle. By the contraction of the latter, its contents are forced through the pulmonary artery into the lungs. The blood having been purified in the lungs, is returned to the left auricle ; thence into the left ventricle ; and it is then forced into the aorta, or large artery which supplies, by its different branches, all parts of the system with blood. Each compartment of the heart is furnished with appropriate valves to cause the blood to be forced *forward* in its regular course, by the muscular contractions of this viscus. These contractions are the result of an inherent and independent power.

The contractions of the heart force the blood into and along the arteries. When this force begins to be spent as the distance from the heart increases, it receives aid from the action of the muscular coat of the arteries themselves, which forces along the blood to their utmost extremities.

The arteries continue to branch off into more and more minute divisions as they recede from the heart, until the tubes are much less in diameter than the finest hair. These, *capillaries* as they are called, open by exceedingly minute mouths in every part of the frame, for the deposition of those secretions from the blood which maintain the vitality and healthy action of the parts, supply the animal waste, &c.

The capillaries, commencing their return toward the heart, constantly reünite, forming larger tubes which are called *veins*, which bring back such portions of the blood carried out by the arteries, as has not been expended in nourishing the system. The blood now deprived of its oxygen, and loaded with carbon, is unfit for farther circulation until re-purified in the lungs. It is of a darker color than the arterial blood. It is no longer urged on by the contractile power of the tubes through which it flows, but by the partial vacuum formed in the right auricle (as at each contraction it forces its contents into the right ventricle,) and by atmospheric pressure. —

THE LUNGS.—The lungs are bodies composed of separate minute air-cells, communicating with the *bronchial tubes*, or subdivisions of the windpipe. They also contain many arteries, and veins. On the delicate membraneous walls of the air-cells the venous blood is carried by innumerable tubes so thin as to permit their contents to be acted upon by the atmospheric air which fills the cells at every inspiration. Here the blood gives off its carbon, and receives oxygen from the air, and thus is prepared for its return to the heart, and to be again sent through the system.

The right lung is somewhat larger than the left, and both fill their respective cavities when inflated. They are entirely free from any attachment to the pleura—the membrane which lines the ribs—when in their natural state. When the animal has been bled to death, the lungs are of a light color ; but if the animal has died with all its blood in it, their color resembles that of the liver. This can, however, be readily distinguished from *hepatization*—the result of certain diseases—as will be hereafter shown. —

THE WINDPIPE, LARYNX, PHARYNX, &c.—The bronchial tubes constantly uniting as they approach the anterior portion or root of each lung, finally form a single large tube, as they make their exit from each lobe, and these, uniting into one, form the windpipe. This is a well known cartilaginous tube which passes out of the chest between the first two ribs, and ascends on the front part of the neck. It unites with the larynx, which continues the air passage from the lungs to the mouth. The œsophagus leaves the chest close beside the windpipe, and ascends the neck on the left side of the latter. It communicates with the pharynx which communicates with the mouth. The food on being swallowed enters the pharynx or food bag, which is directly above the larynx—so that the food traverses the entrance to the latter. It is deterred from entering the windpipe by the *epiglottis*, a triangular lid or valve which projects upward from the floor of the passage, and which closes upon and covers the *glottis*, or entrance into the windpipe, when any substance more dense than air comes in contact with it in its downward passage.

THE THYROID AND PAROTID GLANDS.—The Thyroid glands are located on each side of the trachea. The parotid glands are situated immediately below the ear, behind the angle of the lower jaw. There are certain other glands situated beneath the lower jaw, not necessary here to be referred to.

THE HEAD AND ITS CONTENTS.

Fig. 49.

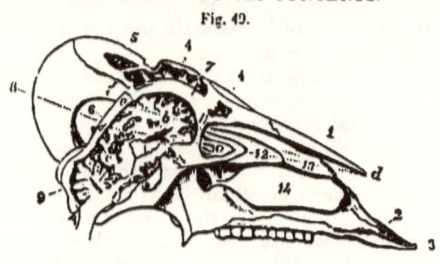

BONES OF THE HEAD.

1. The nasal bone.
2. The upper jaw bone.
3. The intermaxillary bone, which supports the pad which supplies the place of upper front teeth.
4, 4. The frontal sinus.
5. Cavity or sinus of the horn, communicating with the frontal sinus. It is here shown by the removal of a section of the base of the horn.
6. The parietal bone.
7. The frontal bone.
8. Vertical section of the brain.

9. Vertical section of the cerebellum.
   a. The cineritious portion of the brain.
   b. The medullary portion.
10. The ethmoid bone.
11. The cribriform or perforated plate of the ethmoid bone. "It separates the nasal cavity from the brain; it is thin almost as a wafer, and pierced by numerous holes, through which the olfactory nerve penetrates, in order to spread itself over the inner part of the nose."
12. The lower cell of the ethmoid bone.
13. The superior turbinated bone.
14. The inferior turbinated bone.
17. The sphenoid bone.

The above cut, copied from Youatt, gives, with the subjoined explanations, a sufficient description of most of the structures of the head. Some, however, demand a little more particular description.

THE BRAIN.—The brain of the sheep is smaller in proportion than that of Man, but is shaped so nearly like the latter, and so closely resembles it in its general structure and conformation, that it furnishes the medical student with a good substitute for the brain of the human subject! The brain is invested in a membrane called the *pia mater.* The cranium or skull is lined by the *dura mater,* and between this and the former there is a delicate membrane called the *tunica arachnoides.*

THE NERVES.—Ten pair of nerves arise from the brain, and thirty pair from the spinal cord. These supply the sense of seeing, hearing, tasting, smelling, feeling, &c. &c.; and a portion of them, termed nerves of mo tion, communicate that volition of the brain to the different parts of the system, which produces motion. A description of these various nerves, or even an enumeration of them, would be of no *practical* benefit in a mere popular veterinary treatise.

THE TEETH.—The sheep has 24 molar teeth, and eight incisors. The latter are confined to the lower jaw, being opposed to a firm, hard, elastic pad or cushion on the upper jaw. The incisors are *gouge*-shaped—*i. e.*, concave without and convex within—which enables the sheep to crop the herbage closer to the ground than our other domestic ruminant, the ox.

The lamb is born without incisor teeth, or it has but two. In three or four weeks, it has eight small, shortish ones, as represented in fig. 50.—

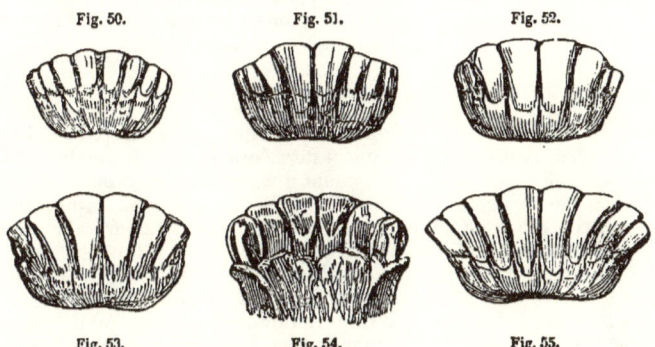

Fig. 50.   Fig. 51.   Fig. 52.

Fig. 53.   Fig. 54.   Fig. 55.

When not far from a year old—though sometimes not until fourteen, fifteen, or even sixteen months old—the two central incisors are shed, and their place is supplied by two longer and broader teeth, as in fig. 51. The sheep is then termed, in this country, a *yearling*, or *yearling past*. Two of the "lamb teeth" continue to be annually shed and their places supplied with the permanent ones until the sheep becomes "*full-mouthed*." Fig. 52 presents the teeth of a two-year-old-past—fig. 53 of a three-year-old-past—fig. 55 of a four-year-old-past. The four-year-old-past is, in reality, nearly or quite five years old, before it obtains its whole number of *fully-grown* permanent teeth. The two-year-old and three-year-old also about reach their next year before their additional incisors are *fully grown*.— Hence, the English writers all speak of two broad teeth (meaning fully-grown ones) as indicating the age of two years; four broad teeth, three years; six broad teeth, four years; and eight broad teeth, or full-mouthed, five years. I prefer the English arrangement, as more accurate, but the other is the common one in the Northern and Eastern States; and, as it is a matter of little practical consequence, it will here be adhered to.

Fig. 54 gives an *inside* view of the incisors of a three-year-old-past—an *outside* view of which is given in fig. 53. The two remaining lamb teeth are here shown, which in the outside view are concealed by the last pair of permanent teeth. From their being thus concealed, the three is often mistaken for the four-year-old-past, by those who do not *count* the permanent teeth.

At six years old, the incisors begin to diminish in breadth. At seven they have lost their fan-like shape, being equilateral, long, and narrow.— At eight, they are still narrower; and this year or the next, reversing the flaring or divergent position in which they are shown in fig. 55, they begin to point in toward the two central ones. Their narrowness and inward direction increases for a year or two more, when they begin to drop out. Sheep fed on turnips or other roots, lose their teeth earlier than those which only receive grain, hay, &c. in winter. At twelve years old, the incisors are usually gone with the exception of one or two loose ones. And here let me remark that when the incisors are reduced to one or two, they should always be twitched out with a pair of nippers. They are useless for the purpose for which they were formed, and they prevent that contact of the lower *gum* with the pad above, which is *now* the only substitute for teeth in cropping grass. When all the incisors are gone, the gums of the lower jaw rapidly harden, and I have known ewes to live for years, keep in fair condition and rear lambs, without an incisor tooth in their heads !

The above remarks are more particularly applicable to the Merino breed. The other breeds, so far as my acquaintance extends, lose their teeth, or become " broken-mouthed " somewhat earlier ; and they dwindle away and die soon after they begin to lose their teeth.

THE LOWER EXTREMITIES.

The BIFLEX CANAL.—The lower extremities of the sheep, including the legs, feet, &c., require no anatomical description. I will simply call attention to the biflex or interdigital canal, the nature and diseases of which have been the subjects of so many errors. It is a small orifice opening externally on the front of each pastern immediately above the cleft between the toes. It bifurcates within, a tube passing down on each side of the inner face of the pastern, winding round and ending in a *cul de sac.*

The use of this canal is a matter of doubt. Mr. Spooner thinks the hair always found in it is " *excreted* from the internal surface," and " from the smallness of the opening it cannot escape, or rather is detained for a useful purpose." He continues :

" The use of this canal, thus stuffed with hair, is self-evident. We have mentioned the great motion possessed by this pastern joint, which is so great as to threaten to chafe the skin by the friction of one side against the other. It is to prevent or ward off this friction that these biflex canals, or rather *hair-stuffed cushions,* are provided."

In my judgment, this is a very far-fetched conclusion, and Mr. Youatt's is little more satisfactory. Diseases originating in this canal are sometimes confounded with hoof-ail ; and the canal, or a portion of it, is often dissected, or rather *mangled* out by ignorant charlatans in pursuit of an imaginary *worm,* which, they induce the credulous farmer to believe, originates the hoof-ail ! The hoof-ail proper has nothing to do with, nor do its characteristic lesions extend to this canal.

FEBRILE DISEASES.

Simple inflammatory, malignant inflammatory, and typhus fevers often devastate the flocks of Europe; but they seem scarcely to be known in the United States, and are included in no American work on the diseases of sheep which has fallen under my eye.

The same remark applies to phrenitis (inflammation of the brain), pleuritis (inflammation of the membrane which lines the thorax), gastritis (inflammation of the stomach), enteretis (inflammation of the intestines), cystitis (inflammation of the bladder), laryngitis (inflammation of the larynx), and several other inflammatory diseases.

OPHTHALMIA.—Ophthalmia, or inflammation of the eye, is not uncommon in our country, but is little noticed, as in most cases it disappears in a few days, or, at worst, is only followed by *cataract*. The cataract being usually confined to one eye does not appreciably affect the value of the animal, and therefore has no influence on its market price. As a remedy for this disease, Mr. Grove recommended blowing pulverized *red chalk* into the inflamed eye! Others squirt into it tobacco juice, from that ever ready reservoir of this nauseous fluid, their mouths! I apprehend that all such prescriptions are far worse than nothing.

Conceiving it a matter of humanity to do *something*, I have in some instances drawn blood from under the eye, bathed the eye in tepid water, and occasionally with a weak solution of the sulphate of zinc combined with tincture of opium. These applications diminish pain and accelerate the cure.

PNEUMONIA.—Pneumonia, or inflammation of the lungs, is not a common disease, in the Northern States, but undoubted cases of it sometimes occur, after sheep have been exposed to sudden cold—particularly when recently shorn. The adhesions occasionally witnessed between the lungs and pleura of slaughtered sheep, betray the former existence of this disease—though in many instances it was so slight as to be mistaken, in the time of it, for a hard cold. The sheep laboring under pneumonia is dull, ceases to ruminate, neglects its food, drinks frequently and largely, and its breathing is rapid and laborious. The eye is clouded—the nose discharges a tenacious, fetid matter—the teeth are ground frequently, so that the sound is audible to some distance. The pulse is at first hard and rapid—sometimes intermittent; but before death it becomes weak. During the hight of the fever, the flanks heave violently. There is a hard, painful cough during the first stages of the disease. This becomes weaker, and seems to be accompanied with more pain as death approaches.

After death, the lungs are found more or less *hepatized, i. e.* permanently condensed, and engorged with blood, so that their structure resembles that of the *hepar*, or liver—and they have so far lost their integrity that they are torn asunder by the slightest force.

It may be well in this place to remark that when sheep die from any cause *with their blood in them*, the lungs have a dark hepatized appearance. But whether actually hepatized or not, can be readily decided by compressing the windpipe, so that air cannot escape through it, and then between such compression and the body of the lungs, in a closely fitting orifice, insert a goose-quill or other tube, and continue to blow until the lungs are inflated so far as they can be. As they inflate, they will become lighter colored, and plainly manifest their cellular structure. If any portions of them cannot be inflated, and retain their dark, liver-like consistency and color, they exhibit hepatization—the result of high inflammatory action—and a state utterly incompatible, in the living animal, with the discharge of the natural functions of the viscus.

With the *treatment* of pneumonia, I have but little personal experience. In the first or inflammatory stages of the disease, bleeding and aperients are clearly called for. Mr. Spooner recommends "early and copious bleeding, repeated, if necessary, in a few hours . . . this followed by aperient medicines, such as 2 oz. of Epsom salts, which may be repeated in smaller doses if the bowels are not sufficiently relaxed. . . . The following sedative may also be given with gruel twice a day :

Nitrate of potash.............................1 drachm.
Digitalis, powdered ....1 scruple.
Tartarized antimony ....1 do.

The few cases I have seen have been of a sub-acute character, and would
not bear treatment so decidedly and *I* think *dangerously* antiphlogistic.
Mr. Youatt remarks :

"Depletion may be of inestimable value during the continuance—the short continuance—
of the febrile state; but excitation like this will soon be followed by corresponding exhaustion, and then the bleeding and the purging would be murderous expedients, and gentian,
ginger, and the spirit of nitrous ether will afford the only hope of cure."

BRONCHITIS.—It would be difficult to suppose that where sheep are subject to pneumonia they would not also be subject to bronchitis—which is
an inflammation of the mucous membrane which lines the bronchial tubes
—the air-passes of the lungs. I have seen no cases, however, which I
have been able to identify as bronchitis, and have examined no subjects,
after death, which exhibited its characteristic lesions. Its symptoms are
those of an ordinary cold, but attended with more fever and a tenderness
of the throat and belly when pressed upon.

Treatment.—Administer salt in doses from 1½ to 2 oz., with 6 or 8 oz.
of lime-water, given in some other part of the day. This is Mr. Youatt's
prescription.

CATARRH.—Catarrh is an inflammation of the mucous membrane which
lines the nasal passages—and it sometimes extends to the larynx and pharynx. In the first instance—where the lining of the nasal passages is
alone and not very violently affected—it is merely accompanied by an increased discharge of mucus, and is rarely attended with much danger. In
this form it is usually termed *snuffles,* and high-bred English mutton sheep,
in this country, are apt to manifest more or less of it, after every sudden
change of weather. When the inflammation extends to the mucous lining
of the larynx and pharynx, some degree of fever usually supervenes, accompanied by cough, and some loss of appetite. At this point the English veterinarians usually recommend bleeding and purging. Catarrh rarely
attacks the American fine-wooled sheep with sufficient violence in *summer,*
to require the exhibition of remedies. I early found that depletion, in
catarrh, in our severe *winter* months, rapidly produced that fatal prostration, from which it is next to impossible to recover the sheep—entirely impossible, without bestowing an amount of time and care on it, costing far
more than the price of any ordinary sheep.

The *best* course is to *prevent* the disease, by judicious precautions. With
that amount of attention which every prudent flock-master should bestow
on his sheep, the hardy American Merino is little subject to it. Good,
comfortable, but *well-ventilated* shelters, constantly accessible to the sheep
in winter, with a sufficiency of food regularly administered, is usually a
sufficient safeguard ; and after some years of experience, during which I
have tried a variety of experiments on this disease, I resort to no other
remedies—in other words, I *do nothing* for those occasional cases of ordinary catarrh which arise in my flock, and they never prove fatal.

MALIGNANT EPIZOOTIC CATARRH.—Essentially differing from the preceding in type and virulence is an epidemic, or, more properly speaking,
an epizoötic, malady, which as often as once in eight or ten years sweeps
over extended sections of the Northern States, destroying more sheep than
all the other diseases put together. It usually makes its appearance in winters characterized by rapid and violent changes of temperature. The
Northern farmers speak of these as the "bad winters" for sheep—frequently without assigning any name to the malady. Others term the lat

ter " *The Distemper*," and others again call it the " *Grub in the Head*," at-
tributing the evil exclusively to the presence of these parasites. The
latter, as I shall hereafter show, is an entirely erroneous hypothesis.

The winter of 1846–7 was one of these " bad winters," and the destruction of sheep in New-York, and some adjoining States, was very ex
tensive. Some flock-masters lost half, others three-quarters, and a few
seven-eighths of their flocks. One individual within a few miles of me lost
five hundred out of eight hundred—another nine hundred out of one
thousand! But these *severe* losses fell mainly on the holders of the delicate Saxon sheep, and perhaps, generally, on those possessing not the best
accommodations, or the greatest degree of energy and skill.

I lost about fifty sheep during this winter, and never having seen any description of the pathology of this disease, its diagnosis, its lesions—or, in
short, any attempt to ascertain its specific character or proper classification in our ovine nosology—I shall attempt to supply some of these omissions. Not dreaming then of a publication of this kind, my notes were
only taken for private reference, and were not as full as they should be for
a veterinary treatise. I *might* supply some of these omissions accurately
from *recollection*, but do not consider it proper thus to *endanger* the accuracy of records, which as far as they go, I think may *now* be *implicitly relied on*. My post-mortem examinations were made at intervals snatched
from other pressing engagements. This fact, and certain preconceived
views—which I subsequently found *erroneous*—prevented me from making
those examinations, and more particularly the *records* of them, as minute
and extended as could be wished. I then sought only to convince *myself*
of the true nature and character of the disease.

In detailing the results of my experience in the premises, I conceive it a
duty to frankly state the *whole facts*. The records of *mismanagement* and
error, are often as useful, nay, more so, than those of successful management, and it is a pitiful pride which prevents any man, who pretends to
communicate information to the public, from giving that public the benefit of his examples which are to be *avoided*, as well as those which are to
be *followed*.

Up to February, my sheep remained apparently perfectly sound, and
they were in good flesh. Each flock had excellent shelters, were fed regularly, etc., and although sheep were beginning to perish about the country, my uniform previous impunity in these " bad winters " led me to entertain no apprehensions of the prevailing epizoötic. About the first of
February, my sheep went into the charge of a new man, hired upon the
highest recommendations. A few days after, I was called away from home
for a week. The weather during my absence was, a part of the time, very
severe. The sheep-house occupied by one flock containing one hundred
sheep, was, with the exception of two doors, as close a room as can be
made by nailing on the wall-boards vertically and without lapping, as is
common on our Northern barns.* One of the doors was always left open,
to permit the free ingress and egress of the sheep, and for necessary ventilation. A half dozen ewes which had been untimely impregnated by a
neighbor's ram, were on the point of lambing, and it being safer to confine
the ewes in a warm room over night, the shepherd, instead of removing
them to such a room, confined the whole flock in the sheep-house every
night, and rendered it warm by closing *both doors !* After two or three
hours, the air must have become excessively impure. On entering the
sheep-house, on my return, I was at once struck with the fetid, highly offensive smell. A change, too, slight but ominous, had taken place in the

* Boards in these cases shrink so as to leave slight cracks between them.

appearance of a part of the flock. They showed no signs of violent colds, I heard no coughing, sneezing, or labored respiration—and the only indication of catarrh which I noticed, was a nasal discharge, by a few sheep. But those having this nasal discharge, and some others, looked dull and drooping; their eyes ran a little—were partially closed, the caruncle and lids looked pale—their movements were languid—and the shepherd complained that they did not eat quite so well as the others. The pulse was nearly natural—though I *thought* a trifle too languid.

Not knowing what the disease was—and fully believing that depletion by bleeding or physic was not called for, let the disease be what it would. I contented myself with thoroughly purifying the sheep house—seeing that the feeding, etc.,* was managed with the greatest regularity—and closely watching the farther symptoms of disease in the flock. In about a week, the above described symptoms were evidently aggravated, and there had been a rapid emaciation, accompanied with debility, in the sheep first attacked. The countenance was exceeding dull and drooping—the eye kept more than half closed—the caruncle, lids, &c. almost bloodless—a gummy yellow secretion below the eye—thick glutinous mucus adhering in and about the nostrils—appetite feeble—pulse languid—and the muscular energy greatly prostrated. Nothing unusual was yet noticed about their stools or urine.

I now had all the diseased sheep removed from the flock, and placed in rooms the temperature of which could be easily regulated.

I commenced giving slight tonics and stimulants, such as gentian, ginger, etc., but apparently with no material effect. They rapidly grew weaker, stumbled and fell as they walked, and soon became unable to rise. The appetite grew feebler—the mucus at the nose, in some instances, tinged with dark grumous blood—the respiration oppressed, and they died within a day or two after they became unable to rise.

I proceeded to make some post-mortem examinations, which I shall here detail, although, as I have before remarked, they are extremely imperfect. I was at first inclined to suspect that the primary disease was one of some of the abdominal or thoracic viscera, and this impression was confirmed by the abnormal condition of these viscera in the first subjects examined. I therefore improperly confined my attention to these, and some of the external tissues, *without any examination of the interior organs of the head and neck.* I shall give my notes verbatim as they were taken down at the time, whether the appearances detailed have, as I *now believe,* any connection with the fatal disease or not.

Case 1st. Old sheep. Much emaciated—mouth and lips covered with yellow froth—yellow waxy matter under eyes—adhesive mucus in and about nostrils. On opening, external tissues appear healthy—two hydatids on omentum of the size of a walnut—gall-bladder enlarged and enormously distended with pale, and apparently not properly eliminated bile—gall-bladder slightly adhering to omentum—mesenteric glands enlarged—other abdominal viscera believed to be normal—fæces in rectum thought to indicate a constipated habit—stomachs rather empty. Thoracic viscera healthy.

Case 2d. Two years old. External appearances as in Case 1st, with the exception of the yellow froth about the mouth. External tissues healthy. Gall-bladder very small and nearly empty—bile pale and uneliminated—mesenteric glands enlarged—schirrous tumor at the junction

---

* They had been fed with bright hay three times a day, and turnips. As those affected as above did not eat their turnips well, I commenced feeding some oats, *in addition* to the turnips. I believed that a generous food was called for, and I gave it

of the cœcum and colon of the size of a butternut. Superior lobe of left lung adherent to pleura costalis—three lobes of right lung ditto, with slight traces of *recent* inflammation. Hydro-pericarditis—the pericardium slightly inflamed and containing something more than a gill of serum.

Case 3d. Old, and in lamb. External appearances and tissues as in Case 2d. Omentum dark-yellowish, or yellowish-brown by deposition of lymph, the result of inflammatory action—gall-bladder precisely as in Caso 2d—tabes mesentrica or enlargement of the mesenteric glands, as in the preceding cases. Middle lobe of right lung slightly hepatized, and adherent to pleura costalis—hydro-pericarditis, (a gill of serum in pericardium.)

Case 4th. Yearling ram. External appearances and tissues as in preceding cases. Two small hydatids on omentum—gall-bladder as in two preceding cases—mesenteric glands as in preceding cases. Traces of diarrhea. Thoracic viscera healthy.

Case 5th. Lamb. External appearance as in preceding cases—omentum as in Case 3d, and small hydatid on it—gall-bladder as in three preceding cases—ditto of mesenteric glands. Thoracic viscera healthy.

Case 6th. Four-year-old ram, killed for examination, in the first stage of the disease. Yet strong, appetite good, in fair condition, and exhibited no particular external indications of disease except running at the eyes, a slight gummy deposition below them—and some mucus about the nostrils. Gall-bladder but little better filled than in preceding cases—mesenteric glands same as in preceding cases. Thoracic viscera healthy.

Remarks on Preceding Cases.—I had started on the supposition that the fatal disease would be found one of the lungs, consequent on catarrh. I thought it *might* prove a species of pneumonia, though some of the characteristic symptoms of that disease seemed to be wanting; but I believed it would rather prove to be phthisis pulmonalis, or pulmonary consumption. To the last disease, when it assumes the form of what is popularly called "quick consumption," it seemed to me to bear several striking analogies. But the post-mortem examinations above detailed, entirely overthrow these suppositions. Except in Case 2d, there were no manifestations of *recent* inflammation of the lungs. The adhesions in Case 3d, were evidently referable to a past date. In the other four cases, the lungs were in a healthy condition—exhibiting not a trace of hepatization, tubercles, ulcers, or other abnormal action! In Case 6th, where the disease was in its first observable and therefore inflammatory stage, *none* of the thoracic viscera presented a particle of inflammation!

Then what *was* the disease? It was evidently the same in the several cases, yet the lesions disclosed by post-mortem examination were very various. Hence, I was led to conclude that these lesions were the results of *symptomatic* disease, and that the *primary* one was not yet discovered. The malady continued to spread. New cases occurred daily—it began to exhibit itself in my other flocks. It had manifestly put on the character of an epizoötic—or, if I may be permitted to coin a word, an en-zoötic. I now gave orders to have every sheep removed from the several flocks, as soon as it should be attacked with disease. I also resolved on more extended post-mortem examinations The following are the notes taken in the immediately succeeding cases.

Case 7th. Yearling. External appearance as in the preceding cases—external tissues normal—mesenteric glands slightly enlarged—gall-bladder of natural size, with good bile, and with the natural discolorations about it. Thoracic viscera healthy, with exception of pericardium, which exhibited traces of recent inflammation and contained a gill of serum.

The thorax also contained considerable fluid, which escaped without admeasurement.

I now examined the bronchial tubes, the lower portions of the windpipe,
œsophagus, &c., and found them all in an apparently healthy condition.
Before tracing these passages to the throat, I removed the upper portion
of the skull and carefully examined the brain and its investing membranes. All seemed in a perfectly normal state. I then made a longitu-
dinal section down through the middle part of the whole head, as is shown
in fig. 49, and *the seat and character of the fatal malady stood at once
revealed !*

The mucous membrane lining the whole nasal cavity, highly congested
and thickened throughout its whole extent, betrayed the most intense inflammation. At the junction of the cellular ethmoid bones with the cribriform plate, (in the ethmoidal cells,) slight ulcers were forming on the mem-
branous lining ! The inflammation also extended to the mucous membrane of the pharynx, and say three inches of the upper portion of the œsophagus. Here it rather abruptly terminated.

Case 8th. Old, in lamb. External appearances as in preceding cases—
abdominal parietes healthy—all the viscera apparently healthy. The inflammation of the mucous membrane lining the nasal cavity, pharynx, and
upper portion of œsophagus, as in Case 7th, only not quite so acute—no
ulcers on the membrane.

Cases 5th and 6th *reviewed.* The heads of these two subjects having
been accidentally preserved, I examined them, and found the inflammatory
action of the mucous membrane same as in cases 7th and 8th. Nor have I
a particle of doubt that the same would have been found the case in *all*
the preceding subjects, had they been examined.

Nosology and Treatment.—I had little difficulty in coming to the conclusion that the primary and main disease was a species of catarrh. It evidently,
however, differed from ordinary catarrh in its diagnosis, and in the extent
of the lesions accompanying both the primary and symptomatic diseases.

In no case, even in the first attack, did I notice anything—the fever—
the accelerated pulse—the redness about the eyes and nostrils—the coughing, etc., accompanying an ordinary severe attack of catarrh. And it was
for this reason that I was misled as to the seat of the malady. From the
very outset, according to my observations, the type of the disease was
typhoid—sinking—rapidly tending to fatal prostration.

How to reduce the local inflammation of the membrane lining the nasal
cavities, I was at a loss to determine. I was satisfied that there was too
much debility to admit of an antiphlogistic course of treatment. Still, to
make myself sure, I bled in three or four cases, and, as I anticipated, it
evidently accelerated the fatal catastrophe. Blistering could not be brought
near to the seat of the inflammation, excepting on the nose, and independent of the extreme difficulty of treating a blister on a spot so constantly
exposed to dirt, the rubbing of hay, etc., in winter feeding, I believed it
could have little effect, on an account of the thick nasal bone intervening
between it and *any portion* of the inflamed membrane. And, moreover,
the greater portion of the inflamed membrane rested on bones detached,
except at one extremity, from all connection with the nasal bone. I blew
Scotch snuff (through paper tubes) up the nostrils of some of the sheep,
for two objects—1, to remove, by sneezing, the mucus, which mechanically, and evidently injuriously, obstructed respiration ; and 2, to produce a
new action, by which an increased mucous secretion would be excited,
and thus the congested membrane relieved. But, farther than this, I re-

sorted to no local or other treatment designed specifically to reach the local inflammation.

The next step was to fix on the constitutional treatment. The liver was evidently in a torpid state. There was a functional derangement in the mesenteric and probably other glands, and a want of activity in the general secretory system. What medicine would stimulate the liver, cause it to secrete the proper quantity as well as quality of bile, change the morbid action of the glands and secretory system, and restore activity and health to the vital functions generally ? In my judgment, nothing promised so well as mercury ; and by its well known effect on the entire secretory system, it would powerfully tend to relieve the congested membranes of the head. In this opinion I was joined by a learned and experienced physician, who, both as a matter of taste and humanity, has given no little attention to veterinary science and practice. The proto-chloride of mercury (calomel) was supposed to possess too much specific gravity to reach the fourth stomach, with any certainty, administered in a liquid ; and if administered as a ball or pill, it would be almost *sure* not to reach that stomach.* The dissolved bi-chloride of mercury (corrosive sublimate) was therefore hit upon. One grain was dissolved in two ounces of water, and one-half ounce of the water (or one-eighth of a grain of corrosive sublimate) was exhibited in a day, in two doses.

As constipation existed in most of the cases, it was thought that the bowels required to be stimulated into action, and slightly evacuated with a mild laxutive. Having noticed in similar cases of debility and torpor of the intestinal canal, that purgation is often followed by a serous diarrhea, difficult to correct, and leading to rapid prostration, and there being no intestinal irritation to suffer exacerbation, I thought that rhubarb—from its well known tendency to give tone to the bowels, and its secondary effect as a mild astringent—was particularly indicated. It was given in a decoction—the equivalent of ten or fifteen grains at a dose—accompanied with the ordinary carminative and stomachic adjuvants, ginger and gentian, in infusion.

To a portion of the sheep I administered the rhubarb and its adjuvants alone ; to others I gave the bi-chloride of mercury *in addition* to the preceding. I employed these courses of treatment in a number of cases, the records of all which have been accidentally destroyed with the exception of the following three.

Case 9th. Ram, three years old. Has been drooping and weak, with feeble appetite, for some time—has been separated from flock. Has eaten his oats irregularly for several days, and refused turnips, bran, etc., altogether—much emaciated—eyes partly closed, with a yellowish deposit below them—caruncle and lids bloodless—nostrils impeded with adhesive yellowish mucus.

March 17th. Weaker than before—would not rise to feed—not seen to eat or ruminate—gait, when helped up, weak and staggering ; eyes nearly closed—stooled dry, hard fæces—urine dark and reddish. Exhibited rhubarb with ginger and gentian in gruel—blew snuff into nostrils. March 18th, morning—Weaker ; refused to eat anything. Exhibited rhubarb, ginger and gentian in gruel. Noon—Urine seemingly bloody : breathing labored : exhibited corrosive sublimate in gruel. Night—Dying. March 19th, morning—Dead.

Post-mortem appearances. Inner edges of both lobes of liver softened about two inches from horizontal fissure : hypropericarditis and hydro-

* For reasons which will be hereafter given under the head of " The Proper Way of Administering Medicine "

thorax—nearly half pint of serum in latter. Othe
normal. Lining of superior portion of œsophagus a
Case 8th.

Case 10th. Three-year-old ewe. Drooping for so
emaciated and weak: cannot rise without help: ap
trils and eyes as in Case 9th: appetite considerable
served. March 17th. Exhibited ginger and gen
snuff in nostrils. Latter produced sneezing and a
18th: Morning. Weaker and would not eat. I
lier: ate hay and grain; exhibited ginger and gen
uations thin: urine of a natural color. 19th. Mori
Exhibited same remedies as before. The same co
three days: the sheep appearing rather to gain, whe
found dead. No post-mortem examination made.

Case 11th. Old ewe. Symptoms precisely as in
occasional grinding of the teeth. March 17th. Treat
9th. Lived three days and appeared to rally a little,
lamb and died. Post-mortem examination. Abdon
—gall-bladder filled with pale bile: liver normal
throughout its entire extent, and pale: portions of it
ganized than others: no parasites in its ducts. The
Sub-acute inflammation of the mucous lining of the no
superior portion of the œsophagus. Slight ulcer in 1

I made various other post-mortem examinations.
in every case were in a more or less abnormal state
same variety in the locality of the diseased action
cases. But so far as the seat and character of the c
concerned, it was *uniform in every case.* The only
tensity, as exhibited by the extent of the lesions.

Not a single sheep recovered after the emaciation
ceeded to any great extent! One such only lingered
Its wool gradually dropped off: it seemed to rally a
and then relapse; and it perished one night in a rai
erality of instances the time from the first observed s
varied from ten to fifteen days. A few died in a sho

In the three cases last detailed, the disease had evi
far to be arrested by *any* treatment. I much regret t
of the other cases, which would throw farther lig
thought that the treatment produced favorable effect:
particularly when resorted to at the commencement o
events, some of the sheep recovered under the treatm
der that including the exhibition of the bi-chloride of
few, if any, recovered without any treatment. Cand
however, that the results of the treatment were far fr
isfactory—that the cases of recovery were much fewe
have merely stated what I believe to be the facts in tl
feel prepared to make *any recommendations.*

The epizoötic gradually abated toward spring, and
ncen in perfect health.

Near spring, many farmers found what seemed to th
ber of grubs in the head (frontal sinuses) of the she
prevailing epizoötic, and therefore they attributed the
and this seems to be the prevailing popular opinion.
cases in my flock, I discovered more or less grubs;
'nstances an unusual number. In other cases wher

toms and the post-mortem appearances were almost identical, no grubs were to be seen. For this reason, and others which I shall assign when treating of grub in the head, I conclude that the popular opinion is erroneous.

THE ROT.—The existence and prevalence of the Rot in the United States have been sufficiently alluded to in Letter XIV. Notwithstanding its comparative rareness here, so far as is known, at present, I think it expedient to give a full description of it. It may be more prevalent hereafter, or it may be found peculiar to localities where sheep have not yet been introduced. And whether so or not, as its existence will often be feared and suspected in diseased flocks, it is proper that the flock-master always have it in his power to clearly identify this terrible destroyer.

The diagnosis of the disease is thus given by Mr. Spooner.*

"The first symptoms attending this disease are by no means strongly marked; there is no loss of condition, but rather apparently the contrary; indeed, sheep intended for the butcher have been purposely *cothed* or rotted in order to increase their fattening properties for a few weeks, a practice which was adopted by the celebrated Bakewell. A want of liveliness and paleness of the membranes generally may be considered as the first symptoms of the disease, to which may be added a yellowness of the caruncle at the corner of the eye. Dr. Harrison observes, 'when in warm, sultry or rainy weather, sheep that are grazing on low and moist lands feed rapidly, and some of them die suddenly, there is fear that they have contracted the rot.' This suspicion will be farther increased if, a few weeks afterward, the sheep begin to shrink and become flaccid about the loins. By pressure about the hips at this time a crackling is perceptible now or soon afterward, the countenance looks pale, and upon parting the fleece the skin is found to have changed its vermilion tint for a pale red, and the wool is easily separated from the pelt; and as the disorder advances the skin becomes dappled with yellow or black spots. To these symptoms succeed increased dullness, loss of condition, greater paleness of the mucous membranes, the eyelids becoming almost white and afterward yellow. This yellowness extends to other parts of the body, and a watery fluid appears under the skin, which becomes loose and flabby, the wool coming off readily. The symptoms of dropsy often extend over the body, and sometimes the sheep becomes *chockered*, as it is termed—a large swelling forms under the jaw, which, from the appearances of the fluid it contains, is in some places called the *watery poke*. The duration of the disease is uncertain; the animal occasionally dies shortly after becoming affected, but more frequently it extends to from three to six months, the sheep gradually losing flesh and pining away, particularly if, as is frequently the case, an obstinate purging supervenes."

Mr. Youatt thus describes the post-mortem appearances: †

"When a rotted sheep is examined after death, the whole cellular tissue is found to be infiltrated, and a yellow serous fluid everywhere follows the knife. The muscles are soft and flabby: they have the appearance of being macerated. The kidneys are pale, flaccid, and infiltrated. The mesenteric glands enlarged, and engorged with yellow serous fluid. The belly is frequently filled with water or purulent matter; the peritoneum is everywhere thickened, and the bowels adhere together by means of an unnatural growth. The heart is enlarged and softened, and the lungs are filled with tubercles. The principal alterations of structure are in the liver. It is pale, livid, and broken down with the slightest pressure; and on being boiled it will almost dissolve away. When the liver is not pale, it is often curiously spotted. In some cases it is speckled like the back of a toad. Nevertheless, some parts of it are hard and schirrous; others are ulcerated, and the biliary ducts are filled with flukes. Here is the decided seat of disease, and it is here that the nature of the malady is to be learned. *It is inflammation of the liver.* . . . The liver attracts the principal attention of the examiner: it displays the evident effects of acute and destructive inflammation; and still more plainly the ravages of the parasite with which its ducts are crowded. Here is plainly the original seat of the disease—the center whence a destructive influence spreads on every side. . . . The Fluke—the *Fasciola* of Linnæus—the *Distoma hepaticum* of Rhodolphi—the *Planaria* of Goese—is found in the biliary ducts of the sheep, the goat, the deer, the ox, the horse, the ass, the hog, the dog, the rabbit, the guinea-pig, and various other animals, and even in the human being. It is from three quarters of an inch to an inch and a quarter in length, and from one-third to half an inch in greatest breadth.

* Spooner, p. 391, et supra.
† Youatt, p. 447, et supra.

Figs. 56 and 58 represent this parasite of its usual size and appearance, and its resemblance to a minute sole, divested of its fins, is very striking. The head is of a pointed form, round above and flat beneath; and the mouth opens laterally instead of vertically.

Fig. 56.   Fig. 57,   Fig. 58.

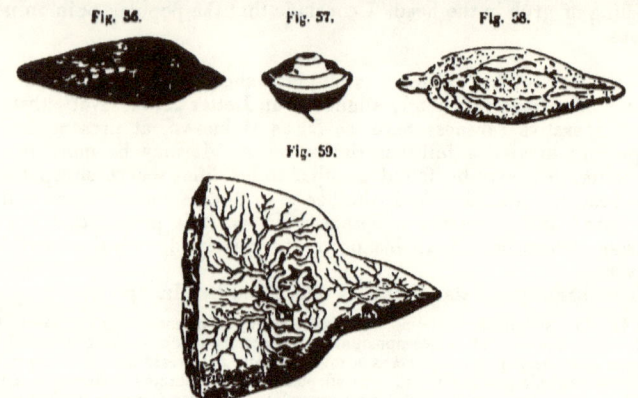

Fig. 59.

THE FLUKE.

There are no barbs or tentacules, as described by some authors. The eyes are placed on the most prominent part of the head, and are very singularly constructed (fig. 57). They have the bony ring of the bird. . . . The anastomoses of the blood-vessels which ramify over the head are plainly seen through a tolerable microscope. The circulating and digestive organs are also evident, and are seated almost immediately below the head. The situation of the heart is seen in fig. 56, and the two main vessels evidently springing from it, and extending through almost the whole length of the fluke. Smaller blood-vessels, if so they may be called, ramify from them on either side. The convolutions of the bowels appear in fig. 59, and the vent, both for the fœces and the ova, and probably for the connection between the sexes, is on the under part, and almost close to the neck. . . . .

In the belly, if so it may be called, are almost invariably a very great number of oval particles, hundreds of which, taken together, are not equal in bulk to a grain of sand. They are of a pale red color, and are supposed to be the spawn or eggs of the parasite. . . . .

There can be no doubt that the eggs are frequently received in the food. Having been discharged with the dung, they remain on the grass or damp spot on which they may fall, retaining their vital principle for an indefinite period of time. . . . They find not always, or they find not at all, a proper nidus in the places in which they are deposited; but taken up with the food, escaping the perils of rumination, and threading every vessel and duct until they arrive at the biliary canal, they burst from their shells, and grow, and probably multiply. . . . .

Leeuwenhoek says that he has taken 870 flukes out of one liver, exclusive of those that were cut to pieces or destroyed in opening the various ducts. In other cases, and where the sheep have died of the rot, there were not found more than ten or twelve. . . .

Then, is the fluke worm the cause or the effect of rot? To a certain degree both. They aggravate the disease; they perpetuate a state of irritability and disorganization, which must necessarily undermine the strength of any animal. . . . . Notwithstanding all this, however, if the fluke follow the analogy of other entozoa and parasites, it is the effect and not the cause of rot. . . . .

The rot in sheep is evidently connected with the soil or state of the pasture. It is confined to wet seasons, or to the feeding on ground moist and marshy at all seasons. It has reference to the evaporation of water, and to the presence and decomposition of moist vegetable matter. It is rarely or almost never seen on dry or sandy soils and in dry seasons; it is rarely wanting on boggy or poachy ground, except when that ground is dried by the heat of the summer's sun, or completely covered by the winter's rain. On the same farm there are certain fields on which no sheep can be turned with impunity. There are others that seldom or never give the rot. The soil of the first is found to be of a pervious nature, on which wet cannot long remain—the second takes a long time to dry, or is rarely or never so. . . . .

Some seasons are far more favorable to the development of the rot than others, and there is no manner of doubt as to the character of those seasons. After a rainy summer or a moist autumn, or during a wet winter, the rot destroys like a pestilence. A return and a continuance of dry weather materially arrests its murderous progress. Most of the sheep that had been already infected die ; but the number of those that are lost soon begins to be materially diminished. It is, therefore sufficiently plain that the rot depends upon, or is caused by, the existence of moisture. A rainy season and a tenacious soil are fruitful or inevitable sources of it. . . . . The mischief is effected with almost incredible rapidity."

Mr. Youatt here gives various instances to prove that rot is engendered in a few hours and even minutes. * He farther says :

"It is an old observation that all pasture that is suspected to he unsound, the sheep should be folded early in the evening, before the first dews begin to fall, and should not be released from the fold until the dew is partly evaporated. . . . .
Then the mode of prevention—that with which the farmer will have most to do, for the sheep having become once decidedly rotten, neither medicine nor management will have much power in arresting the evil—consists in altering the character of as much of the dangerous ground as he can, and keeping his sheep from those pastures which defy all his attempts to improve them. . . . . If all unnecessary moisture is removed from the soil, or if the access of air is cut off by the flooding of the pasture, no poisonous gas has existence, and the sheep continue sound. . . . .
The account of the *treatment* of rot must, to a considerable extent, be very unsatisfactory."

Mr. Youatt proceeds to recommend the *sale* of sheep to the *butcher* when they are found to be rotted ! Rot hastens for a short period the accumulation of fat. Bakewell—a man whose name is associated with the exhibition of prodigious abilities in the improvement of stock, but, in my mind tarnished also by an equal exhibition of selfishness and absolute meanness—displayed a characteristic sagacity in *purposely* rotting his sheep to avail himself of the above circumstance !† It is with pain I make the following quotation from Youatt—the only thing of such a character I remember to have noticed in his voluminous works :

· It is one of the characters of the rot to hasten, and that to a strange degree, the accumulation of flesh and fat. Let not the farmer, however, push this experiment too far. Let him carefully overlook every sheep daily, and dispose of those which cease to make progress, or which seem beginning to retrograde. It has already been stated that the meat of the rotted sheep, in the early stage of the disease, *is not like that of the sound one; it is* pale and not so firm; but it is not unwholesome (!) and it is coveted by certain epicures, *who, perhaps, are not altogether aware of the real state of the animal (!!)* All this is *matter of calculation*, and must be left to the owner of the sheep; except that, if the breed is not of very considerable value, and the disease has not proceeded to emaciation or other fearful symptoms, the first loss will probably be the least; and if the owner can get anything lil o a tolerable price for them, the sooner they are sent to the butcher, or *consumed at home*, the better. Supposing, however, that their appearance *is beginning to tell tales about them*, and they are too far gone to be disposed of in the market or consumed at home, are they to be abandoned to their fate ? No : far from it."

Conceding to Mr. Youatt the whole benefit of that saving clause about " consumption at home," the above sentence is one which I could well wish stricken from his valuable work. The sale of the meat of diseased animals, for human consumption, is abhorrent to decency and propriety, and there is not a respectable American family which would not revolt at the idea of either selling or consuming such meat.

Of the treatment of rot, Mr. Youatt continues :

" If it is suited to the convenience of the farmer, and such ground were at all within his reach, the sheep should be sent to a salt-marsh in preference to the best pasture on the best farm. There it will feed on the salt incrusted on the herbage, and pervading the pores of every blade of grass. A healthy salt-marsh permits not the sheep to become rotten which graze upon it; and if the disease is not considerably advanced, it cures those which are sent upon it with the rot. . . . Are there any indications of fever—heated month, heaving flanks, or failing appetite ? Is the general inflammation beginning to have a determination to that part on which the disease usually expends its chiefest virulence ? Is there yellow-

* Youatt, p. 453.
† So say both Spooner and Youaz.

ness of the lips and of the mouth, of the eyes, and of the skin? At the same time, are there no indications of weakness and decay? Nothing to show that the constitution is fatally undermined? Bleed—abstract, according to the circumstances of the case, eight, ten, or twelve ounces of blood. There is no disease of an inflammatory character at its commencement which is not benefited by early bleeding. To this let a dose of physic succeed—two or three ounces of Epsom salts, administered in the cautious manner so frequently recommended; and to these means let a change of diet be immediately added—good hay in the field, and hay, straw, or chaff, in the straw-yard.

The physic having operated, or an additional dose, perchance, having been administered in order to quicken the action of the first, the farmer will look out for farther means and appliances. . . . . Two or three grains of calomel may be given daily, but mixed with half the quantity of opium, in order to secure its beneficial, and ward off its injurious effects on the ruminant. To this should be added—a simple and cheap medicine, but that which is the sheet-anchor of the practitioner here—common salt. . . . . In the first place, it is a purgative inferior to few, when given in a full dose; and it is a tonic as well as a purgative. . . . A mild tonic, as well as an aperient, is plainly indicated soon after the commencement of rot. The doses should be from two to three drachms, repeated morning and night. When the inflammatory stage is clearly passed, stronger tonics may be added to the salt, and there are none superior to the gentian and ginger roots; from one to two drachms of each, finely powdered, may be added to each dose of the salt. . . . . The sheep having a little recovered from the disease, should still continue on the best and driest pasture on the farm, and should always have salt within their reach. . . . . The rot is not infectious."

———

DIARRHŒA.—This disease is often more properly a *nervous* than a *febrile* one—in the former case, a morbid increase of the peristaltic motion of the bowels—in the latter, an inflammation of the *mucous* coat of the smaller intestines. But for the purpose of viewing it in connection with dysentery, to which it is sometimes closely allied, and into which it often runs—and which is clearly a febrile disease—it will be described here.

Common diarrhœa, purging, or scours, manifests itself simply by the copiousness and fluidity of the alvine evacuations. It is brought on by a sudden change from dry feed to green, or by the introduction of improper substances into the stomach. It is important to clearly distinguish this disease from dysentery. In diarrhœa there is no apparent general fever; the appetite remains good; the stools are thin and watery, but unaccompanied with slime (mucus) and blood; the odor of the fæces is far less offensive than in dysentery; the general condition of the animal is but little changed.

Treatment.—Confinement to dry food for a day or two, and a gradual return to it, oftentimes suffice. I have rarely administered anything to grown sheep, and never have lost one from this disease. To lambs, especially if attacked in the fall, the disease is more serious. If the purging is severe, and especially if any mucus is observed with the fæces, the feculent matter should be removed from the bowels by a gentle cathartic—as half a drachm of rhubarb, or an ounce of linseed-oil, or half an ounce of Epsom salts to a lamb. This should always be followed by an astringent, and in nine cases out of ten, the latter will serve in the first instance. I generally administer, say, $\frac{1}{4}$ oz. of prepared chalk in half a pint of tepid milk, once a day for two or three days, at the end of which, and frequently after the first dose, the purging will have ordinarily abated or entirely ceased.

The following is the formula of the English "sheep's cordial" usually prescribed in cases of diarrhœa by the English veterinarians, and there can be no doubt it is a safe and excellent remedy—better probably than simple chalk and milk, in severe cases: Take of prepared chalk one ounce, powdered catechu half an ounce, powdered ginger two drachms, and powdered opium half a drachm; mix them with half a pint of peppermint water—give two or three table-spoonsfull morning and night to a grown sheep, and half that quantity to a lamb

DYSENTERY.—Dysentery is caused by an inflammation of the mucous or inner coat of the larger intestines, causing a preternatural increase in their secretions, and a morbid alteration in the character of those secretions. It is frequently consequent on that form of diarrhea which is caused by an inflammation of the mucous coat of the smaller intestines. The inflammation extends throughout the whole alimentary canal, increases in virulence, and it becomes dysentery—a disease frequently dangerous and obstinate in its character, but fortunately not common among sheep in this part of the United States. Its diagnosis differs from that of diarrhea in several readily observed particulars. There is evident fever; the appetite is capricious, ordinarily very feeble; the stools are as thin or even thinner than in diarrhea, but much more adhesive in consequence of the presence of large quantities of mucus. As the erosion of the intestines advances, the fæces are tinged with blood; their odor is intolerably offensive; and the animal rapidly wastes away. The course of the disease extends from a few days to several weeks.

Treatment.—I have seen but a few well-defined cases of dysentery, and in the half-dozen instances which have occurred in my own flock, I have usually administered a couple of purges of linseed-oil, followed by chalk and milk as in diarrhea (only doubling the dose of chalk), and a few drops of laudanum, say twenty or thirty—with ginger and gentian. According to my recollection, about one-third of the cases have proved fatal, but they have usually been old and feeble sheep.

Farther inquiry satisfies me that moderate bleeding should be resorted to in the first or inflammatory stage of the disease, or whenever decided febrile symptoms are found to be present.

Mr. Youatt prescribes bleeding, cathartics, mashes, gruel, &c. He says:

"Two doses of physic having been administered, the practitioner will probably have recourse to astringents. The sheep's cordial will probably supply him with the best; and to this, tonics may soon begin to be added—an additional quantity of ginger may enter into the composition of the cordial, and gentian powder will be a useful auxiliary. With this—as an excellent stimulus to cause the sphincter of the anus to contract, and also the mouths of the innumerable secretory and exhalent vessels which open on the inner surface of the intestine—half grain of strychnine may be combined. . . . . Smaller doses should be given for three or four days."

GARGET—Is an inflammation of the udder, with or without general inflammation. Where simply an inflammation of the udder, it is usually caused by a too great accumulation of milk in the latter prior to lambing, or in consequence of the death of the lamb. It is not the serious malady, here, described by the English veterinarians.

Treatment.—Drawing the milk partly from the bag so that the hungry lamb will butt and work at it an unusual time in pursuit of its food, and bathing it a few times in cold* water, usually suffices. If the lamb is dead, the milk should be drawn a few times, at increasing intervals, washing the udder for some time in cold water at each milking. In cases of obdurate induration, the udder should be anointed with iodine ointment. If there is general fever in the system, an ounce of Epsom salts may be given.

NERVOUS DISEASES.

APOPLEXY.—Soon after the sheep are turned to grass in the spring, one of the best conditioned sheep in the flock is sometimes suddenly found dead.

---

* The English veterinarians recommended *warm* fomentations.

The symptoms which precede the catastrophe are occasionally noted. The sheep leaps frantically into the air two or three times, dashes itself on the ground and suddenly rises, and dies in a few moments. Such cases occur but now and then, and none have ever occurred in my flock to my knowledge. I have therefore had no opportunity of observing the diagnosis, or making dissections. There can be little doubt, however, that the disease is apoplexy.

Desirous to raise the condition of a poorish flock (the poorest sheep culled from my other flocks) somewhat too rapidly, perhaps, some winters since, in addition to good hay three times a day, I ordered them fed a gill of oats per head; and as rapidly as it could be done without bringing on scours, I had them fed a liberal allowance of Swedish turnips—about as much as they would eat up clean. They gained perceptibly. One day a sheep was reported to me as having become suddenly blind and motionless. I immediately examined it. It was in good fair condition. It stood with its head a little down—its eyes were glassy and staring—it was stone blind! The evening before nothing unusual had been perceived about it. I bled it at the inner angle of each eye, and the blood had scarcely started before its sight began to return. In less than a minute it walked off among its companions. It had no relapse. Another case was soon reported; I treated it in the same way, and with the same apparent effect. The symptoms soon returned, however, and I bled again. This appeared to produce but a partial restoration of the sight. The sheep would not follow its companions into and out of the sheep-house. When approached, it would run about knocking its head against fences, &c. It lost condition, finally became unable to rise, and died. Another one, after being bled, fed regularly, but its sight was never restored. It lived along thus for three or four weeks, and then fell into a hole containing water, and perished. Another apparently recovered, all but sight, and continued in my flock for more than a year afterward. The eye was bright and *clear*, as in *gutta serena*, and the blindness would not be suspected, unless the sheep was cornered up. Then, if the catchers remained momentarily *still*, it would as soon run into their arms or against the fence, as in any other direction. Perhaps fifteen cases occurred. In three or four instances the *blind* sheep, when they moved, constantly traveled round in a circle. In about as many cases, they twisted themselves about without progressing, the head was drawn round toward one side, they fell, ground their teeth, and their mouths were covered with a frothy mucus. In neither of the latter description of cases did bleeding at the inner angles of the eyes afford anything more than temporary relief. They all proved fatal.

At the time these things occurred, I regret to say that I had paid but very little attention to veterinary science, and had never made a dissection. I did nothing but bleed at the inner angles of the eyes, and made no post-mortem examinations.

Taking into consideration the feed and the symptoms, there can be but little doubt, I think, that all these cases were referable to a *determination of blood to the brain*. The sheep were not *fat*, but the secretions of blood were rapidly and powerfully increased by rich and abundant food.

Treatment.—If the eyes are prominent and fixed, the membranes of the mouth and nose highly florid, the nostrils highly dilated, and the respiration labored and stertorous, the veins of the head turgid, the pulse strong and rather slow, and these symptoms attended by a partial or entire loss of sight and hearing, it is one of those decided cases of apoplexy which require immediate and decided treatment. As the good effects of venesection, in all cases, and especially in this, depend not only upon the amount of blood abstracted, but also upon the rapidity with which it is drawn from the veins, the eye-veins are not the proper ones to open. They are so small that the blood flows slowly, and if cut directly *across*, as is usually done, they soon contract, and the flow of blood is arrested before a sufficient quantity has been abstracted. It is better to have recourse at once to the jugular vein. The animal should be bled until an obvious constitutional effect is produced—the pulse lowered and the rigidity of the muscles relaxed. An aperient should at once follow bleeding, and if the animal is strong and plethoric, a sheep of the size of the Merino would require at least two ounces of Epsom salts, and one of the large mutton sheep more. If this should fail to open the bowels, half an ounce of the salts should be be given, say, twice a day.

In the milder cases which I have mentioned as occurring in my own flock, I think had I bled more thoroughly, in the very first attack, and given a mild aperient of Epsom salts, most of the sheep would have recovered.

PHRENITIS, TETANUS, EPILEPSY, PALSY, RABIES.—I never have seen a well-defined case of either of these maladies among our sheep, though, in a few instances, something which struck me at the time as somewhat analogous to paralysis or palsy. Palsy is a diminution or entire loss of the powers of motion in some part of the body. I have occasionally seen, in the winter, poor lambs, or poor pregnant ewes, or poor feeble ewes immediately after yeaning in the spring, lose the power of walking or standing rather too suddenly to have it satisfactorily referable to increasing debility. The animal seems to have lost all strength in its loins, and the hind-quarters are powerless. It makes ineffectual attempts to rise, and cannot stand if placed upon its feet.

Treatment.—Warmth, gentle stimulants, and good nursing, might raise the patient, but in nineteen cases out of twenty it would be more economical and equally humane, to at once deprive it of life.

COLIC.—Sheep are occasionally seen, particularly in the winter, lying down an , rising every moment or two, and constantly stretching their fore and hind legs so far apart that their bellies almost touch the ground. They appear to be in much pain, refuse all food, and not unfrequently die, unless relieved. This disease is popularly known as the " *stretches,*" and is erroneously attributed to introsusception of an intestine. Some farmers worry the sheep with a dog, and others hold it up by the hind legs, to effect a cure ! I consider it a sort of flatulent colic induced by costiveness.

Treatment.—Half an ounce of Epsom salts, a drachm of ginger, and sixty drops of essence of peppermint. The salts alone, however, will affect the cure, as will an equivalent dose of linseed-oil, or even hog's lard.

LETTER XVI.

DISEASES AND THEIR TREATMENT—(Continued.)

Cachectic Diseases...Hydatid on the Brain—diagnosis—common methods of treating it—treatment of French and English veterinarians...The Pelt Rot...Local diseases...Grub in the head—the nature of the disease, if one—erroneous popular opinions—location of the grub—description of the fly (*Œstrus ovis*)—method of attacking the sheep—conduct of the sheep—appearance of the larva—its habits—the chrysalis—the larva found in the heads of healthy sheep—not believed to be the cause or source of fatal disease—Mr. Bracy Clark's and Mr. Youatt's opinion—method of preventing and of expelling the grub... Scab—nature of it—habits of the *acari*—description of them—contagiousness of the disease—post-mortem appearances—treatment...Erysipelatous scab—treatment...Disease of Biflex Canal—nature and treatment...Hoofail—first indications—erroneous statements of foreign veterinarians—of Mr. Youatt—author's experience with it—diagnosis—chronic hoofail—can it be cured?—difficulties—preparation of the foot—ordinary treatment—proper treatment—cost of curing a flock—cheap partial remedies—suggestions—contagiousness of the disease—how communicated...Fouls—cause and treatment...Broncho-cele or goitre—diagnosis—treatment...Miscellaneous diseases...Poison from eating Laurel—symptoms—treatment...Sore Face—cause and treatment...Loss of cud—not a disease...Hoove—cause—symptoms—cure...Obstruction of Gullet, or choking—treatment...Fractures—treatment, &c...Method of administering medicine into the stomach...Method of bleeding...The place of feeling the pulse...List of medicines employed in treating the diseases of sheep...Ale...Aloes...Alum...Antimony...Arsenic...Blue Vitriol...Camphor....Carraway seeds....Catechu....Chalk...Corrosive Sublimate...Digitalis...Epsom Salts...Gentian...Ginger...Iodine...Lard...Lime, carbonate of...Lime, chloride of....Linseed Oil... Mercury...Muriatic Acid...Nitrate of Potash...Nitrate of Silver...Nitric Acid...Opium...Pepper... Pimento...Rhubarb...Salt...Sulphate of Iron...Sulphur....Sulphuric Acid....Spirit of Tar....Tar... Tobacco...Turpentine...Verdigris...Zinc.

CACHECTIC DISEASES.

HYDATID ON THE BRAIN.—This disease, known as turnsick, sturdy, staggers, etc., is spoken of by Chancellor Livingston, and other writers of reputation, as having occurred in this country within their own observation. I have never seen a case of it, and shall be obliged, therefore, to make use of the descriptions of others. Mr. Spooner says :

" The symptoms are a dull, moping appearance, the sheep separating from the flock, a wandering and blue appearance to the eye, and sometimes partial or total blindness ; the sheep appears unsteady in its walk, will sometimes stop suddenly and fall down, at others gallop across the field, and after the disease has existed for some time will almost constantly move round in a circle—there seems, indeed, to be an aberration of the intellect of the animal. These symptoms, though rarely all present in the same subject, are yet sufficiently marked to prevent the disease being mistaken for any other. On examining the brain of sturdied sheep, we find what appears to be a watery bladder, termed a hydatid, which may be either small or of the size of a hen's egg. This hydatid, one of the class of entozoöns has been termed by naturalists the *hydatis polycephalus cerebralis*, which signifies the *many-headed hydatid of the brain;* these heads being irregularly distributed on the surface of the bladder, and on the front part of each head there is a mouth surrounded by minute sharp hooks within a ring of sucking disks. These disks serve as the means of attachment by forming a vacuum, and bring the mouth in contact with the surface, and thus by the aid of the hooks the parasite is nourished. The coats of the hydatid are disposed in several layers, one of which appears to possess a muscular power. These facts are developed by the microscope, which also discovers numerous little bodies adhering to the internal membrane. The fluid in the bladder is usually clear, but occasionally turbid, and then it has been found to contain a number of minute worms. "

According to Mr. Youatt, this disease attacks many of the weakly lambs in the English flocks. It usually appears, he remarks, " during the first year of the animal's life, and when he is about or under six months old." It succeeds a " a severe winter and a cold, wet spring."— He says :

" If there is only one parasite inhabiting the brain of a sturdied sheep, its situation is very uncertain. It is mostly found beneath the pia-mater, lying upon the brain, and in or upon the scissure between the two hemispheres. If it is within the brain, it is generally in one of the ventricles, but occasionally in the substance of the brain, and, in a few instances, in that of the cerebellum . . . .

. This is a singular disease;' but it is a sadly prevalent and fatal one in wet and moorish districts. . . . . It is much more fatal in France than in Great Britain. It is supposed that nearly a million of sheep are destroyed in France every year by this pest of the ovine race. . . . .

The means of cure are exceedingly limited. They are confined to the removal or destruction of the vesicle. Medicine is altogether out of the question here."

Many barbarous methods have been adopted to rupture the hydatid, which I will not disgust you by repeating. Mr. James Hogg thrust a wire up the nostrils of the sheep, and through the plate of the ethmoid bone *into the brain*, and thus, as he assures us, punctured the hydatid and " cured many a sheep!"* This practice, which I cannot characterize otherwise than as atrocious, is justly condemned by Mr. Youatt. The dotted lines *d, e*, and *d, d*, in fig. 49, show how limited a portion of the brain could be reached with a wire or trochar by piercing the plate of the ethmoid bone—the only portion of the walls of the skull thin enough to be so pierced by a trochar introduced at the nostrils.

Mr. Parkinson " pulled the ears very hard for some time," and then cut them off close to the head ! †

Where the hydatid is not imbedded in the brain, its constant pressure, singularly enough, causes a portion of the cranium to be absorbed, and finally the part immediately over the hydatid becomes thin and soft enough to yield under the pressure of the finger. When such a spot is discovered, the English veterinarians usually dissect back the muscular integuments, remove a portion of the bone, carefully divide the investing membranes of the brain, and then, if possible, remove the hydatid whole —or, failing to do this, remove its fluid contents. The membranes and integuments are then restored to their position, and an adhesive plaster placed over the whole. The French veterinarians usually simply puncture the cranium and the cist with a trochar, and laying the sheep on its back, permit the fluid to run out through the orifice thus made. A common awl would answer every purpose for such a puncture. The puncture would be the preferable method for the unskilled practitioner. But when we take into consideration the hazard and cruelty attending the operation at best, and the conceded liability of a return of the malady—the growth of new hydatids—it becomes apparent that, in this country, it would not be worth while, unless in the case of uncommonly valuable sheep, to resort to any other remedy than depriving the miserable animal of life.

—

PELT ROT—Is classified as a disease by Mr. Livingston, and various other American writers. Mr. Livingston says :

" This is often mistaken for the scab, but it is in fact a different and less dangerous disease ; in this the wool will fall off, and leave the sheep nearly naked ; but it is attended with no soreness, though a white crust will cover the skin from the wool which has dropped. It generally arises from hard keeping and much exposure to cold and wet, and, in fact, the animal often dies in severe weather from the cold it suffers by the loss of its coat. The remedy is full feeding, and a warm stall, and anointing the hard part of the skin with tar, oil, and butter." ‡

I have seen frequent cases of the pelt rot, but never have done any thing for it, scarcely considering it a disease. If the condition of a poor sheep is raised as suddenly as practicable, by generous keep in the winter, the wool is very apt to drop off, and if yet cold, the sheep will require warm shelter.

* Hogg on Sheep, p. 59.
† Parkinson on Sheep, vol. 1, p. 412.
‡ Livingston on Sheep, Appendix, p. 179.

LOCAL DISEASES.

"GRUB IN THE HEAD."—If the "grubs" found in the frontal and maxillary sinuses of the sheep actually, in any case, produce *disease*, it must be, in my judgment, by the irritation and inflammation which they induce in the mucous membrane which lines those cavities. The popular theory that the grub causes death by boring through the bony walls which surround the brain, and attacking the substance of the brain itself, is, it seems to me, utterly absurd. The only part of the skull where it could even be fancied that such a perforation would be practicable, is the cribriform plate of the ethmoid bone (11 of fig. 49,) which is very thin and is pierced with numerous small holes for the passage of nerves. But an inspection of the same figure will show that the sinus where the parasite is generally found lodged, is not in immediate juxtaposition with the cribriform plate, and that a passage from the former to the brain, would lead directly through the frontal bone—the thickest one of the whole cranium. I never saw but one grub in the cells of the ethmoid bone near the cribriform plate, and that, I judged at the time, was thrown there accidentally by the violence attending the opening of the head.* But if the grub actually penetrates to the brain, the fact would readily be disclosed after death. The full-grown grub would necessarily leave an orifice of considerable diameter through the skull. Who has seen any such orifice in the cribriform plate or elsewhere? Who has seen *any* orifice but the natural ones of the cribriform plate, *filled* with the nerves which pass through them? The farmer splits open the head of a sheep with an ax, cutting, mangling and scattering its contents, by the repeated blows necessary to effect his purpose.—Under such circumstances grubs are sometimes found scattered through all the nasal cavities—over and among the brains—and *on the ground.*—The proof is just as strong, here, that prior to opening the head, some of the grubs were *on the ground*, as that they were in the brain!

The "grub" of popular parlance is the larva of the *Œstrus ovis*, or gad-fly of the sheep. The latter is represented of the natural size in figures 60 and 61. It is composed of five rings. It is tiger-colored on the back and belly, sprinkled with spots and patches of brown. The wings are striped. The comparative proportions of the head, corslet, wings, etc. are sufficiently seen in the cuts. He who desires a full, scientific description of these insects, or who would fully investigate their habits and economy, will do well to consult the excellent monograph of them by Mr. Bracy Clark, the celebrated veterinarian.

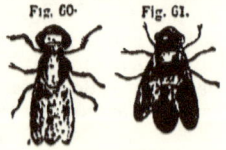

Fig. 60.　　　Fig. 61.

SHEEP GAD-FLY.

The sheep gad-fly is led by instinct to deposit its eggs within the nostrils of the sheep. Its attempts to do this, most common in July and August, are always indicated by the sheep, which collect in close clumps with their heads inward and their noses thrust close to the ground, and *into it*, if any loose dirt or sand is within their reach. If the fly succeeds in depositing its egg, it is immediately hatched by the warmth and moisture of the part, and the young grubs, or larvæ, crawl up the nose, finding their devious way to the sinuses, where, by means of their tentaculæ, they attach themselves to the mucous membrane lining those cavities. During the ascent of the larvæ, the sheep stamps, tosses its head violently, and often dashes away from its companions wildly over the field. The larvæ re-

---

* The head was cloven whl. an *ax!* It is proper to say, however, that various writers speak of having found the grubs in the ethmoid cells, and indeed in all the nasal cavities.

main in the sinuses feeding on the mucus secreted by the membrane, and apparently creating no farther annoyance, until ready to assume their pupa form in the succeeding spring.   Figures 62 and 63 give the shape and an upper and under view of the full-grown larva.

Fig. 62.

Fig. 63.

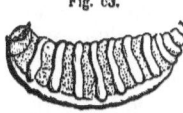

THE "GRUB" OR LARVA OF THE SHEEP GAD-FLY.

The body consists of eleven rings, colorless in the young grub, but the elevated portions growing darker with age, and becoming a dark brown when the full size is attained.   There are round spots of a still darker color on each of these bands.   At the edges of the rings are a few short hairs, and lower down some round darkish spots, as shown in fig. 62.— Small red spines, as shown in fig. 63, cover the space between the rings on the belly.   The remainder of the body (with the exception of the posterior stigmata) is white.   The tentaculæ, as well as certain appendages on each side of the anus, the purposes of which have not been discovered, are seen in fig. 63.

The larva having remained in the sinuses through the fall and winter, abandons them as the warm weather advances in the latter part of spring. It crawls down the nose, creating even greater irritation and excitement than when it originally ascended, drops on the ground, and rapidly burrows into it.   In a few hours its skin has contracted, become of a dark brown color, and it has assumed the form of a chrysalis, as seen in fig. 64. Or rather, this figure exhibits the *shell* of the chrysalis, after the escape of the fly ; and fig. 65 shows the upper extremity or head of the pupa, detached by the fly in its escape.

Fig. 64.   Fig. 65.

SHELL
OF CHRYSALIS.

The experiments of Valisnieri go to show that the *Œstrus ovis never eats*—and this is the received opinion.— The male, after impregnating two or three females, dies, and the latter having deposited their ova in the nostrils of the sheep, also soon perish.

The larva in the heads of sheep may, and probably do add to the irritation of those inflammatory diseases, such as catarrh, which attack the membraneous lining of the nasal cavities ; and they are, as we have seen, a powerful source of momentary irritation in the first instance, when ascending to and descending from their lodging-place in the head. But in the interval between these events—extending over a period of several months—not a movement of the sheep indicates the least annoyance at their presence, or reveals to the veterinarian whether they exist in the sinuses or not.   It would be very difficult to believe that all the local irritation which these parasites could cause, would be sufficient to terminate life, and, so far as my observation has extended, post-mortem examination discloses no lesions which would in anywise sanction such belief. The larvæ, moreover, are found, at the proper season, in the heads of nearly all sheep—the healthy as well as the diseased—and I never have been able to ascertain that the number of them is greater, on the average, in the heads of those sheep which were supposed to have fallen victims to their attacks, than in the heads of *perfectly healthy sheep* slaughtered for the table.   And to prove that the popular ideas on the subject are but vague and crude—not the result of that long and close comparison of symptoms, results, and post-mortem appearances, which would give weight to the opinions of the most unerudite—we have but to notice a few of the cases popularly referred to the "grub in the head." A sheep in the highest condition and apparent health leaps into the air two or three times, and suddenly dies, and if a grub can be found in the cavities of the head, that is the undoubted destroyer. Another wastes away for months and dies lingeringly, a mere skeleton, and the same proof establishes the same fact. Whether there has been fever or no fever—whether there has been obstinate constipation, or equally obstinate dysentery—whether one viscus or another exhibit traces of abnormal action—whether the disease has been acute or chronic—in a word, whatever the form or character of the malady—however diametrically different the diagnosis and the lesions, it is a clear case of "grub in the head," if two or three of those parasites are found there !

Mr. Bracy Clark and Mr. Youatt, so far from regarding the larva of the *Œstrus ovis* as the cause of a fatal disease, suggest that they may even *promote* the *health* of the sheep by diminishing the tendency to cerebral disease—especially determinations of blood—by establishing counter irritation ! Mr. Spooner does not speak of their producing fatal effects in any instances, nor am I aware that any late scientific veterinarians do.

Treatment.—Though the presence of the grub constitutes no disease, some think it well to diminish their number by all convenient means.— One simple way of effecting this is by turning up with a plow a furrow of earth in the sheep pasture. Into this the sheep will thrust their noses on the approach of the *Œstrus*, and thus many of them escape its attacks.— Some farmers smear the noses of their sheep with tar occasionally, during the proper season—the odor of which is believed to repel the fly. Others compel the sheep to smear their own noses every week or two, by feeding them their salt sprinkled over tar. Blacklock says that the larvæ may be dislodged even from the sinuses, by blowing tobacco smoke for some moments through the tail of a pipe into each nostril. I have never tried the experiment.

———

THE SCAB.—The scab is a cutaneous disease, analogous to the mange in horses and the itch in men. It is caused and propagated by a minute insect, the *acarus*. M. Walz, a German veterinarian, who has thrown great light on the habits of these parasites, says :

"If one or more female acari are placed on the wool of a sound sheep, they quickly travel to the root of it, and bury themselves in the skin, the place at which they penetrated being scarcely visible, or only distinguished by a minute red point. On the tenth or twelfth day a little swelling may be detected with the finger, and the skin changes its color, and has a greenish blue tint. The pustule is now rapidly formed, and about the sixteenth day breaks, and the mothers again appear, with their little ones attached to their feet, and covered by a portion of the shell of the egg from which they have just escaped. These little ones immediately set to work, and penetrate the neighboring skin, and bury themselves beneath it, and find their proper nourishment, and grow and propagate, until the poor animal has myriads of them to prey on him, and it is not wonderful that he should speedily sink. Some of the male acari were placed on the sound skin of a sheep, and they too burrowed their way and disappeared for a while, and the pustule in due time arose ; but the itching and the scab soon disappeared without the employment of any remedy.

The figures on the next page are copied from M. Walz's work :
The female acarus brings forth from eight to fifteen young at a litter.
The scab is often produced spontaneously in England by mismanagement of various kinds, such as "bad keep, starvation, hasty driving, dogging, and exposure afterward to cold and wet ;" and it spreads rapidly

by contagion.   It is very prevalent there, and annually causes an immense loss in the wool and flesh of the British flocks.   In the United States it is comparatively little known, and so far as I am able to learn, never origin-ates spontaneously.   It is a singular fact that short-wooled sheep, like the

Fig. 66.   Fig. 67.   Fig. 68.

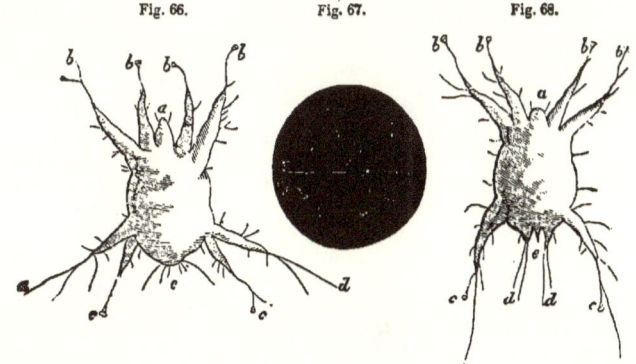

THE ACARUS WHICH CAUSES SCAB.

Fig. 67.—The acari of their natural size on a dark ground.
Fig. 66.—The female of 366 times the natural size, larger than the male, of an oval form, and provided with eight feet, four before and four behind.
 a.—The sucker.
 b. b. b. b.—The four anterior feet, with their trumpet-like appendices.
 c. c.—The two interior hind feet.
 d. d.—The two outward feet, the extremities of which are provided with some long hairs, and on the other parts of the legs are shorter hairs.   To these hairs the young ones adhere, when they first escape from the pustule.
 e.—The tail, containing the anus and vulva, garnished with some short hairs.
Fig. 68.—The male on its back, and seen by the same magnifying power.
 a.—The sucker.
 b. b. b. b.—The fore-legs with their trumpet-like appendices, as seen in the female.
 c. c.—The two hind-legs, with the same appendices and hairs.
 d.—The rudiments of the abdominal feet.
 e.—The tail.

Merino, are much less subject to its attacks, and this is probably one reason for its little comparative prevalence in the United States.   Mr. Youatt observes:

"The old and unhealthy sheep are first attacked, and long-wooled sheep in preference to the short; a healthy short-wooled sheep will long bid defiance to the contagion, or probably escape it altogether."

It spreads from individual to individual and from flock to flock, not only by means of direct contact, but by the acari left on posts, stones, and other substances against which diseased sheep have rubbed themselves.   Healthy sheep are therefore liable to contract the malady if turned on pastures pre-viously occupied by scabby sheep, though some considerable time may have elapsed since the departure of the latter.

The sheep laboring under the scab is exceedingly restless.   It rubs it-self with violence against trees, stones, fences, &c.   It scratches itself with its feet, and bites its sores and tears off its wool with its teeth.   As the pustules are broken, their matter escapes, and forms scabs covering ed, inflamed sores.   The sores constantly extend, increasing the misery of the tortured animal.   If unrelieved, he pines away and soon perishes.

I have never had an opportunity to observe the post-mortem appear-nces.   Mr. Youatt says:

"The post-mortem appearances are very uncertain and inconclusive.   There is generally chronic inflammation of the intestines, with the presence of a great number of worms.   The liver is occasionally schirrous, and the spleen enlarged; and there are frequently serous effusions in the belly, and sometimes in the chest. There has been evident sympathy between the digestive and the cutaneous systems."

Treatment.—About twelve years since, I purchased 150 fine-wooled sheep just driven into the county from a considerable distance. I placed them on a farm then owned by me, in another town, and did not see them for about three weeks. One of my men then reported to me that the sheep were amiss—that they were shedding off their wool—sore spots were beginning to show on them—and that they rubbed themselves against the fence-corners, &c. Though I had never seen the scab, I took it for granted that this was the disease. No time was to be lost, as I had 700 other sheep on the farm—though fortunately, thus far, the new comers had been kept entirely separate from them. Barely looking into Mr. Livingston's work for a remedy, I provided myself with an ample supply of tobacco and set out. The sheep had been shorn, and their backs were covered with scabs and sores. They evidently had the scab. I had a large potash kettle sunk partly in the ground as an extempore vat, and an unweighed quantity of tobacco put to boiling in several other kettles. The only care was to have *enough* of the decoction, as it was rapidly wasted, and to have it *strong enough*. A little spirits of turpentine was occasionally thrown on the decoction, say to every third or fourth sheep dipped. It was necessary to use it sparingly, as, not mixing with the fluid and floating on the surface, too much of it otherwise came in contact with the sheep. Not attending to this at first, two or three of the sheep are thrown into great agony, and appeared to be on the point of dying. I had each sheep caught and its *scabs scoured off*, by two men who rubbed them with stiff shoe-brushes, dipped in a suds of tobacco-water, and soft soap. The two men then dipped the sheep all over in the large kettle of tobacco-water, rubbing and kneading the sore spots with their hands while immersed in the fluid. The decoction was so strong that many of the sheep appeared to be sickened either by immersion or by its fumes; and one of the men who dipped, though a tobacco-chewer, vomited, and became so sick that his place had to be supplied by another.

The effect on the sheep was almost magical! The sores rapidly healed, the sheep gained in condition, the new wool immediately started, and I never had a more perfectly healthy flock on my farm. Though administered with little reference to economy, the remedy was a decisive one.—With a vat like fig. 27, (Letter XII,) this would not necessarily be a very expensive method, with sheep recently sheared. But the assaults of the scab usually come on in the spring before shearing time, and it would require an immense quantity of the tobacco decoction to dip sheep with their fleeces on, however carefully it might be pressed out.

The following is the remedy recommended by Chancellor Livingston:

"First, I separate the sheep (for it is very infectious); I then cut off the wool as far as the skin feels hard to the finger; the scab is then washed with soap-suds, and rubbed hard with a shoe-brush, so as to cleanse and break the scab. I always keep for this use a decoction of tobacco, to which I add one-third by measure of the lye of wood ashes, as much hog's-lard as will be dissolved by the lye, a small quantity of tar from the tar-bucket, which contains grease, and about one-eighth of the whole by measure of spirits of turpentine. This liquor is rubbed upon the part infected, and spread to a little distance round it, in three washings, with an interval of three days each. I have never failed in this way to effect a cure when the disorder was only partial. . . . I cannot say whether it would cure a sheep infected so as to lose half its fleece."[*]

The following remedies are much used in Great Britain:

No. 1.—Dip the sheep in an infusion of arsenic, in the proportion of

---

* Livingston's Essay. Appendix p. 177.

ienic to twelve gallons of water. The sheep should pre-
in soap and water. The infusion must not be per-
mouth or nostrils.

)mmon .mercurial ointment, for bad cases, rub it own
3 weight of lard—for ordinary cases, five times its weight
tle of this ointment into the head of the sheep. Part tho
se the skin in a line from the head to the tail, and then
) ointment with the finger the whole way. Make a sim-
olication, on each side, four inches from the first, and so
3 body. The quantity of ointment (after being com-
lard) should not exceed two ounces, and considerably
suffice. A lamb requires but one-third as much as a
is will generally cure, but if the sheep should continue
ter application of the same should be made in ten days.
lard or palm oil 2 lbs., oil of tar ½ lb., sulphur 1 lb.—
last two, then rub down the compound with the first.—
way as No. 2.
corrosive sublimate ½ lb., white hellebore, powdered, ¾
oil 6 gallons, rosin 2 lbs., tallow 2 lbs. " The first two
. little of the oil, and the rest being melted together, the
illy mixed." This is a powerful preparation and must
freely.
es the preference to No. 1, as least troublesome ; Mr.
and the author of the Mountain Shepherd's Manual to
certainly prefer No. 3, if it is, as it is asserted to be,
ir the reason that it contains no poisonous or dangerous

3 scab, or erysipelas, attended with considerable itch-
.cks the English flocks, but I have heard of no cases of
ld be classified as a febrile disease. It is treated with
2, venesection, and oil or lard applied to the sores.

BIFLEX CANAL.—From the introduction of foreign bod
anal, or from other causes, it occasionally becomes the
on. This is sometimes confounded with the hoof-ail,
'e entirely distinct and different from each other. In-
biflex canal causes an enlargement and redness of the
y about the external orifice of the canal. The toes are
by the tumor. I never have known it to attack more
never have allowed it to go to the point of ulceration,
do if neglected. There is none of that soreness and
.ween the back part of the toes—and none of that pecu-
tinguishes the hoof-ail. I never have found it anything
sease as it is described tc be by the English veterina-

ave always scarified the coronet, making one or two
the principal swelling around the mouth of the canal
with tar—and paid no more attention to it.

) first symptom of this troublesome malady, which is or-
1 lameness of one or both of the fore feet. But on daily
of a flock which have the disease among them, it will be
he lesions manifest themselves for several days before
with lameness. Scarcely any English writer whom I

have read, describes with respectable accuracy the first appearances of the hoof-ail as it exhibits itself *in this country, and among the fine-wooled sheep.* Mr. Youatt says:

"The foot will be found hot and tender, the horn softer than usual, and there will be enlargement about the coronet, and a slight separation of the hoof from it, with portions of the horn worn away, and ulcers formed below, and a discharge of their fetid matter. The ulcers, if neglected, continue to increase; they throw out fungous granulations, they separate the hoof more and more from the parts beneath, until at length it drops off."

The above is *not* a description of the consecutive symptoms of the hoof-ail as *I* have seen them. The hoof, instead of being softened, is perceptibly *hardened*, I think, by the presence of the disease. There is occasionally an enlargement about the coronet, but this is not common in the outset; and so far from the horn first separating from the foot at that point, it is the last place where it usually adheres when the soles are eaten away by the ulcerous matter, and the mere outside shell remains. I never have known a hoof to drop off, entire, in the sense in which I understand the closing part of Mr. Youatt's remark.

My first introduction to this disease was by its breaking out in its most malignant form in a flock of eight hundred sheep, with which I had placed, early in the preceding spring, a few valuable sheep received from abroad which were infected with the hoof-ail, without my having the slightest suspicion of the fact. The disease, when of long standing, and well kept under, shows itself but very little during the winter and spring, unless the foot is directly examined. Every sheep in that eight hundred took the disease, sometimes first in one foot, then in another, then in a third, and when the fourth one was attacked, perhaps it was again bursting out in one of the *cured feet!* I considered the sheep valuable, had much of the *esprit du corps* of a young flock-master, and was *determined* to conquer the malady at any cost and at all hazards. I have little doubt that every sheep in the flock was "doctored" on the average ten times each, and it was very rarely that I permitted *any other person than myself to cut away the horn and prepare the foot of a single sheep for the application of the remedies!* When I look back to that period—the sheep on some remote pastures—not a shed on them to shelter myself or assistants from the burning August sun as we bent ten or twelve hours a day over our task—our only "operating room" a yard in the corner of two fields—blood and pus encrusting hands and garments, and occasionally by an unlucky stroke of the knife showered over *face* and *bosom*—the crawling maggots—the intolerable fetor :—I hardly know whether to take credit to myself for or to laugh at the stanchness of my zeal. But, worst of all, with all my labor, I had

——"scotch'd the snake, not killed it!"

The disease appeared in my flock, though in a much mitigated form, the next summer. I think I then cured it—but I was not allowed to escape thus. In the succeeding summer, accident again brought it among my sheep. In a word, I have first and last served a five years' apprenticeship to combating the hoof-ail. Having seen it in every possible phase —having experimented with almost every recommended remedy not obviously empirical—I shall be excused if I speak my own opinions with a de-

* As I have before stated, when discussing "the most profitable breed for the South," the hoof of the Merino and that of the English Long-Wooled sheep, is essentially different. The latter usually retains its natural shape and thickness, and although the side-crust sometimes turns under, it is but a comparatively thin slip of horn, which is subsequently worn or broken off—or it is easily removed by the knife. The hoof of the Merino grows rapidly, *especially when the animal has the hoof-ail.* The horny soles will sometimes become nearly an inch thick, and the toes will elongate and turn up in front like horns, to the length of three and even four inches. The weight of the Merino is much less than that of the Long-Wool. Take these facts into consideration, together with some of the other circumstances detailed in the introductory remarks to Letter XIV, and perhaps it sufficiently accounts for some differences in the diagnosis of the disease between the two countries.

gree of confidence, even if they chance to conflict with those of professed and eminent veterinarians.

As all are aware, the horny covering of the sheep's foot extends up, gradually thinning out, some way between the toes or divisions of the hoof, and above these horny walls the "cleft" is lined with skin. When the points of the toes are spread apart, this skin is shown in front, covered with short, soft hair. The back part of the toes, or the "heels," can be separated only to a little distance, and the skin in the cleft above them is naked. In a healthy foot, the skin throughout the whole cleft is as firm, sound, *dry* and uneroded, as on any other part of the animal.

The first symptom of hoof-ail is a slight erosion, accompanied with inflammation and heat of the naked skin in the *back part* of the cleft, immediately above the heels. The skin assumes a *macerated* appearance, and is kept moist by the presence of a sanious discharge from the ulcerated surface. As the inflammation extends, the friction of the parts causes pain and the sheep limps. At this stage the foot *externally*, in a great majority of cases, exhibits not the least trace of disease, with the exception of a slight redness, and sometimes the appearance of a small sore at the upper edge of the cleft, when viewed from behind.

The ulceration of the surface rapidly extends. The thin upper edges of the inner walls of the hoof are disorganized, and an ulceration is established between the hoof and the fleshy sole. A purulent fetid matter is exuded from the cavity. The extent of the separation daily increases, and the ulcers also form sinuses deep into the fleshy sole. The bottom of the hoof disappears, eaten away by the acrid matter, and the outer walls, entirely separated from the flesh, hang only by their attachments at the coronet. The whole fleshy sole is now entirely disorganized, and the entire foot is a mass of black, putrid ulceration; or, as it more commonly happens, the fly has struck it, and a dense mass of writhing maggots cover the surface, and burrow in every cavity. The fore-feet are generally first attacked, and most usually but one of them. The animal at first manifests but little constitutional disturbance. It eats as usual. By the time that any considerable disorganization of the structures has taken place in the first foot—sometimes sooner—the other fore-foot is attacked. That becoming as lame as the first, the miserable animal seeks its food on its knees, and if forced to rise, its strange, hobbling gait betrays the intense agony occasioned by bringing its feet in contact with the ground. There is a bare spot under the brisket of the size of the palm of a man's hand, which looks red and inflamed. There is a degree of general fever—and the appetite is dull. The animal rapidly loses condition. The appearance of the maggot soon closes the scene. Where the rotten foot is brought in contact with the side in lying down, the filthy ulcerous matter adheres to and saturates the short wool, (it being but a month and a half or two months after shearing,) and maggots are either carried there by the foot, or they are soon generated there. A black crust is soon formed round the spot. It is the decomposition of the surrounding structures, and innumerable maggots are at work below, burrowing into the integuments and muscles and eating up the miserable animal alive. The black festering mass rapidly spreads, and the poor sufferer perishes, we cannot suppose otherwise than in tortures the most excruciating.

Sometimes but one fore-foot is attacked, and subsequently one or both hind ones. There is no uniformity in this particular, and it is a singular fact that when two or even three of the feet are dreadfully diseased, the fourth may be entirely sound. So also one foot may be cured, while every other one is laboring under the malady.

The highly offensive odor of the ulcerated feet is so peculiar that strictly pathognomonic of the disease—and would reveal its charact one familiar with it, in the darkest night.

When the disease has been well kept under during the first seas its attack, but not entirely eradicated, it will almost or entirely disap as cold weather approaches, and does not manifest itself until the weather of the succeeding summer. It then assumes a mitigated fo the sheep are not rapidly and simultaneously attacked—there seems less inflammatory action, constitutionally, and in the diseased parts course of the disease is less malignant and more tardy, and it more ly yields to treatment. If well kept under the second summer, it i milder the third. A sheep will occasionally be seen to limp, but its dition will scarcely be affected, and dangerous symptoms will rarel pervene. One or two applications made during the summer, in s way, as I shall presently describe, that one thousand sheep can be mitted to the treatment in half a day—with but a trifle of labor and pense—will now suffice to keep the disease under. At this point a vigor in the treatment will entirely extinguish the disease.

With all its fearful array of symptoms, can the hoof-ail be cured *first* attack on a flock? The worst case can be promptly cured, as I by repeated experiments. Take a single sheep, put it by itself, an minister the remedies *daily* after the English fashion, or as I presently prescribe, and there is not an ovine disease which more s yields to treatment. But as already remarked, in a preceding Lett this country, where sheep are so cheap, and labor in the summer m so dear, it would out of the question for an extensive flock-master tempt to keep each sheep by itself, or to make a *daily* application of edies. There is not a flock-master within my knowledge who has pretended to apply his remedies oftener than once a week, or *regula* often as that, and not one in ten makes any separation between the eased and healthy sheep of a flock into which the malady has been on introduced. The consequence necessarily is that though you may cu sheep *now diseased*, it has infected or inoculated others—and these in scatter the contagion, before they are cured. There is not a partic doubt—nay, I know, by repeated observation, that a sheep once en cured may again contract the disease, and thus the malady performs a petual circle in the flock. Fortunately, however, the susceptibility to tract the disease diminishes, according to my observation, with every ceeding attack; and fortunately also, as already stated, succeeding att *cœteris paribus*, become less and less virulent.

What course shall then be pursued? Shall the flock-master sac his sheep—shall he take the ordinary half-way course—or shall he ex more on the sheep than they are worth in attempting to cure them? ther. The course I would advise him to pursue, will appear as I the experiments I have made.

Treatment.—The *preparation of the foot*, where any separate indiv treatment is resolved upon—and this is always necessary, at least in cases—is a subject of no dispute. But the labor can be prodigi economized by attention to a few not very commonly observed partic Sheep should be yarded for the operation immediately after a rain, if ticable, as then the hoofs can be readily cut. In a dry time, and af night which has left no dew on the grass, their hoofs are almost as t as horn. They must be driven through no mud, or soft dung, on way to the yard, which would double the labor of cleaning their fe The yard m be small, so they can be easily caught, and it must be

well littered down, so they shall not fill their feet with their own excrement. If the straw is wetted, their hoofs will not of course dry and harden
us rapidly as in dry straw. Could the yard be built over a shallow, grav-
elly-bottomed brook,* it would be an admirable arrangement. The hoofs
would be kept so soft that the greatest and most unpleasant part of the la
bor, as ordinarily performed, would be in a great measure saved, and they
would be kept free from that dung which by any other arrangement will,
more or less, get into their clefts.

The principal operator or foreman seats himself in a chair—a couple of
good knives, a whetstone, the powerful toe-nippers (fig. 21, Letter XII,)
a bucket of water with a couple of linen rags in it, and such medicines as
he chooses to employ, within his reach. The assistant catches a sheep and
lays it partly on its back and rump, between the legs of the foreman, the
head coming up about to his middle. The assistant then kneels on some
straw or seats himself on a low stool at the hinder extremity of the sheep.
If the hoofs are long, and especially if they are dry and tough, the assistant presents each foot to the foreman, who shortens the hoof with the toe-
nippers. If there is any filth between the toes, each man takes his rag
from the bucket of water, and draws it between the toes and rinses it, until the filth is removed. Each then seize their knives, and the process of
paring away the horn commences. And *on the effectual performance of
this, all else depends.* A glance at the foot will show whether it is the seat
of the diseased action. The least experience cannot fail in properly set-
tling this question. An experienced *finger*, placed on the back of the
pastern close above the heel, would at once detect the local inflammation
(by its heat) *in the dark.*

If the disease is in the first stage—*i. e.* there is merely an erosion and ulceration of the cuticle and flesh in the cleft *above* the walls of the hoof, no
paring is *necessary.* But if ulceration has established itself between the
hoof and the fleshy sole, the ulcerated parts, be they more or less extensive, MUST BE ENTIRELY DENUDED OF THEIR HORNY COVERING, cost what it
may of time and care. It is better not to wound the sole so as to cause it
to bleed freely, as the running blood will wash off the subsequent application, but no fear of wounding the sole must prevent a full compliance with
the rule above laid down. At the worst, the blood will stop flowing after
a little while, during which time no application need be made to the foot.

If the foot is in the third stage—a mass of rottenness and filled with
maggots—in the first place pour a little spirits of turpentine (a bottle of it,
with a quill through the cork, should be always ready,) on the maggots
and most of them will immediately decamp, and the others can be removed with a probe or small stick. Then *remove every particle of loose
horn, though it should take the entire hoof*—and it *will* generally take the
whole hoof in such cases. The foot should be now cleansed with a solution of chloride of lime, in the proportion of one pound of chloride to one
gallon of water. If this is not at hand, plunging the foot repeatedly in
water, just short of scalding hot, will answer every purpose. The great
object is to clean the foot *thoroughly.* If there are any considerable fungous granulations, ("proud-flesh,") they should be excised with a pair of
scissors, or the actual cautery (hot iron.)

And now comes the important question *what constitutes the best remedy?*
The recommended prescriptions are innumerable. The following are
some of the most popular ones.† 1. 4 oz. blue vitriol, 2 oz. of verdigris.

* A portion of any little brook might be prepared by *planking* the bottom, and widening it if desirable
† The first three are given in the American Shepherd, pp. 379-80.

to a junk-bottle of wine. 2. Spirits turpentine, tar and verdigris in equal parts. 3. 3 quarts of alcohol, 1 pint spirits of turpentine, 1 pint of strong vinegar, 1 lb. blue vitriol, 1 lb. copperas, 1½ lbs. verdigris, 1 lb. alum, 1 lb. of saltpetre, pounded fine : mix in a close bottle, shake every day, and let it stand six or eight days before using : also mix 2 pounds of honey and 2 quarts of tar, which must be applied after the previous compound. " Two applications will entirely remove the disease," says this recipe, which was once, I believe, hawked about the country as a patent cure—being sold at five dollars to each purchaser, he giving a promise of inviolable secrecy ! 4. Apply diluted aquafortis (nitric acid) with a feather to the ulcerated surface. 5. Apply diluted oil of vitriol (sulphuric acid) in the same way. 6. Same of muriatic acid. 7. Dip the foot in tar nearly at the boiling point, &c.

After a thorough trial of the above and a multitude of other prescriptions,* I have come to the conclusion that in the first and second stages of the disease—before the ulcers have formed sinuses into the sole, and wholly or partly destroyed its structure—that no application, simple or compound, is preferable to a saturated solution of blue vitriol, (sulphate of copper.) In my judgment, no beneficial addition can be made to it as a remedy. Of the manner of applying it I shall speak presently.

In the third stage, when the foot is a festering mass of corruption, after it has been cleansed as already directed, it requires some strong caustic to remove the unhealthy granulations—the dead muscular structures—and to restore healthy action. Lunar caustic I think preferable to any other application, but it is too expensive. Mr. Youatt gives a decided preference to chloride of antimony, and I think him correct. This is frequently not attainable in the country drug-stores, and muriatic acid may be resorted to, or even nitric or sulphuric acid. The diseased surface is touched with the caustic (applied with a swab formed by fastening a little tow on the end of a stick,) until the objects above pointed out are obtained. I have then usually treated the foot with the solution of blue vitriol, and subsequently coated it over with tar which has been boiled, and is properly cooled. The last protects the raw wound from dirt, flies, &c. Sheep in this stage of the disease should certainly be separated from the main flock, and looked to as often as once in three days. With this degree of attention, their cure will be rapid, and it is astonishing with what celerity the obliterated structures of the foot will be restored.

The ordinary method of using the solution of blue vitriol is to pour it from a bottle with a quill in the cork, into the foot, when the animal lies on its back between the operators, as already described. In this way a few cents' worth of vitriol will serve for a large number of sheep. But the method is imperfect, because, without remarkable care, there will almost always be some slight ulcerations not uncovered by the knife—the passages to them will be devious, and perhaps nearly or quite closed—and the solution will not reach them. Thus the disease will only be temporarily suppressed, not cured.

I had a flock of sheep a few years since which were in the second season of the disease. They had been but little looked to during the summer, and as cold weather was setting in, many of them were considerably lame—some of them quite so. The snow fell and they were brought into the yards, limping and hobbling about deplorably. This sight, so disgraceful to me as a farmer, roused me into activity. I bought a quantity

*Many of them resorted to "against the stomach of my sense," to give myself and others indisputable ocular proof of their inutility—or that they were no better than cheaper, simpler, and more easily attainable medicines

of blue vitriol—made the necessary arrangements—and once more took the chair as principal operator! Never were the feet of a flock more thoroughly pared. Into a large washing tub, in which two sheep could stand conveniently, I poured a saturated solution of blue vitriol and water, *as hot as could be endured by the hand even for a moment.* The liquid was about four inches deep on the bottom of the tub, and was kept at about that depth by frequent additions of *hot* solution. As soon as a sheep's feet were pared, it was placed in the tub and held there by the neck, by an assistant. A second one was prepared and placed beside it. When the third one was ready, the *first* was taken out, and so on. Two sheep were thus constantly in the tub, and each remained in it about five minutes.— The cure was *perfect!* There was not a lame sheep in the flock during the winter or the next summer! The hot liquid penetrated to every cavity of the foot, and doubtless had a far more decisive effect *even on the uncovered ulcers,* than would have been produced by merely wetting them. Perhaps the lateness of the season was also favorable, as in cold weather the ulcers of ordinary virulence discharge no matter to inoculate the healthy feet, and thus, at the time of applying the remedy, there are no cases where there has been inoculation not yet followed by those lesions which admit of cure. *Whether so thorough a soaking would destroy the virus in the inoculated foot,* I cannot pretend to decide.

I think that the vitriol required for the above one hundred sheep was about twelve pounds, and that it cost me fifteen cents per pound. The account then would stand thus:

```
12 lbs. of vitriol at 15 cents..............................$1,80
Labor of 3 men one day each..............................  2,25

     Total.............................................$4,05
```

or about *four cents* per sheep. I have not a doubt that three such applications at intervals of a week, would effectually cure the disease, as every new case would be arrested and cured before it had time to inoculate others. I have no doubt that it would do this at any time of year, and even during the first and most malignant prevalence of the contagion, PROVIDING THE PARING WAS SUFFICIENTLY THOROUGH. The second and third parings would be a mere trifle, and the liquid left at the first and second applications could again be used. Thus sheep could be cured at about twelve cents per head. This is vastly cheaper in the long run than the ordinary temporizing method—where people count the cost of a few pounds of blue vitriol, but not their time, and who thus 'keep the disease lingering in their flocks for years. Indeed, if partial and temporizing treatment is all that is aimed at,—if the flockmaster is content to simply keep the disease under—I can point out methods quite as efficacious as the common one by paring and applying washes from a bottle—*as ordinarily performed*—and not costing a tithe as much.

Between the corners of two sheep-pastures (1, 2, of fig. 69,) construct the dividing fence as represented in the cut. A narrow passage is thus left from one field to another. This passage should be about 2 or 2½ feet wide and 12 feet long. The fence on each side of the passage should be an upright board fence, so that the space can be entirely filled on the bottom with a flat trough, (the bottom formed of a plank) with side and end boards about five inches high. ·In this trough place say a bushel and a half or two bushels of un

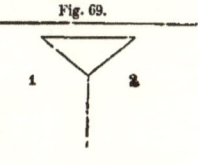

Fig. 69.

slacked lime,* slack it, and then fill the trough nearly full of water.—
Through this drive the flock several times from one field to the other—until the lame ones manifest much suffering. Repeat this once a week the
first summer that the disease appears, putting in fresh lime each time.—
This does not appear to *cure* the hoof-ail, but it keeps it under; the sheep
keep their condition, and show little lameness. The second or third summer of the disease, three or four such applications usually answer for the
entire season. Some use dry slacked lime, as the same trough-full will
then answer for several applications. The trough in this case must have
a roof over it. I never have tried the last method. If the dry lime will
get sufficiently between the toes—and it is said to—it will answer the
purpose where it touches more effectually than even the liquid, but it
would not be so likely to penetrate into cavities. Some who use the lime
remedy, pare the feet once pretty thoroughly prior to the first application,
but afterward neglect them. Others neglect paring entirely, i. e. beyond
shortening the toes once a year, as is practiced with all fine-wooled flocks.

Fig. 70 is an improvement on the
more common arrangement exhibited in fig. 69. The dotted lines
enclose good-sized yards in the corners of two adjoining pastures.—
Two drivers can yard the sheep in
one of these, and drive the sheep
from one to the other any number
of times, without chasing them
about a large field. The labor can

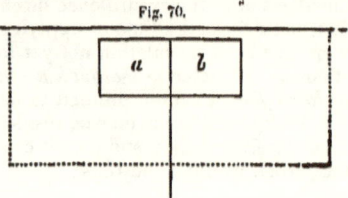

Fig. 70.

therefore be performed much more rapidly, and it requires less force. A
couple of active fellows would yard and submit a flock of two or three
hundred sheep to the process in less than an hour. When the sheep are
first yarded, if there are any very lame ones, draw them out and place
them in one of the small pens (a, b.) Their feet can be examined, and if
necessary a little extra pains taken with them, by paring, cauterizing, etc.
Each sheep as treated is put into the other small pen, where it can be retained until the flock is discharged, and then removed to a separate pasture from the others, if considered desirable.

Where two yards are constructed, as in fig. 70, it is obvious that the arrangement can be made elsewhere as well as in the corner of two fields;
though if the sheep are wild, it may require a few rods of wing fence (in
the place of the dividing one between the fields, as seen in figures 69 and
70,) for the more convenient *cornering* of the sheep to yard them. Thus
one such apparatus might be made to conveniently answer for a whole
farm, though thousands of diseased sheep were scattered in different flocks
over it, and may be placed at a spot where water, etc. are convenient.

Where lime and water are used, the sheep must be driven through the
trough slowly and quietly—as otherwise the lime will be scattered over
their wool, into their eyes, &c. If the lime is fresh burned and highly
caustic, it would be likely to destroy their eyes. Indeed, pure fresh-burned
lime sometimes will take the hair off from their pasterns and shanks. It is
better, therefore, to use it when somewhat re-carbonized by exposure to
the air.

Wood ashes are said to produce the same effect with lime. It is claimed
that sheep kept on lands where the timber has been recently burned,
(" new clearings,") will recover from the hoof-ail. *Query :* If this be true

* To be added to, from time to time, if the number of sheep run through is large enough to waste it materially, before they are sufficiently wetted.

might not the lye of ashes, of the proper strength, make au adequate sub stitute for lime and water ?

Some Northern farmers drive their sheep over dusty roads as a *remedy* for *hoof-ail !* Opposed as it would seem to be to sound theory—sadly as it is at variance with the practice of foreign veterinarians who employ "low-pledgets," "gaiter boots," etc., to exclude all dirt from the diseased surface, it does actually seem in cases of ordinary virulence—especially where the disease is chronic—to *dry up the ulcers and keep the malady under!*

There is an important point to be regarded in exhibiting remedies for the hoof-ail, the mention of which I have reserved until now, as it concerns *all remedies equally.* Many farmers select rainy weather to "doctor" the sheep. Their feet are then soft, and it is therefore on all accounts good economy, when the feet are to be pared, and each separately treated, *provided* they can be kept in sheep-houses, or under shelters of any kind, until the rain is over and the grass again dry. If immediately let out in wet grass of any length, the vitriol or other application is measurably washed away. This is avoided by many, by dipping the feet in warm tar —an excellent plan under such circumstances. The tar is probably a good application at any time, but I do not consider it necessary, in ordinary cases, unless the sheep must be turned out into wet grass.

A flock of sheep which have been cured of the hoof-ail, are considered more valuable than one which has never had it. They are far less liable to contract the disease from any casual exposure—and its ravages are far less violent and general among them.

I am strongly disposed to believe that hoof-ail is propagated in this country only by *inoculation*—the contact of the matter of a diseased foot with the integuments lining the bifurcation of a healthy foot. That it is propagated in some of those ways classed under the ordinary designation of *contagion* is certain. I could indisputably authenticate more than a hundred cases, where the sheep on a farm, indeed through a neighborhood, had been notoriously exempt from hoof-ail from the first settlement of the country—so that the inhabitants did not even know what the disease was—until some diseased flock was introduced from abroad. It was so in the region where I live, and I well recollect when a flock of Saxons, driven from a neighboring county, first introduced it among our sheep. There has not been a diseased flock in the county which could not trace it back to that flock. And the contagion was spread by them as readily on our dry hill-farms as on low and moist ones.

That it may be propagated by *inoculation* I *know* by direct experiment. I have placed the matter of diseased feet on the skin lining the cleft of a healthy foot under a variety of circumstances—sometimes when that skin was in its ordinary and natural state—sometimes after a very slight scarification—sometimes when macerated by moisture. The disease has been communicated under each of these circumstances, and in a majority of all the instances, amounting to sixteen or seventeen.

That there is not even a supposed or pretended case, to my knowledge, on record where the disease has originated spontaneously, in the Northern States, I have already asserted.* I regard Professor Dick's statements of the manner in which the disease originates, which I have quoted,† as wholly inapplicable to *our country with its present breeds of sheep,* and I cannot sufficiently express my surprise that this eminent veterinarian should nave adopted—what I deem so unqualified an absurdity—the non-contagion theory.

I have been disposed to trace the propagation of the disease exclusive-

*In the beginning of Letter XIV.   †Ib.

ly to *inoculation*, from having observed on my own farm and elsewhere, that healthy flocks have occupied with impunity fields adjoining those occupied by diseased ones—an open board or rail fence only separating them.   I have drawn the same inference also from the manner in which the disease attacks flocks.   The whole, or any considerable number, though sometimes rapidly, are never *simultaneously* attacked, as we should expect among animals so gregarious, if the disease could be communicated by simple contact, inhaling the breath or other effluvium.   But not having positive and demonstrative proof of the correctness of the proposition, I would advise no man to incur any risks, unnecessarily, founded on this assumption, without first *satisfying himself* on the point.

The matter of diseased feet is left on grass, straw, and other substances, and thus is brought in contact with the inner surfaces of healthy feet.— Sheep therefore contract the disease from being driven over the pastures, yarded on the straw, &c., where diseased sheep have been, perhaps even days before.   The matter would probably continue to inoculate until dried up by the air and heat, or washed away by the rains.   The stiff upright stems of closely mown grass (as on meadows,) are almost as well calculated to receive the matter of diseased feet, and deposit it in the clefts of healthy ones, as any means which could be devised artificially.   I do not consider it *entirely safe* to drive healthy sheep over *roads*, and especially into *washing-yards* or *sheep-houses*, where diseased sheep have been, until rain has fallen, or time has elapsed for the matter to dry up.   On the moist bottom of a washing-yard, and particularly in houses or sheds, kept from sun and wind, and rain, this matter might be preserved for some time in a condition to inoculate.

FOULS.—Sheep are much less subject to this disease than cattle, but are subject to it if kept in wet, filthy yards, or on moist, poachy ground.   It is an irritation of the integument in the cleft of the foot, slightly resembles incipient hoof-ail, and produces lameness.   But it produces no serious structural disorganization—disappears without treatment—is not contagious—and appears in the wet weather of spring and fall, instead of the dry, hot period of summer when the hoof-ail rages most.   A little solution of blue vitriol, or a little spirits of turpentine, either followed by a coating of warm tar, promptly cures it.

GOITRE OR BRONCHOCELE.—I never have seen this classed among the diseases of sheep, but the "swelled neck" in lambs is, like the goitre, an enlargement of the thyroid glands, and it is strikingly analogous to, if not identical with, that disease.   It is congenital.   The glands at birth are from the size of a pigeon's to that of a hen's egg—though more elongated and flattened than an egg in their form.   The lamb is exceedingly feeble, and often perishes almost without an effort to suck.   Many even make no effort to rise, and die as soon as they are dropped.   It is rare that one lives —though three or four years since, a lamb in my flock having *one* of the thyroid glands enlarged, grew up a large, healthy sheep.   At a year old, when disposed of, the enlarged gland was of the size of a goose-egg.

No inconsiderable number of lambs annually perish from this disease.— It does not appear to be an epizoötic, though I think it more prevalent some seasons than others.   It does not seem to depend upon the water, or any other natural circumstances of a region, (as goitre is usually supposed to,) as it may not prevail in the same flock or on the same farm once in ten years.   I never have been able to trace it to any particular kind of food.   That when it does appear, it is induced by some common local or alimentary cause, I am induced to infer from the fact that its attacks are rarely isolated. When there are any instances of it in a flock, there are usually a number of them. I have lost lambs by it two seasons—from six to ten per cent. of the whole number. Francis Rotch, Esq. of Louisville, Otsego county, lost a much heavier per centage than this (my impression would now be nearly *fifty* per cent.) of his choice South-Down lambs, a few years since. I am acquainted with various other instances where the loss has ranged from ten to twenty per centum.

When congenital goitre has thus appeared among my lambs, the ewes *have been in unusually high condition.* The same was true of Mr. Rotch's ewes, as he wrote me at the time. Whether this coëxistence implies causality, I do not pretend to decide. High condition in the ewe *may* be *one* of the inducing causes.

Treatment.—I know of no treatment which will reach the case. Indeed, the lamb is dying, almost, when born—and remedies are out of the question. Should one having the disease chance to live, it would scarcely be worth while to attempt reducing the enlargements of the glands. Perhaps keeping the breeding ewes uniformly in fair, plump, but not *high* condition, would be as effectual a preventive as any.

MISCELLANEOUS DISEASES.

POISON FROM EATING LAUREL.—I often hear of this from our drovers, who take sheep in the spring to the Philadelphia and New-Jersey markets, through Northern Pennsylvania, on the Old Red Sandstone formation of which the beautiful *Kalmia angustifolia* is abundant. The following description of the effects on the sheep of eating this plant, and the proper remedial treatment, though, I confess, not very *satisfactory* to *me,* I extract entire from the " American Shepherd," * as I have no experience whatever in the premises, and no better account within my reach :

" Sheep and calves will often, in the winter or spring of the year, eat greedily of the low Laurel *(Kalmia angustifolia).* The animal appears to be dull and stupid, swells a little, and is constantly gulping up a greenish fluid which it swallows down ; a part of it will trickle out of its mouth, and discolor its lips. The plant probably brings on a fermentation in the stomach, and Nature endeavors to throw off the poison herb by retching or vomiting.

Treatment.—In the early stages, if the greenish fluid be suffered to escape from the stomach, the animal most generally recovers. To effect this, gag the sheep, which may be done in this manner : Take a stick of the size of your wrist and six inches long—place it in the animal's mouth—tie a string to one end of it, pass it over the head and down to the other end, and there make it fast. The fluid will then run from the mouth as fast as thrown up from the stomach. In addition to this, give roasted onions and sweetened milk freely."

I have somewhere, I think, seen drenches of milk and castor-oil prescribed for sheep poisoned with laurel ; and I should, without farther knowledge of the subject, consider it treatment promising better results than the preceding.

SORE FACE.—Sheep feeding on pastures infested with John's wort *(Hypericum perforatum)* not unfrequently exhibit an irritation of the skin about the nose and face, which causes the hair to drop off from the parts The irritation sometimes extends over the whole body, though no such case has fallen under my observation. Mr. Morrel says : † " If eaten in too large quantities, it produces violent inflammation of the bowels, and is frequently fatal to lambs, and sometimes to adults."

Treatment.—Rub a little sulphur and lard on the irritated surface. If there are symptoms of inflammation of the bowels, Mr. Morrell prescribes

---

* American Shepherd, p. 361.   † Ib. 374.

tar—"putting it into the mouth of the sheep with a flattened stick."
Abundance of salt is considered, and probably truly, a preventive. I
have a sheep pasture considerably infested with this difficultly exterminated weed, and I do not recollect an instance of a sheep exhibiting the
effects of eating it, in several years. It is certain that my sheep have
plenty of salt, whether this is the preventive or not.

SORE MOUTH.—The lips of sheep sometimes become suddenly sore it.
the winter, and swell to the thickness of a man's hand. The malady
occasionally attacks whole flocks, and becomes quite fatal. No cases of
it having been brought under my observation, I am unable to state
whether, in accordance with the popular description, the lesions are confined to the lips. I should *presume not*. It is usually attributed to
noxious weeds cut with the hay.

Treatment.—Mr. Morrell states that he has had the disease in his flock,
and has cured it immediately by smearing the diseased lips with tar.*

LOSS OF CUD.—The "loss of the cud" ranks as an important disease in
the nosology of the "Cattle Doctor," and frequently calls forth all the skill
of that functionary to manufacture a *new cud*, which is placed in the
mouth of the animal as a *substitute for the one which was lost !* That
person must be little versed in the physiology of ruminants who needs to
be told that the accidental loss of one of the cuds, in the process of re-
mastication, would be a matter of no sort of consequence. The sheep,
as well as the cow, not unfrequently nearly or entirely *ceases to ruminate*
but this is the *result*, not the *cause*, of disease. It is diagnostic of all
important diseases, and when observed, its warning should never go un-
heeded.

HOOVE.—This is not common, to any dangerous degree, among sheep,
but if turned upon clover when their stomachs are empty, it will sometimes ensue. It is a distention of the paunch by gas extricated from
the fermentation of its vegetable contents, and evolved more rapidly, or in
larger quantities, than can be neutralized by the natural alkaline secretions
of the stomach. When the distention is great, the blood is prevented
from circulating in the vessels of the rumen, and is determined to the
head. The diaphragm is mechanically obstructed from making its ordinary contractions, and respiration, therefore, becomes difficult and imperfect. Death soon supervenes. In ordinary cases, gentle but prolonged
driving will effect a cure. Where the animal appears swelled almost to
bursting, and is disinclined to move, it is better to at once open the
paunch. At the most protuberant point of the swelling, on the left side,
a little below the hip bone, plunge a trochar or knife, sharp at the point
and dull on the edge, into the stomach. The gas will rapidly escape, carrying with it some of the liquid and solid contents of the stomach. If no
measures are taken to prevent it, the peristaltic motion, as well as the
collapse of the stomach, will soon cause the orifices through the abdomen
and paunch not to coincide, and thus portions of the contents of the former
will escape into the cavity of the latter. However perfect the cure of
hoove, these substances in the belly will ultimately produce fatal irritation.
To prevent this, a canula or little tube should be inserted through both
orifices as soon as the puncture is made. Where the case is not imminent,
alkalies have been sometimes successfully administered, which combine

* American Shepherd, p. 375.

with the carbonic acid gas, and thus at once reduce its volume. A flexible probang—or, in default of it, a rattan or grape-vine—with a knob on the end, may be gently forced down the gullet, and thus the gas permitted to escape.

OBSTRUCTION OF THE GULLET OR "CHOKING."—After pouring a little oil in the throat, the obstructing substance can be frequently moved up or down by external manipulation. If not, it may usually be forced down with a flexible rod, the head of which is guarded by a knob or a little bag of flax-seed. The latter having been dipped in hot water for a minute or two, is partly converted into mucilage, which constantly exudes through the cloth, and protects the œsophagus from laceration. But little force must be used, and the whole operation conducted with the utmost care and gentleness, or the œsophagus will be so far lacerated as to produce death, although the obstruction is removed.

FRACTURES.—Of these Mr. Blacklock concisely says :

"If there be no wound of the soft parts, the bone being simply broken, the treatment is extremely easy. Apply a piece of wet leather, taking care to ease the limb when swelling supervenes. When the swelling is considerable, and fever present, you can do no better than open a vein of the head or neck, allowing a quantity of blood to escape, proportioned to the size and condition of the animal, and the urgency of the symptoms. Purgatives in such cases should never be neglected. Epsom salts, in ounce doses, given either, as a gruel or a drench, will be found to answer the purpose well. If the broken bones are kept steady, the cure will be complete in from three to four weeks, the process of reunion always proceeding faster in a young than in an old sheep. Should the soft parts be injured to any extent, or the ends of the bone protruding, recovery is very uncertain, and it will become a question whether it would not be better at once to convert the animal into mutton."

TREATMENT.

METHOD OF ADMINISTERING MEDICINE INTO THE STOMACH.—The stomach into which we wish to administer medicines, is the fourth, or digesting stomach. The comparatively insensible walls of the rumen are but slightly acted upon, excepting by doses of very improper magnitude For the reasons given when the course of the food through the stomachs was described, medicine to reach the fourth stomach should be given in a state as near approaching fluidity as may be. And even then it may be given in such a *manner* as to defeat our object. Mr. Youatt says :

"If the animal forcibly gulps fluids down, or if they are given hastily and bodily by the medical attendant, they will fall on the canal at the base of the gullet with considerable momentum, and force asunder the pillars and enter the rumen; if they are drank more slowly, or administered gently, they will trickle down the throat and glide over these pillars, and pass on through the maniplus to the true stomach."

METHOD OF BLEEDING.—Bleeding from the ears or tail, as is commonly practised, rarely extracts a quantity of blood sufficient to do any good where bleeding is indicated. To bleed from the eye-vein, the point of a knife is usually inserted near the lower extremity of the pouch below the eye, pressed down, and then a cut made inward toward the middle of the face. Daubenton recommends bleeding from the angular or cheek vein,

"—— in the lower part of the cheek, at the spot where the root of the fourth tooth is placed, which is the thickest part of the cheek, and is marked on the external surface of the bone of the upper jaw by a tubercle, sufficiently prominent to be very sensible to the finger when the skin of the cheek is touched. This tubercle is a certain index to the angular vein which is placed below. . . . . The shepherd takes the sheep between his legs; his left hand more advanced than his right, which he places under the head, and grasps

2 M

the under jaw near to the hinder extremity, in order to press the angular vein, which passes in that place, to make it swell; he touches the right cheek at the spot nearly equidistant from the eye and mouth, and there finds the tubercle which is to guide him, and also feels the angular vein swelled below this tubercle; he then makes the incision from below upward, half a finger's breadth below the middle of the tubercle."

When the vein is no longer pressed upon, the bleeding will ordinarily cease. If not, a pin may be passed through the lips of the orifice, and a lock of wool tied round them

For thorough bleeding, the jugular vein is generally to be preferred. The sheep should be firmly held by the head by an assistant, and the body confined between his knees, with its rump against a wall. Some of the wool is then cut away from the middle of the neck over the jugular vein, and a ligature, brought in contact with the neck by opening the wool, is tied around it below the shorn spot near the shoulder. The vein will soon rise. The orifice may be secured, after bleeding, as described in the preceding method.

As once before remarked, the good effects of bleeding depend almost as much on the *rapidity* with which the blood is abstracted, as on the *amount* taken. This is especially true in acute disorders. Blacklock tersely remarks : " *Either bleed rapidly or bleed not at all.*" The orifice in the vein, therefore, should be of some length, and I need not inform the least experienced practitioner that it should be made lengthwise with the vein. A lancet is by far the best implement, and even a short-pointed penknife is preferable to the bungling fleam.

Another important rule in venesection is that, where indicated at all, it should always be resorted to as nearly as possible to the *commencement* of the malady.

The amount of blood drawn should never be determined by *admeasurement*, but by constitutional effect—the lowering of the pulse, and indications of weakness. In urgent cases as, for example, apoplexy or cerebral inflammation, it would be proper to bleed until the sheep staggers or falls.

The amount of blood in the sheep is less, in comparison, than that in the horse or ox. The blood of the horse constitutes about one-eighteenth part of his weight, that of the ox at least one-twentieth, while the sheep, in ordinary condition, is one-twenty-second. For this reason, we should be more cautious in bleeding the latter, especially in frequently resorting to it. Otherwise, the vital powers will be rapidly and fatally prostrated. Many a sheep is destroyed by bleeding freely in disorders not requiring it, and in disorders which 'did require it at the commencement, but of which the inflammatory stage has passed.

THE PLACE OF FEELING THE PULSE.—The number of pulsations can be determined by feeling the heart beat on the left side. The femoral artery passes in an oblique direction across the inside of the thigh, and about the middle of the thigh its pulsations and the character of the pulse can be most readily noted. The pulsations per minute in a healthy adult sheep are set down by Gasparin at 65, by Youatt at 70, and by Hurtrel d'Arboval at 75. My own observations accord most nearly with those of Gasparin.

LIST OF MEDICINES EMPLOYED IN TREATING THE DISEASES OF SHEEP

ALE.—In cases of debility, unaccompanied with fever, a small amount of ale is sometimes found a good stimulant. It may be given to feeble

·sheep which have become unable to stand from having been too long cast —especially if they have laid on the snow, or on damp cold ground. It is sometimes given in addition to other medicine, in the place of the ordinary stimulants.

ALOES—Are occasionally used as a purgative in sheep medicine by farmers, but their use is justly condemned by all veterinarians.

ALUM—Used as an astringent, but is inferior to many others.

ANTIMONY (*The chloride or butyr of)*—the best caustic to remove fungous granulations, dead muscular structures, etc., in the last and worst stage of hoof-ail—applied with a swab or feather.

ARSENIC—Employed in the proportion of half a pound to twelve gallons of water, to cure scab. An infusion of it is also used to kill ticks, &c. From its liability to adhere to vessels, or to come in contact with substances which may be subsequently eaten, it is a dangerous remedy, and one which I would never have employed on my farm.

BLUE-VITRIOL (*Sulphate of Copper)*—Used internally as a strong tonic, but inferior to others. Dissolved in hot water, and applied to morbid sores, an astringent, alterative, and mild caustic, of the most admirable character. It is superior to all other applications in ordinary cases of hoof-ail.

CAMPHOR—Used with oil as an external stimulant on swellings, &c.

CARRAWAY-SEEDS—Given favorably in doses of two or three drachms, as a stomachic with other medicines.

CATECHU—A valuable astringent, in doses of half a drachm. It is one of the ingredients of the celebrated " sheep's cordial," spoken of under the head of " diarrhea."

CHALK, *Prepared*, by its alkaline properties, neutralizes the acidity of the stomach, and thus checks diarrhea. It is a very valuable remedy in doses from half an ounce to an ounce, exhibited as directed under the head of " diarrhea."

CORROSIVE SUBLIMATE (*Bi-chloride of Mercury)*—The most convenient form in which mercury can be exhibited internally. The *proto-chloride*, or calomel, from its great gravity, could not, with any certainty, be made to reach the fourth stomach. It would seem that mercury should be a useful remedy in several of the diseases of sheep. I have administered it only in the cases specified under the head of " malignant epizoötic catarrh," and then apparently with some benefit. It would be well if a series of careful experiments could be instituted of its value in the appropriate ovine diseases. It is very little used by veterinariaus, in this country or Europe.· A solution of corrosive sublimate is used for the destruction of ticks, &c., and sometimes as a wash in the scab, but its use for these purposes is liable to the same objections with that of arsenic.

DIGITALIS (*Foxglove)*—A sedative employed in most of the fever medicines of the English veterinarians. Dose, one scruple.

EPSOM SALTS (*Sulphate of Magnesia)*—In doses from half an ounce to one, and in some few cases two ounces, the best purgative which can, in almost every disease, be administered to sheep.

GENTIAN—Decidedly the best vegetable tonic in use. Dose, from one to two drachms.

GINGER—A stomachic and tonic, given with almost every aperient, in doses of from half a drachm to a drachm. It prevents griping.

IODINE.—The hydriodate of potash in the proportion of one part to seven or eight parts, by weight, of lard, constitutes an ointment which is a powerful stimulant to the absorbent vessels, and therefore is an excellen application to glandular swellings, or to indurated tumors. It is a goo application to the swelled udder *(q. v.)* in garget.

LARD—A mild and gentle purgative in doses of two ounces. The basis of most ointments, and applied externally in almost every case as an emollient and lubricant in the place of oils.

LIME, *Carbonate of*—Used as a caustic to run flocks of sheep through, in the "hoof-ail," *quem vide.*

LIME, *Chloride of*—An excellent antiseptic and disinfectant, and a good application to foul ulcers.

LINSEED-OIL—A good purgative in two ounce doses. Preferable to Epsom salts in cases of great intestinal irritation, but not otherwise.

MERCURY.—The common mercurial ointment, rubbed down with five parts of lard, for severe cases, and seven parts for ordinary cases, of scab, is an effectual cure,

MURIATIC ACID *(Spirit of Salt)*—Next to chloride of antimony, the best caustic in the worst stage of hoof-ail.

NITRATE OF POTASH *(Nitre or Saltpetre)*—In doses one drachm, a cooling diuretic.

NITRATE OF SILVER *(Lunar Caustic)*—Superior to all other caustics, but too expensive for general use. For poisonous wounds, and particularly for the bite of a mad dog, it has no substitute.

NITRIC ACID *(Aquafortis)*—Sometimes used as a substitute for chloride of antimony, or muriatic acid, as a caustic in hoof-ail. Used by drovers, also, to harden the soles of feet which have become thin and tender by driving. It is touched over the sole with a feather.

OPIUM—An invaluable sedative, and anti-spasmodic, and is employed in nearly all prescriptions for diarrhea and dysentary, and also in colic drinks. It is an important part of the "sheeps cordial." It is commonly used in the form of a tincture, or laudanum. Dose, one drachm.

PEPPER, *Black*—Given in small quantities in milk, to new-born lambs, when chilled.

PIMENTO *(Allspice)*—A substitute for ginger, in the same doses, but not so valuable.

RHUBARB—Unites the properties of a cathartic and subsequent astringent. In small doses it is a tonic and stomachic, invigorating the digestion. When the bowels are relaxed and torpid, and the stomach in a feeble state, it would seem the most appropriate purgative, when a purgative is indicated.

SALT *(Muriate of Soda)*—An ounce constitutes a purgative ; in small quantities a tonic and stomachic. The necessity of keeping sheep freely supplied with salt has been referred to under Summer and Winter Management.

SULPHATE OF IRON *(Copperas, or Green Vitriol)*—Used in washes for the hoof-ail, but superseded by sulphate of copper. Internally, a tonic.

SULPHUR, *Flower of*—In doses of from one to two ounces, a good aperient. It is the basis of various ointments.

SULPHURIC ACID *(Oil of Vitriol)*—A powerful caustic used as a substitute for the acids already alluded to, in the worst stage of hoof-ail.

SPIRIT OF TAR—Destroys maggots, and repels the attack of flies. Flies will not approach a part over which it has been smeared.

TAR—Is a valuable application to the feet, nose, back of the horns, &c., under the various circumstances detailed in Summer Management, and in the treatment of grub in the head, hoof-ail, &c.

TOBACCO—An infusion of it destroys vermin, and also is a cure for scab, *quem vide.*

TURPENTINE, *Spirits of*—Prevents the attack of flies, and drives away maggots. It is a useful application to old sores, wounds, &c.

VERDIGRIS *(Acetate of Copper)*—Used in hoof-ail; but adds nothing, 1 think, to the good effects of the sulphate of copper.

ZINC, *Carbonate of*—Mixed with lard, constitutes a valuable emollient and healing ointment. It is mixed in the proportion of one part of the carbonate, by weight, to eight of the lard.

LETTER XVII.

SHEEP-DOGS, WOOL DEPOTS, &c.

The estimation in which dogs have been held by different nations, &c...The Sheep-Dog—Buffon's description of him...The Spanish Sheep-Dog—Origin—Introduction into the United States—Value—Arrogante—his history...The Hungarian Sheep-Dog—Mr. Paget's description of—probable origin—The Mexican Sheep-Dog—Mr. Lyman's description of—Mr. Kendall's...South American Sheep-Dogs—Darwin's description of...The English Sheep-Dog—Mr. Gates's description of...Mr. Colman's...The Scotch Sheep-Dog—Mr. Hogg's account of...Mr. Peters's...Necessity of accustoming Sheep to a dog...Wool Depots—Mr. Blanchard's account of their origin—Letter from Mr. Peters, describing their object, methods of doing business, and advantages—Utility of these depots—their especial utility to the South...A correction—Mr. Rulin...Note in relation to Australia—Statistics of its Wool Trade brought down to 1846.

Dear Sir :—In all ages of the world, and among nearly all nations, savage and civilized,* the dog has been the friend and cherished companion of man. The Egyptians placed him among their gods. The Greeks held him in the highest estimation. His figure mingles with that of warriors and demi-gods on their friezes; and Argus, the dog of Ulysses, lives as immortal in the Odyssey, (*vide Book XVII., p.* 344 *to* 400) as his sagacious master, or the faithful Penelope. Alexander the Great founded a city in honor of a dog! The Romans treated him with similar respect. His skin covered the statues of the sacred Lares; his figure, as the emblem of care and vigilance, stood at the feet of these household gods—venerated and loved as the tutelary manes of departed ancestors. Horace in his Ode to Cassius Severus (*Book V., Ode VI.,*) compares himself to the Molossian, or the tawny Spartan dog, which defends the flocks, and with ears erect, pursues the wild beast through the deep snows. Virgil, in the delightful Georgics, admonishes the Roman shepherds not to neglect the care of their dogs :

> " Nec tibi cura canum fuerit postrema : sed unà
> Velocas Spartæ catulos, acremque Molossum,
> Pasce sero pingui : nunquam, custodibus illis,
> Nocturnum stabulis furem, incursusque luporum,
> Aut impacatos à tergo horrebis Iberos. "
>
> [*Georg. Liber III., commencing at line* 404.

Thus translated by Sotheby :

> Nor slight thy dogs ; on whey the mastiffs feed,
> Molossian race, and hounds of Spartan breed ;
> Beneath their care, nor wolves, nor thieves by night,
> Nor wild Iberian shall thy fear excite.

These " Spartan hounds," I may remark, par parenthesis, are the ones spoken of by Shakspeare, in that glorious description of the music of a pack in full cry, and of the points of a hound, in *Midsummer-Night's Dream :*

> *Hippolita.*—I was with Hercules, and Cadmus, once,
> • When in a wood of Crete they bayed the bear
> With hounds of Sparta : never did I hear
> Such gallant chiding ; for, besides the groves,
> The skies, the fountains, every region near
> Seemed all one mutual cry : I never heard
> So musical a discord, such sweet thunder.

> *Theseus*—My hounds are bred out of the Spartan kind,
> So flewed, so sanded ; and their heads are hung
> With ears that sweep away the morning dew ;
> Crook-kneed, and dew-lapped, like Thessalian bulls ;
> Slow in pursuit, but matched in mouth like bells,
> Each under each. A cry more tunable
> Was never hallo'd to, nor cheered with horn,
> In Crete, in Sparta, nor in Thessaly.

* The only exceptions which now occur to me are the Jews, the Hindoos, and the Mahommedan nations and tribes

Arrian, Pliny, Oppian, Ælian, and a host of other writers of the Empire descant on the praises of the dog, or give anecdotes of his courage, strength, and fidelity.

In the chivalric ages, he was the companion of knights and princes—the soul of the manly field-sports of those times. Even prelates followed him to the chase. The abbots of St. Hubert *bred* a celebrated race of hounds St. Hubert himself, St. Eustace, and many others on the canonized calen dar, were keen hunters. " *Whereupon,*" says the author of the "Noble Art of Venerie," &c., published in 1611, " *we may conceive that (by the grace of God) all good huntsmen shall follow them into Paradise !*" Truly, a consoling religious *sequitur !*

Scott, in his beautifully descriptive poetry, and still more poetical prose, has given us a whole picture gallery of dogs, from the Middle Ages down The few which start up first in memory, (in *my* memory,) because, proba bly, linked with the most interesting associations, are Fangs—a genuine Saxon—gaunt and unkempt, but stanch as his master, Gurth, the son of Beowulph ; the noble hound of Sir Kenneth ; the " two dogs of black Saint Hubert's breed," that with Fitz-James pursued their quarry into the wild pass of the Trosachs ; the faithful little terrier, which,

> —— " on the dark brow of the mighty Helvellyn,
> The much-loved remains of her master defended,
> And chased the hill fox and the raven away ; "

and last, not least, Hector McIntyre's bitch Juno, which stole the butter, and broke the "lachramatory from Clochmaben," of the glorious old Antiquary. They stand out on the canvas like Landseer's pictures. We pause to hear them *bark !* It has often occurred to me that Scott omitted a fine opportunity, indeed, made a *hiatus vale deflendus*, in not introducing one or more of the Alpine spaniels—or dogs of Mount St. Bernard—into his Anne of Geierstein, providing it could be done, (on which point I am uninstructed,) without a violent anachronism. When Arthur clung dizzy and stupefied to the trunk of the tree which hung over the beetling verge of the precipice—when the cry of the Swiss maiden announced approach ing succor, should it not have had for its accompaniment the baying of one of those great dogs of the Alps—the deep and far-heard reverbera tions of which so often calls help to the perishing traveler, for miles, through the howling storm ? Should not the dog of Donnerhugel, on the night-watch of Graffs-lust, have been of the same breed—huge, shaggy, and daring as himself ? The portrait of Barry, a Bernardine dog which saved the lives of forty persons, and finally perished in an avalanche in guid ing some travelers to St. Pierre, is to be found in every print-shop. It represents him carrying a child on his shoulders—clinging by his shaggy hair,—which he found in the Glacier of Balsore, and rescued from approaching death.

Scott is not the only modern poet who has admired and sung the praises of the dog. And I do not recollect the instance of one, who has mentioned him, that is, the *well-bred* dog, who has not praised him, except Byron 'n these moody lines :

> " Perchance my dog will whine in vain,
> Till fed by stranger hands ;
> But long ere I come back again
> Would tear me where he stands. "

In his epitaph on his Newfoundland dog, the noble poet retracted this ungenerous libel, and pays one of the warmest tributes to the fidelity of the dog, on record.

Volumes of anecdotes of canine sagacity might be easily compiled.

Reasoning powers the dog undoubtedly possesses, quite on a par with ordinary humanity, if we may believe scores of these writers. But it is probable that the grandsires of some of them "drew good *long-bows* at Hastings," and they, like Hubert, may lay claim to a hereditary knowledge of the weapon. It is to be feared that dog-stories will soon be sunk to a par with *fish-stories !* The truth is, the dog knows enough, and there are authenticated cases enough of his wonderful sagacity, without having an air of discredit thrown over the whole of them, by fanciful exaggerations.

The comparative intelligence, and the comparative value to man, of the different species of the dog, would be very differently estimated by those who have been placed in situations to be particularly benefited by the peculiar instincts of this race or that. Nearly every species has some traits, some uses, where it is unequaled by the others; and each in its place is valuable. I do not, however, mean these remarks, or any others which I have made in favor of the dog, to apply to the mongrel tribe of curs. That there have been valuable individuals from this disreputable stock, all must admit; but the miserable, cowardly and thievish character of the mass of them has been proverbial in all time. Far too many of them are kept by our farmers in the place of noble and serviceable animals and multitudes of them, owned by idlers and vagabonds, infest the country and do ten times more mischief to our flocks than diseases and beasts of prey.

———

THE SHEEP-DOG.—Buffon thus eloquently describes the sheep-dog,[*] and compares his sagacity and value to man, with other races[†] :

" This animal, faithful to Man, will always preserve a portion of his empire and a degree of superiority over other beings. He reigns at the head of his flock, and makes himself better understood than the voice of the shepherd. Safety, order, and discipline are the fruits of his vigilance and animal. They are a people submitted to his management, whom he conducts and protects, and against whom he never applies force but for the preservation of good order. . . . If we consider that this animal, notwithstanding his ugliness, and his wild and melancholy look, is superior in instinct to all others; that he has a decided character in which education has comparatively little share; that he is the only animal born perfectly trained for the service of others ; that, guided by natural powers alone, he applies himself to the care of our flocks, a duty which he executes with singular assiduity, vigilance, and fidelity ; that he conducts them with an admirable intelligence, which is a part and portion of himself ; that his sagacity astonishes at the same time that it gives repose to his master, while it requires great time and trouble to instruct other dogs for the purposes to which they are destined ; if we reflect on these facts, we shall be confirmed in the opinion that the shepherd's dog is the true dog of Nature, the stock and model of the whole species."

I shall call attention to but a few of the most distinguished varieties of the sheep-dog.

THE SPANISH SHEEP-DOG.—Of the origin of this celebrated race, I do not recollect to have seen anything. I have observed them several times spoken of, latterly, in newspapers and agricultural publications, as the same variety with the Alpine Spaniel, or Bernardine dog. This, I think, must be an error, though there may be a general resemblance between the two species. Arrogante, on the next page, though a dog of prodigious power, decidedly lacks the massive proportions, both in body and limbs, of several Bernardine dogs, which I have seen, of unquestiona-

———

[*] I stated near the close of Letter V. that there are no shepherd dogs large and powerful enough to encounter and kill wolves and vagrant dogs, except the great Sheep-dog of Spain, and that he is so ferocious that he might frequently bring his owner into difficulty, and even endanger human life.—I was mistaken. Crosses between this and other species seems to have mitigated the ferocity of the Spanish dog, and still left it within the power of two to overcome a wolf, as will appear from what follows.
[†] Buffon's Natural History, vol. v., pp. 306, 318.

ble lineage. The *temper* and disposition of the two species, too, seems to me to be essentially different.

Mr. Trimmer, and various other foreign writers, speak in warm terms of the value of the Spanish sheep-dog, for guarding the migratory flocks of that country from the attacks of wolves—staying behind to protect feeble and lagging sheep, &c. In the Memoirs of the Pennsylvania Agricultural Society, there is a communication from the well-known John Hare Powell, Esq., of Philadelphia, from which the following are extracts :—

" The first importations of Merino sheep were accompanied by some of the large and powerful dogs of Spain, possessing all the valuable characteristics of the English shepherd's dog, with sagacity, fidelity and strength peculiar to themselves. . . . . Their ferocity, when aronsed by any intruder, their attachment to their own flock, and devotion to their master, would, in the uncultivated parts of America, make them an acquisition of infinite value, by affording a defence against wolves, which they readily kill, and vagrant cur dogs, by which our flocks are often destroyed. The force of their instinctive attachment to sheep, and their resolution in attacking every dog which passes near to their charge, have been forcibly evinced upon my farm. '

Fig. 71.

HOWLAND SC

ARROGANTE—A SPANISH SHEEP-DOG

Arrogante, whose portrait is above given with admirable fidelity, was imported from Spain with a flock of Merinos, a number of years since, by a gentleman residing near Bristol, England. His subsequent owner, Francis Rotch, Esq., of this State, thus describes him in a letter to me, which, though not intended for publication, I will venture to make a few extracts from :

" I have, as you desired, made you a sketch of the Spanish sheep-dog Arrogante, and a villainous looking rascal he is. A worse countenance I hardly ever saw on a dog. His small blood-shot eyes, set close together, give him that sinister, wolfish look, which is most unattractive ; but his countenance is indicative of his character. There was nothing affectionate or joyous about him. He never forgave an injury or an insult: offend him, and it was for life. I have often been struck with his resemblance to his nation. He was proud and reserved in the extreme, but not quarrelsome. Every little cur would fly out at him, as at some strange animal; and I have seen them fasten for a moment on his heavy, bushy tail, and yet he would stride on, never breaking his long, ' loping,' shambling trot. Once I saw him

2 N

turn, and the retribution was awful ! It was upon a large, powerful mastiff we kept as a night-guard in the Bank. He then put forth his strength, which proved tremendous ! He coat hung about him in thick, loose, matted folds, dirty and uncared-for,—so that I presume a dog never got hold of anything about him deeper than his thick, tough skin, which was twice too large to fit him anywhere, and especially around the neck and shoulders. The only other evidence of his uncommon strength which I had observed, was the perfect ease with which he threw himself over a high wall or paling, which often drew my attention, because he seemed to me wanting in that particular physical development which we are accustomed to consider as necessary to muscular power. He was flat-chested, and flat-sided, with a somewhat long back and narrow loin. (My drawing foreshortens his length.) His neck, forearm and thigh certainly indicated strength. If the Spanish wolf and the dog ever cohabit, he most assuredly had in him such a cross; the very effluvia of the animal betrayed it. In all in which he differed from the beautiful Spanish shepherd-dog, he was wolfish both in form and habits." But, though no parlor beauty, Arrogante was unquestionably a dog of immense value to the mountain-shepherd. Several times, he had met the large wolf of the Appenines, and without aid slain his antagonist. The shepherds who bred him said it was an affair of no doubtful issue, when he encountered a wolf single-handed. His history, after reaching England, you know."

Some portions of that history I cannot resist the temptation of narrating, as illustrative of the character of this interesting breed, and commemorative of the virtues of the stern, but honest and dauntless Arrogante. If his courage was tinctured with ferocity, and sometimes instigated by a revenge, going a little beyond the canon which permits bad debts to be paid in kind, he did everything *openly !* He made no sneakish, cur-like attacks, on the heels of his foe. By him, as by Robin Hood and his merry men—commemorated by Drayton—

    " Who struck below the knee [was] not counted then a man ;"

and his spring was always at the *throat* of his quarry. But he made not that deadly spring until he gave " warning fair and true," and never without provocation.†

Soon after Arrogante's arrival in England, a ewe under his charge chanced to get cast in a ditch, during the temporary absence of the Spanish shepherd who had accompanied the flock and dog at their importation. An English shepherd, in a spirit of vaunting, insisted on relieving the fallen sheep, in preference to having the absent shepherd called, though warned by his companions to desist. The stern stranger dog met him at the gate and also warned him with sullen growls, growing more menacing as he approached the sheep. The shepherd was a powerful and bold man, and felt that it was too late now to retract with credit. On reaching the sheep, he bent carefully forward, with his eyes on the dog, which instantly made a spring at his throat. A quick forward movement of his arm saved his throat, but the arm was so dreadfully lacerated that immediate amputation became necessary. To save the dog, which had but done his duty, *as he had been taught it,* from the popular excitement, he was shipped in a vessel which sailed that very afternoon, from Bristol for America. He was sent to Francis Rotch, Esq., then a resident of New-Bedford.

For a long time Arrogante would not pay the least attention to his new master ; the voice of the latter would scarcely arrest him for a moment. After attempting in vain, for several weeks, to obtain some recognition of mastership from him, Mr. Rotch chained him securely to a tree, punished him severely, and then, with not a few misgivings, released him. But he submitted, for he well knew that the punishment came from his master, and afterward gave a cold, haughty obedience to all required of him.

---

* I never have supposed, from the several conversations which I have had with Mr. Rotch on the subject, that Arrogante was anything less than a thorough-bred Spanish shepherd-dog. Mr. Rotch here means that he was an ill-favored individual of the family—and he thinks that this *may* be owing to a bar-sinister on his escutcheon, left there by some wolfish gallant. His temper was even *less ferocious* than Mr. Powel describes that of *his* Spanish dogs.
† Was there anything *wolf-like* in all of this ?

Stupid and apparently sleeping much of the day, nothing, however, escaped his observation, or was subsequently erased from his memory. If led round a building, or enclosure, or even an open space, at night-fall, in a manner to evince particular design, during the entire night like a senti- nel he traversed some part of the guarded ring, permitting neither man nor beast to pass *in* or *out* from it.

Arrogante was a "temperance man," of the straightest sect—an out-and out *teetotaler*—and if tolerant of deviations from his *creed*, he could bear none, from the sobriety of his *practice*. Never would he confess acquain- tance with a drunken man—though the hand of that man fed him. The bailiff, who usually fed Arrogante, used occasionally to come home late in the evening a little "*fou*,"and never could he in this condition get his foot on the premises! The old man has plead guilty to more than one night's lodgings on the ground, in consequence of Arrogante's temperance scru- ples.

On one occasion a couple of sailors, to take advantage of the tide, came unexpectedly, and without giving any notice, on the farm, at 3 A. M., to take away some potatoes they had purchased. Arrogante thought it was not so "nominated in the bond;" he forced them to clamber into an empty cart, and there he kept them until morning. They tried the expe-- riment of putting a leg over the side once or twice, but were admonished in too unequivocal a manner to keep quiet, to need any farther hints. They lost the tide, and were in great tribulation, but, like honest fellows, confessed the fault was their own.

I might, did limits allow, recount many more anecdotes displaying the iron determination and fixed precision with which this noble dog obeyed his instructions in guarding sheep or other property committed to his charge. He was a decided "strict constructionist," swerving not from the letter of his commission, and woe to him who attempted to countervail the tenor of that commission!

Drunkenness was destined to prove as fatal as it was detestable, to Arro- gante. A gentleman occupied a cottage orné by the sea-side, the lane to which ran along the farm, and near the stable which Arrogante made his head-quarters, when not on particular duty. The gentleman was reg- ularly *introduced* to him, and warned against ever provoking him. Re- turning him home late one Saturday evening on horseback, from a *conviv- ial* meeting, as he galloped through the lane, he met the dog, and wan- tonly struck him or struck at him with a hunting-whip. He was a large man, and rode a tall, powerful horse, and being under speed, he escaped before the astonished dog recovered from his surprise. But the insulted blood of Castile rushed in boiling currents through the veins of the mad- dened Arrogante. He felt, like his countryman De Lerma, in Epes Sar- gent's tragedy of Velasco—

> "Struck like a menial! buffeted! degraded!
> Spare not my life, if mercy thou would show,
> Thou givest me back only what thou hast made
> A burden, a disgrace, a misery!"

But Arrogante felt both the power and will to avenge himself, and he resolved on a bloody retribution.

The next morning the gentleman was on his way to church, mounted as before. The dog heard and knew the tread of his horse, rose from his lair in the stable, walked to the road-side, and stood grimly awaiting his in- sulter. When the latter had approached within a few yards, Arrogante, like a missile projected from a catapult, met him *in the air*, in a deadly spring at his throat. The sudden jump and swerve of the frightened and

very active horse, saved the rider's throat and his life—but so narrowly
had he escaped, that he felt the gnashing teeth of the frenzied brute
scrape down his dress, where they came in contact with and closed upon
his watch, tearing it away with the adjacent clothing.   The horseman fled
for his life, while the baffled dog vented his rage on the gold watch
which he had captured, by chewing it *into atoms !*   The cause of this terrible enset not being disclosed at the time, Mr. R., though convinced from
the character of the dog that he had not been the aggressor, felt constrained to give orders to have him shot.

THE HUNGARIAN SHEEP-DOG.—The following description of the Hungarian Sheep-Dog, occurs in Paget's " Hungary and Transylvania :"*

" It would be unjust to quit the subject of the Puszta Shepherd without making due and
honorable mention of his constant companion and friend, the jubász-hutya—the Hungarian
shepherd dog.   The shepherd dog is commonly white, sometimes inclined to a reddish
brown, and about the size of our Newfoundland dog.   His sharp nose, short erect ears,
shaggy coat, and bushy tail give him much the appearance of a wolf; indeed, so great is the
resemblance, that I have known a Hungarian gentleman mistake a wolf for one of his own
dogs.   Except to their masters, they are so savage that it is unsafe for a stranger to enter the
court-yard of a Hungarian cottage, without arms. I speak from experience ; for as I was walking through the yard of a post-house, where some of these dogs were lying about, apparently
asleep, one of them crept after me, and inflicted a severe wound on my leg, of which I still
bear the marks.   Before I could turn round, the dog was already far off ; for, like the wolf,
they bite by snapping, but never hang to the object like the bull-dog or mastiff.   Their sagacity in driving and guarding the sheep and cattle, and their courage in protecting them from
wolves and robbers, are highly praised ; and the shepherd is so well aware of the value of a
good one, that it is difficult to induce him to part with it."

I have little doubt that the Hungarian dogs above described are the
descendants of the Spanish ones, introduced into Hungary with the Merino sheep, though possibly they may be somewhat crossed by interbreeding
with the dogs of the country.

THE MEXICAN SHEEP-DOG.—The following account of these noble dogs
appears as a communication from Mr. J. H. Lyman, in the third volume of
the American Agriculturist :†

"Although Mr. Kendall and some other writers have described this wonderful animal as
a cross of the Newfoundland dog, such, I think, cannot be the fact ; on the contrary, I have
no doubt he is a genuine descendant of the Alpine mastiff, or more properly, Spanish shepherd dog introduced by them at the time of the Conquest. He is only to be found in the
sheep-raising districts of New Mexico.   The other Mexican dogs, which number more than
a thousand to one of these noble animals, are the results of a cross of everything under the
sun having any affinity to the canine race, and even of a still nobler class of animals if Mexican stories are to be credited.   It is believed in Mexico, that the countless mongrels of that
country owe their origin to the assistance of the various kinds of wolves, mountain cats,
lynxes, and to almost if not every class of four-footed carnivorous animals.   Be this as it may,
those who have not seen them can believe as much as they like ; but eye-witnesses can assert,
that there never was a country *blessed* with a greater and more abundant variety of misera-
rable, snarling, cowardly packs, than the mongrel dogs of Mexico.   That country of a surety
would be the plague-spot of this beautiful world, were it not for the redeeming character of
the truly noble shepherd dog, endowed as it is with almost human intellect.   I have often
thought, when observing the sagacity of this animal, that if very many of the human race
possessed one half of the power of inductive reasoning which *seems* to be the gift of this
animal, that it would be far better for themselves and for their fellow-creatures.
   The peculiar education of these dogs is one of the most important and interesting steps
pursued by the shepherd.   His method is to select from a multitude of pups a few of the
healthiest and finest-looking, and to put them to a sucking ewe, first depriving her of her
own lamb.   By force, as well as from a natural desire she has to be relieved of the contents of her udder, she soon learns to look upon the little interlopers with all the affection
she would manifest for her own natural offspring.   For the first few days the pups are kept
in the hut, the ewe suckling them morning and evening only ; but gradually, as she be-

comes accustomed to their sight, she is allowed to run in a small enclosure with them until she becomes so perfectly familiar with their appearance as to take the entire charge of them. After this they are folded with the whole flock for a fortnight or so, they then run about during the day with the flock, which after a while becomes so accustomed to them, as to be able to distinguish them from other dogs—even from those of the same litter which have not been nursed among them. The shepherds usually allow the slut to keep one of a litter for her own particular benefit ; the balance are generally destroyed.

After the pups are weaned, they never leave the particular drove among which they have been reared. Not even the voice of their master can entice them beyond sight of the flock; neither hunger or thirst can do it. I have been credibly informed of an instance where a single dog having charge of a small flock of sheep, was allowed to wander with them about the mountains, while the shepherd returned to his village for a few days, having perfect confidence in the ability of his dog to look after the flock during his absence, but with a strange want of foresight as to the provision of the dog for his food. Upon his return to the flock, he found it several miles from where left, but *on the road leading to the village*, and the poor faithful animal in the agonies of death, dying of *starvation*, even in the midst *of plenty;* yet the flock had not been harmed by him. A reciprocal affection exists between them which may put to blush many of the human family. The poor dog recognized them only as brothers and dearly loved friends; he was ready at all times to lay down his life for them ; to attack not only wolves and mountain-cats, with the confidence of victory, but even the bear, when there could be no hope. Of late years, when the shepherds of New Mexico have suffered so much from Indian marauders, instances have frequently occurred where the dog has not hesitated to attack his human foes, and although transfixed with arrows, his indomitable courage and faithfulness have been such as to compel his assailants to pin him to the earth with spears, and hold him there until dispatched with stones.

In the above instance the starving dog could have helped himself to one of his *little bro ther* lambs, or could have deserted the sheep, and very soon have reached the settlements where there was food for him. But faithful even unto death, he would neither leave nor molest them, but followed the promptings of his instinct to lead into the settlement; their unconsciousness of his wants and slow motions in traveling were too much for his exhausting strength.

These shepherds are very nomadic in character. They are constantly moving about their camp equipage consisting merely of a kettle and a bag of meal ; their lodges are made in a few minutes, of branches, &c., thrown against cross-sticks. They very seldom go out in the day-time with their flocks, intrusting them entirely with their dogs, which faithfully return them at night, never permitting any stragglers behind or lost. Sometimes different flocks *are* brought into the same neighborhood owing to scarcity of grass, when the wonderful instincts of the shepherds' dogs are most beautifully displayed ; and to my astonishment, who have been an eye-witness of such scenes, if two flocks approach within a few yards of each other, their respective proprietors will place themselves in the space between them, and as is very naturally the case, if any adventurous sheep should endeavor to cross over to visit her neighbors, her dog protector kindly but firmly leads her back, and it sometimes happens, if many make a rush and succeed in joining the other flock, the dogs under whose charge they are, go over and bring them all out, but, strange to say, under such circumstances they *are never opposed by the other dogs.* They approach the strange sheep only to prevent their own from leaving the flock, though they offer no assistance in expelling the other sheep But they *never permit* sheep not under canine protection, nor dogs not in charge of sheep, to approach them. Even the same dogs which are so freely permitted to enter their flocks in search of their own, are driven away with ignominy if they presume to approach them without that laudable object in view.

Many anecdotes could be related of the wonderful instinct of these dogs. I very much doubt if there are shepherd dogs in any other part of the world except Spain, equal to those of New-Mexico in value. The famed Scotch and English dogs sink into insignificance by the side of them. Their superiority may be owing to the peculiar mode of rearing them, but they are certainly very noble animals, naturally of large size, and highly deserving to be introduced into the United States. A pair of them will easily kill a wolf, and flocks under their care need not fear any common enemy to be found in our country.

J. H. LYMAN."

Mr. Kendall* speaks of meeting, on the Grand Prairie,

—a flock numbering seventeen thousand, which immense herd was guarded by a very few men, assisted by a large number of noble dogs, which appeared gifted with the faculty of keeping them together. There was no running about, no barking or biting in their system of tactics ; or the contrary, they were continually walking up and down, like faithful sentinels, on the outer side of the flock, and should any sheep chance to stray from its fellows the dog on duty at that particular post, would walk gently up, take him carefully by the ear

* Vol I., p. 268.

and lead him back to the flock. Not the least fear did the sheep manifest at the approach of these dogs, and there was no occasion for it.

These noble animals seem, according to these and various other corresponding accounts I have seen of them, to leave nothing to desire in the way of a sheep-dog, either for guarding or managing flocks. They would be invaluable in our Southern States, to protect the flocks from the cur-dog which so often attack them, and from the occasional wolves. I hope efforts will be made to introduce them into our country, and then they should be bred in the utmost purity.

—

SOUTH AMERICAN SHEEP-DOG.—Similar to the preceding in character and habits, are the sheep-dogs to be found in various parts of South America. They, too, are undoubtedly an offshoot from the Spanish stem. The following interesting account of them is from Darwin's Journal :

"While staying at this estancia (in Banda Oriental), I was amused with what I saw and heard of the shepherd dogs of the country. When riding, it is a common thing to meet a large flock of sheep guarded by one or two dogs, at the distance of some miles from any house or man. I often wondered how so firm a friendship had been established. The method of education consists in separating the puppy, when very young, from the bitch, and in accustoming it to its future companions. A ewe is held three or four times a day for the little thing to suck, and a nest of wool is made for it in the sheep-pen.—At no time is it allowed to associate with other dogs, or with the children of the family. The puppy, moreover, is generally castrated : so that when grown up, it can scarcely have any feelings in common with the rest of its kind. From this education it has no wish to leave the flock, and just as another dog will defend its master, man, so will these the sheep. It is amusing to observe, when approaching a flock, how the dog immediately advances barking—and the sheep all close in his rear as if round the oldest ram. These dogs are also easily taught to bring home the flock at a certain time in the evening. Their most troublesome fault when young is their desire of playing with the sheep, for in their play, they sometimes gallop their poor subjects most unmercifully. The shepherd dog comes to the house every day for some meat, and immediately it is given to him he skulks away as if ashamed of himself. On these occasions the house dogs are very tyrannical, and the least of them will attack and pursue the stranger. The minute, however, the latter has reached the flock, he turns round and begins to bark, and then all the house dogs take very quickly to their heels. In a similar manner a whole pack of hungry wild dogs will scarcely ever (and I was told by some, never), venture to attack a flock guarded even by one of these faithful shepherds. The whole account appears to me a curious instance of the pliability of the affections of the dog race ; and yet, whether wild, or however educated, with a mutual feeling of respect and fear for those that are fulfilling their instinct of association. For we can understand on no principle the wild dogs being driven away by the single one with its flock, except that they consider, from some confused notion, that the one thus associated gains power, as if in company with its own kind. F. Cuvier has observed that all animals which enter into domestication consider Man as a member of their own society, and thus they fulfil their instinct of association. In the above case the shepherd dogs rank the sheep as their brethren ; and the wild dogs, though knowing that the individual sheep are not dogs, but are good to eat, yet partly consent to this view, when seeing them in a flock, with a shepherd dog at their head."

—

OTHER LARGE RACES OF SHEEP-DOGS.—There are one or two fine species in France, as those of Brie, and Auvergne. In a letter from G. W. Lafayette, to John S. Skinner, Esq., the latter are pronounced equal to the Spanish dogs.* Large powerful races, everywhere possessing the same general characteristics, are to be found in almost every country excepting our own, where the fine-wooled breeds of sheep have been extensively introduced. With a commerce extending to all the maritime nations of the world, singular it is that so little pains have been taken to introduce them.

—

THE ENGLISH SHEEP-DOG.—The following are portraits of a Drover's dog

* See Farmers' Library, Vol. i., p. 465.

and a Scotch Colley slut, imported by B. Gates, of Gap Grove **Lee Co**
Illinois.   They are taken from The Farmers' Library.*

Fig. 72.

DROVER'S DOG, AND COLLEY SLUT.

The Drover's dog, or English sheep-dog, or Butcher's dog—for by all
of these names is he known—is considerably smaller than the species or
families heretofore described, but he is a larger and more powerful dog
than the Colley.   Mr. Gates, in the communication accompanying the
portraits, remarks :

"Much has already been written on the intelligence of the Scotch Colley.   My opinion is
that the English "Butcher's dog" is no way lacking on that point.   Any reader who has
visited Smithfield market in London, on Monday or Friday, will, no doubt, have formed the
same opinion.   There you have an opportunity of seeing a number of these useful animals
at their work.   It would, in fact, be almost impossible to conduct this market without their
aid.   There a vast number of different animals are brought for sale from all parts of the
country, to supply this great metropolis, and are collected in the smallest possible space.
The difficulty of keeping them from mingling with others falls principally on the dog.   If
one slips away, or a particular one is wished to be caught, it is pointed out to him, and is returned back, or held till the owner takes it—the dog always holding them by the side of the
head, so as not to bruise the body.   By a word or motion of the hand, they will run over
the backs of the sheep, to stop them or turn them in a different direction.   I have often
admired, with astonishment, their quick and intelligent actions.   They appear to read the
thoughts of their master by his countenance, for their eye is continually on his, or the flock.
Nothing else can attract his attention when he has work to perform, and at times I have
thought he acted with more judgment than the owner. . . . The breed of "Boxer," (whose
portrait is above given,) is sometimes called the Drover's or Tailless breed."

Mr. Colmar, in one of his Reports, says :

' For a week or more before the tryst, the roads leading to Falkink will be found crowded
with successive droves of cattle and sheep, proceeding to this central point ; and it is ex
uemely curious on the field to see with what skill and care the different parties and herds
are kept together by themselves.   In this matter the shepherds are generally assisted by

their dogs, which appear endowed with a sagacity almost human, and almost to know every individual belonging to their charge. They are sure, with an inflexible pertinacity, to bring back a deserter to the flock."

Mr T. C. Peters, (now of Buffalo, N. Y.,) on his return from Europe, a few years since, brought over a Drover and a Colley. His testimony to their extraordinary value will be found in the American Agriculturist, vol. iii., page 76.

Fig. 73

THE COLLEY.

THE SCOTCH SHEEP-DOG OR COLLEY.—The light, active, sagacious Colley admits of no superior—scarcely of an equal—where it is his business merely to manage his flock, and not to defend them from beasts larger than himself. Mr. Hogg says that "a single shepherd and his dog will accomplish more in gathering a flock of sheep from a Highland farm than twenty shepherds could do without dogs. Neither hunger, fatigue, nor the worst treatment will drive him from his master's side, and he will follow him through every hardship without murmur or repining."

The same well-known writer, in a letter in Blackwood's Magazine, gives a most glowing description of the qualities of his Colley, "Sirrah." One night a flock of lambs, under his care, frightened at something, made what we call in America a regular *stampede*, scattering over the hills in several different bodies, "Sirrah," exclaimed Hogg in despair, "they're a' awa!" The dog dashed off through the darkness. After spending, with his assistants, the whole night in a fruitless search after the fugitives, Mr. Hogg commenced his return to his master's house. Coming to a deep ravine, they found Sirrah in charge, as they at first supposed, of one of the scattered divisions, but what was their joyful surprise to find that not a lamb of the whole flock was missing!

Of the stanch devotedness of the Colley, under any and all circum stances, Mr. Peters gives, in the American Agriculturist, the foll.)wing charactoristic illustration, copied from a Scotch paper:

" The master of the bitch purchased at a fair some eighty sheep, and having occasion to tay a day longer, sent them forward and directed his faithful Colley to drive them home, a distance of about 17 miles. The poor bitch, when a few miles on the road, dropped two whelps; but faithful to her charge, she drove the sheep on a mile or two farther—then allowing them to stop, she returned for her pups, which she carried some two miles in advance of the sheep, and thus she continued to do, alternately carrying her own young ones, and taking charge of the flock, till she reached home. The manner of her acting on this occasion was gathered by the shepherd from various persons who had observed her on the road. "

The Colleys are not now uncommon in the Northern States, and I have often seen proofs of their singular sagacity in collecting, driving, and guarding sheep, and in catching out one from the flock when directed by their masters. I have often seen one drive a flock of fifty or sixty sheep through a crowed street, encountering teams, pedestrians, and other dogs at every step—without the slightest assistance.

ACCUSTOMING THE SHEEP TO THE DOG.—It is a mistake to suppose that a trained sheep-dog will manage any strange flock, however wild and unaccustomed to such company. The sheep must be gradually made acquainted with, and accustomed to the dog. They must know—and they will readily learn it—that he is their friend, their guardian and protector, instead of that hereditary enemy which their instinct teaches them to fly from. A want of knowledge of this fact has frequently led to disappointment and disgust, to a giving up of the valuable dog which it has cost pains and money to procure. Mr. Skinner relates a ludicrous incident of Mr. Jefferson, arising from his not being apprized of this fact. A thoroughly broken sheep-dog had been sent him from abroad, and the great Sage of Monticello, after having held forth *ore rotundo* to some visitors, on the value of these dogs, and their immense convenience—nay, their indispensability in managing flocks, led forth his guests to give a practical exemplification of the qualities of his dog. At the word, the latter made for the sheep. The terrified animals fled in all directions, some of them dashing themselves over precipices and breaking their necks. The dog either shared the same fate, or, mortified at his failure, felt his pride too deeply wounded to return. Mr. Jefferson never recovered him !

WOOL DEPOTS.

Commission merchants who confine their operations exclusively to the sale of Wool, have opened large stores or " Dépôts," at three or four points in the Northern States. Of the origin of this system, Mr. H. Blanchard, of Kinderhook, N. Y., thus spoke at the Agricultural meeting at the Assembly Chamber, Albany, Feb. 3, 1848 :

" From facts that were ascertained by Hon. J. P. Beekman, (then President of the N. Y State Agricultural Society,) at the State Fair held in Poughkeepsie, in 1844, he became con vinced that the growers of Dutchess county, by reason of the superior facilities afforded there for the sale of their fine wools, were procuring from six to eight cents per pound more than many wool-growers in other sections of the State who produced the same quality of wool. The large quantity of fine wool grown in that county, offered great inducements for manufacturers and purchasers of fine wool to make that a place of resort to obtain their sup plies, and thus a fair competition was awakened, which resulted in a just appreciation of the relative value of their wools, and remunerating prices to the fine wool grower. Soon after Dr B.'s return, the evils consequent upon the system of selling wools in our county, as wel: as elsewhere, became a matter of discussion between him and other wool-growers in our vicinity and myself, the result of which was a request from them that I would open what we now term a " Wool Dépôt." The *principles* involved in the dépôt system are not new u being conducted upon those of a commission business; but it is only the details and appli

2 O

cation of these principles to wool when received direct from the grower, that had never before in this country been applied in the same discriminating manner, and with as little expense as by this system."

The objects and advantages of the system, and the method of conducting these establishments are clearly set forth in the following letter from my friend Mr. Peters, to whom, as a keeper of one of these Dépôts, and a gentleman of conceded ability—as well as skill, energy and success in this and in his other business operations—I thought it appropriate to apply for this information.

H. S. RANDALL, Esq.   BUFFALO, N. Y., Dec. 16, 1847.

My Dear Sir: Your kind favor of the 12th inst., making inquiries relative to the Wool Dépôt system, is before me.

It will give me pleasure to answer your queries, not that by so doing I can add anything to the exceeding great value of your Letters to the whole country, and especially to the South and West—yet from my own experience as a wool-grower, and in the management of a Wool Depôt which I established at this place last spring, I may give some information that will be useful to your readers, and may they be millions. In so doing, I will give you 1st, An account of the object; 2d, The method of doing business; and 3d, The advantages of the Wool Depôt system.

THE OBJECT.—Upon no sheep is the wool exactly alike over the whole body; nor is the wool exactly alike upon any single flock. In most flocks there is a great diversity—greater than there should be for the farmer's profit. There is, then, a variety of grades of wool in every flock, and in every section of the country where wool is grown.

Manufacturers first grade the wool; that is, sort the fleeces, making from five to eight or nine different grades. Each fleece is then opened, and stapled, or sorted into the various grades of the factory. Some manufactories use only the finest, others only the coarsest, and others again use only one kind of the intermediate sorts, so that from a single flock, I sold this year wool to five different manufacturers, no one wanting or working the kind tint the other wanted.

The object of the Wool Depôt is to sort and arrange the wool, that the manufacturer can readily obtain the particular kind adapted to his machinery, and to obtain for each sort its fair market value.

METHOD OF DOING BUSINESS.—The system originated with Mr. H. Blanchard, at Kinderhook, some three years ago. Last year, we sent our wool to Mr. Blanchard, and during the winter I visited his establishment, and was so well satisfied with the operation of it, and of the vital importance of the system to the wool-growers everywhere, that I at once made arrangements to open one at this point. I accordingly commenced operations in the spring, and have been successful beyond my most sanguine expectations.

I have a competent and experienced sorter, and when wool is sent in, it is at once sorted in the fleece, each sort weighed, and entered in a book under the name of the person sending it.

I have adopted Mr. Blanchard's method of sorting, as experience has shown that to be the best, under all circumstances. I make, then, five sorts, taking full-blooded Merino for No. 1, and grading down to coarse common wool, which is No. 5. Saxony I grade into Extra, Prime 1, and Prime 2. Then there is a kind of wool which is admirable for combing, and another kind that is wanted for De Laines;—these form five more sorts, making thus ten sorts. But as there is such a difference in the condition of wool when brought into the Dépôt, I usually make two sorts of each number. Thus I have No. 2, and No. 2 A. No. 2 is usually good, but No. 2 A is of the same grade, but is in better con lition, every way a choice article, but still not fine enough to go into a higher grade. The wool is actually worth two or three cents per lb. more than the other number to which it belongs, and but for making this distinction, would not bring its full value. When the wool is properly sorted, it is piled up in a manner that will enable the purchaser to see it at a good advantage;—insured, and held until the market requires it. I make all my sales here, and for cash When the sales are closed, an account is made out and sent to those who have sent me their wool; usually, an account is rendered as fast as any part of a man's wool is sold. I have often been asked, how I could tell whether any man's wool was sold, unless the whole of a sort was sold at a time. It is very easy. Suppose A. has 100 lbs. of No. 1, and I have sold 20,000 lbs. out of 40,000 lbs.—that being the whole amount in the Depôt I have sold one-half of each man's No. 1, and I turn to A.'s account and give him credit for 50 lbs. sold, and so go through and credit each man with his proportion of that number sold.

The charges are, for receiving, sorting, and selling, one cent per lb., and the insurance— which is usually about 36 cts. on $100, for three months. Cartage from the dock is usually three cents per bale The sacks are returned or sold at the option of the owner. They are usually worth about fifty cents, more or less, according to their condition. Each man's wool is carefully examined; if put up in bad order, it is so noted, and a deduction made by the sorter, to make it as it should be. So that it is no object for a man to send to the Dépôt wool in a bad condition.

THE ADVANTAGES.—The foregoing facts would seem to be so plain that it cannot be necessary to refer to the advantages. No man, however, is more at the mercy of the speculator, than the wool-grower. The very fact that he has so many kinds of wool in his clip, prevents him from ascertaining the market value of the whole, for being in comparatively small quantities, he has not enough, if ever so well sorted, to make it an object for the different manufacturers to visit him. He is therefore compelled to sell his whole clip at the price of his present quality, and at prices from five to fifteen cents per lb. under the real market value of his wool. Allow me to illustrate by an example. A farmer has his wheat, corn, oats, and barley, all mixed, and carries it to market in this condition. Will anybody give him the value of each kind of grain? On the contrary, they would not be willing to pay even the value of the cheapest kind. And yet each kind by itself has a market value. Precisely in the same situation is the wool-grower, except that he has no means of knowing the value of the various sorts of wool, except from the speculator.

The advantage of the Dépôt system, then, is, that there his wool is properly sorted. The wheat is separated from the corn, the corn from the oats, and the oats from the barley, and each is made to bring its fair market value. By having an extensive correspondence among the manufacturers throughout the country, I am kept constantly advised of the market; and knowing the price of cloth in the cities, I am enabled to know to a certainty what the price of the various grades of wool should be.

When the manufacturer can get the kind of wool he wants, and in large quantities, he is willing to pay, and does pay a better price than when he has to buy that which he does not want, to get the right sort. It also equalizes the market, and brings the producer and the manufacturer together, without being compelled to pay agents or speculators, and prevents that fluctuation of the market which is always produced by speculation.

But there is another very great advantage growing out of the system. It enables the wool-growers in the various sections of the country to compare wool, and to know who has really the best and most profitable kinds of sheep. It has been strikingly manifest with me this season. For I have been enabled to point out to people in different States West, where they could find the most profitable sheep, by the wool which had been sent me. And in one instance men had been over five hundred miles after sheep, and paid high prices, when there were sheep in their own town worth double the money.

There is no difficulty in sending wool here from any part of the country bordering upon the Ohio, or its tributaries. The expense of transportation will range from one to one and a half cents per lb.—depending much upon the bargain made with the clipper. I have received wool this year from all the Western States, in some instances as far West as the Mississippi River, and the average cost for freight has been about one cent per lb.

It was urged by many last spring that this city was not a good point, inasmuch as it was not sufficiently central in its location. For nothing is more certain, than that a wool Dépôt, to be successful, must be so located as to command a large amount of wool. The larger amount you can concentrate at a point, the more rapid and sure will be your sales. To this city the products of the West naturally tend, and to this point the producer can calculate with great certainty when, and at what expense it will arrive. But after its trans-shipment here, expenses accumulate, without any corresponding benefit. And it is peculiarly so, in regard to wool, coming as it often does in bad order, sacks torn, broken, and wet.

But I have made my letter already longer than I intended, and in speaking of my own Dépôt have perhaps gone more into detail than is necessary.

This much I must be permitted to say to every wool-grower, that the Wool Dépôt system, properly conducted and patronized, is indispensable to ultimate and profitable success.

I remain, my dear sir,

Very sincerely yours,

T. C. PETERS.

Messrs. Perkins and Brown have a Dépôt at Springfield, Mass.; and I believe the establishment of two or three others is in contemplation, by companies or individuals.

Conducted with skill and fidelity, there can be but little doubt that these establishments are alike beneficial to the wool-grower and manufacturer. That Mr. Blanchard's and Mr. Peters's have thus far been so conducted, there is not the least doubt. Of the other I know nothing, though report speaks well of it. The design was not regarded with much favor, in the outset, by many of our most extensive wool-growers. They preferred to "do their own business," and not "pay the wages of an intermediate agent.' But the advantages derived from selling the wool in sorted lots, have been found to far more than overbalance the one cent per pound paid to the "agent" or Dépôt keeper, and the system is rapidly gaining favor. Many of our most experienced wool growers in this State—men the most competent to favorably dispose of their wool—have sent their wool to Messrs. Blanchard and Peters, and I have yet to see or hear of the first person who has been disappointed in the result.

If wool Dépôts are beneficial in the North, where the agents of different manufacturers, and "speculators," visit every man's barn to bid on his wool—and among a class of growers, too, who, from long experience, are familiar with the qualities and comparative values of the staple—how much more beneficial would they be to regions in which the growers are so scattered that they are rarely visited by traveling agents—or if so, not in numbers sufficient to produce that competition which would *compel* them to offer the fair market value of the article : and where, perhaps, in many cases, the growers themselves have not sufficient experience to determine the exact grade of their own clips, even supposing them correctly notified from time to time from abroad, of the market value of the several grades. *The Dépôt system, in my judgment, removes the great and only serious obstacle to successful wool-growing in the South.*

It is not necessary that Dépôts be established *in the Southern States*, to have those States reap the full benefit of the system. For the present, and for some time to come, at least, the North will furnish the *best home market* for fine wools. The wool therefore must, until some changes take place, come to the North before it is sold ; and the transportation must be equally subtracted from the avails, whether the sale is effected at home or at a Northern wool Dépôt. Indeed, it would be *better* to store it in a Dépôt at Kinderhook or Buffalo, than at Charleston or Nashville. And this is for the reason that the two former are much nearer to, and can be more speedily visited by the principal woolen manufacturers of the United States, than the latter. The New-York or New-England manufacturer would be little likely to send an agent to Charleston or Nashville, if he could supply his wants equally cheaply (with the addition of cost of transportation), from Buffalo, Kinderhook, or Springfield. And if supplied any more cheaply at the former places (price of transportation excepted), be it remembered, it would be so much *unnecessarily taken out of the pocket* of the grower.

Should the South at any future day find it more for her interest to ship her wools to Europe, the above considerations will cease to be valid. She would then want Dépôts as much as now, for far more gain, proportionably, is made by *sorting wool* for the foreign, than the American markets. But in that event, the Dépôts would assume a different character, and they would be most appropriately located at the port whence the wools were shipped.

A CORRECTION.—MR. RUFFIN.

In the beginning of Letter VI., I made the following remark in relation to Hon. Edmund Ruffin—" He seems to think lime, of itself, adequate to the full and permanent amelioration of the tertiary soils."—This remark was made on a somewhat too hasty inspection of some of Mr. Ruffin's positions in the Agricultural Survey of South Carolina. Since writing it, I have had the pleasure of reading for the first time Mr. R.'s highly valuable work on Calcareous Manures, and find that I was in error in the statement above made.

Since the preceding Letters were completed, the exceedingly interesting
article from the (English) Farmers' Magazine, which is published below,
has met my eye. It will be seen from it that the conclusions arrived at
by me (see page 123,) in relation to the vast increase in the trans-Atlantic
demand for wool and woolens, are in a rapid course of verification. I
wrote from statistics extending down to 1840. In that year the English
import of wool was forty-six millions of pounds. In 1845, according to
the subjoined authority, it was seventy-six millions of pounds. And this
rapid increase took place, notwithstanding the vast extension in the woolen
manufactures in other nations, particularly in Germany, France, Spain and
Belgium. The extract given from Waterton's "Cyclopædia of Commerce," asserting the improbability of a much greater extension of the
English woolen manufactures, "unless new markets shall be opened," may
be true. But new markets are yet to spring up in Central and Northern
Asia, and even in Northern Europe, which will, in the aggregate, require
an increase of woolen manufactures which the boldest calculator now
scarcely dreams of. For the reasons for this opinion, see page 123.—
Whether England is to supply a greater or less portion of this increasing
demand remains to be seen. If she continues as well prepared as she
now is to compete with other manufacturing nations, doubtless she will
contribute her full share to that supply.
It will also be seen, from the annexed paper, that what I predicted (see
page 121) in relation to the prospective competition (from the year 1840)
in wool-growing, between the densely populated countries of Western
Europe and those in newer settled regions, where land is cheap and population comparatively sparse, has already come to pass. Spain, and even
Germany, which in 1840 supplied England with nearly twenty-two million
pounds of wool—nearly half of the whole import of the latter—have now
been driven almost entirely out of the English market! But, says the
Sydney Herald, Germany, Spain, etc., have renewed the contest in another
form : they have extended their manufacturing operations, and now manufacture their own wool. Admit this : but if German wools cannot compete with others in the English market, which are brought from fifty times
the distance, they cannot compete with them even in the German market,
unless the latter are kept out by duties. The German manufacturer, then,
in working up home wools, pays more for his raw material than the English manufacturer, and he cannot, therefore, compete with him in foreign
markets, nor even in the home one, without a protective Tariff which
would raise the price of the English to that of the German article.—
Tariffs *materially* enhancing the cost of the necessaries of life will not long
be tolerated by the consuming millions, in regions where civilization has
penetrated.
It seems that Australia and Van Diemen's Land are the successful com
petitors which have driven Germany and Spain from the English wool
market. The views set forth by me in Letter IX. in relation to the advantages of the former for wool-growing compared with those of Hungary,
Southern Russia, North and South America, remain the same ; indeed a
careful review of my positions has served to farther convince me of their
correctness. The *character of the population*, and the better commercial
regulations of Australia, have given her a present advantage over new rivals in the Old World ; and America has not yet entered the field of competition. When the Anglo-Saxon of North America enters the lists with
the Anglo-Saxon of Australia, natural advantages will not, as now, be

overbalanced by superior energy and enterprise. The Anglo-Australian will, to say the least of it, meet his full equal in these particulars. And, on the other hand, there is not a rational doubt that the *natural and other present advantages of all kinds are on the side of the Anglo-American.* The portion of North America included in the proper wool-growing zone is immensely greater than in Australia; our climate, all things considered—considering the occasional terrible drouths of Australia—is the best; our lands are cheaper, and will certainly average as good, including our whole Atlantic coast, and including only our territory between the Apalachians and the Rocky Mountains, our land will average by far the best; labor is not dearer among us; we are not a quarter as distant from the English markets; the wool from all parts of our immense interior, instead of being dragged long and expensive journeys in "bullock drays," is already whirled along by steam, or boated on canals or rivers to the seaboard, at a comparatively trifling expense. It would be difficult to name a particular, excepting in the two-cent duty, in which large portions of the United States have not the advantage over Australia for supplying the English wool market, and in other European markets we have perhaps *every* advantage over that Colony.

The Australian Wool Trade*—[By *Wm. Westgarth*, Esq.]—The importance at present assumed by the Australian wool trade in the lists of British Commerce, demands some degree of attention in the history of an Australian settlement. I shall, therefore, devote the present chapter to a short account of this branch of Commerce, in its capacity both of an export from the Australian Colonies and an import into the British market.

In the year 1836, the quantity of wool exported from Sydney amounted to 3,700,000 lbs. weight. The proportion for the Port Philip district, included in this amount, could not, at so early a period of her existence, have exceeded 60,000 lbs. weight. Five years afterward the annual produce had attained to 1,578,000 lbs.; and the lapse of a similar period, bringing us down to the year 1846, exhibits the astonishing quantity of 7,400,000 lbs.† During this interval of ten years the quantity of wool exported from Sydney, exclusive of any from Australia Felix, had increased from three and a half millions to nearly twelve millions of pounds weight.

The importation of wool into the British market appears, indeed—like the rise of the Australian Colonies—to be but a business of yesterday, and one, among numerous other instances, of the wonderful extension of Modern Commerce. In 1820, the quantity imported was under ten millions of pounds weight; in 1845, it had risen to seventy-six millions. The proportion from the Australian Colonies in the former year was the one-hundredth part; it now forms nearly one-half of the whole importation‡, and at the steady and rapid ratio of the present increase of Australian wool, the lapse of a few years will exhibit a quantity far greater than the united total of the wool at present imported into Britain from every quarter of the world. The following Table exhibits the respective averages, in round numbers, for each period of five years from 1826 to 1845; the numbers representing millions of pounds weight:

Average of years.	Foreign Wool.	Colonial Wool.	Total
1826–30	25	2	27
1831–35	34	4	38
1836–40	44	10	54
1841–45	36	22	58
1846	34	30	64

This Table illustrates the extraordinary progress of the colonial production, three-fourths of which are derived from Australia and Van Diemen's Land.

The periodical public sales of colonial wool, which now occupy so important a position among the commercial occurrences of the British Capital, date their origin only so lately as the year 1817. The prices at that time, and for some subsequent period, were only from 2d. to 3d. per lb.; and it was not until twelve or fourteen years afterward that any important advance took place in the value of this commodity. The fine quality of the Australian wool

* From a new work in the press, on Port Philip.
† The wools occasionally sent from Port Philip by way of Sydney, and appearing in the Customs' returns as Sydney exports, are here allowed for. The season or year is taken as ending on the 10th October, as the usual date of 31st December falls in the midst of the wool shipments, and cannot fairly represent the quantities and ratio of progress of each year.
‡ In 1846, the relative quantities imported into Britain were, in round numbers, thirty-four millions of pounds of foreign wool and thirty millions of colonial. For the present year the colonial may be safely assumed at somewhat more than half the importation.

began soon after to attract notice, and in 1835 and 1836 to excite the attention even of foreign manufacturers. From very small beginnings the extent of the periodical auction sales gradually increased. An unprecedented number of 750 bales was announced for one series of sales in 1825 ; and for some years afterward 400 bales were considered to form a very extensive sale. But in July, 1835, 8,746 bales were brought forward, realizing for the better qualities the considerable rates of from 2s. 6d. to 3s. 8d. per pound ; and at the sales of the same month in 1844, there were exposed no less than 31,358 bales.*

The celebrated wools of Australia are derived from two principal breeds of sheep, the Merino and the Saxon. The former is the finest in quality, but it may be doubted if an adequate price has been hitherto derived to compensate for the lighter weight of the fleece. In the Sydney district, attention was chiefly bestowed on the Merino ; in Van Diemen's Land, on the Saxon ; the Port Philip district received a share of both, as the colonists from either locality transported their flocks to her pastures. This mixture of breeds was still farther increased by occasional crosses with the Leicester and South-Down. In fact, from the numbers of inexperienced persons who entered on the occupation of sheep farming in this new settlement, and, without any fixed principles, carried on a mere random system of breeding, the greater portion of the wool consists of every shade of quality that natural accidents could produce. The abundant pasturage of Port Philip appears also to affect the pure Merino wool of the Sydney district, which in the former locality acquires a more open appearance, loses somewhat of its fineness, and increases about a quarter or half a pound in the weight of the fleece. The average weight of the good qualities of Port Philip fleece, after washing, is from 2½ to 2¾ lbs. ; of the Sydney fleeces about 2¼ lbs. There has been for several years a desire to introduce a greater uniformity of quality in the fleeces of each particular grower, and on the whole an inclination to adhere to the production of the finer qualities of wool.

[Here follow details of the Australian method of washing and other preparations for shearing, which are omitted, as they conform in every important particular to the directions laid down in these Letters for those processes.]

- The wool is now ready to be packed and dispatched to the port of shipment. Each fleece is cleared of the locks and clippings or other unseemly portions, and is usually tied with a piece of string, and tightly squeezed into bags containing about one hundred each, or from two hundred and fifty to three hundred pounds weight. The ponderous bullock dray is now yoked to its team of eight or a dozen oxen, and charged with an ample load of the golden fleece, is dispatched from the station on its annual and protracted mission to the port of shipment.

The wool on its arrival in town is now generally classed and re-packed at an establishment for that purpose, unless this process has already been competently performed at the station.— The classification distinguishes only the entire fleece ; it is not attempted to proceed to a minuter distinction of qualities by breaking the fleeces. When intended for sale, the wool, on its arrival in town, is conveniently exhibited in the various bins of the sorting establish ment, and its quality and condition are fairly ascertained. At the establishment of the Messrs. Bakewell, in Melbourne, the wool is assorted first into the two leading divisions of clothing and combing, and each of these descriptions is run out into five qualities, the fifth or lowest being the coarse Leicester breeds. Extra fine lots are classed by themselves ; super-greasy, or kempy, or other defective fleeces, are also classed apart. The charge for sorting is ½d. per pound. The usual charge for hand-washing is 1d. per pound on the weight returned, and for scouring 1d. to 1½d. per pound. The system of re-packing is also of use in exposing any wet or damp that the wool may have acquired on the way from the interior, in which condition it is in danger of heating and even of originating fire in the hold of a vessel during a lengthened voyage.

The shipping season for the Australian staple commences toward the end of October ; but only a few solitary drays have succeeded in reaching town during that month. Considerable quantities have arrived by the end of November ; and during the two succeeding months there is a continuous succession of vehicles pouring into town with their voluminous loads into the various ports of the district.† These arrivals begin to fall off in February ; but during that and the two succeeding months considerable quantities continue to be shipped, including the later shorn fleeces of the young lambs. The shipment of other exports, which are comparatively of un-important amount, terminates with that of the wool. A solitary vessel may linger till July or August, when the transactions of the season are finally closed.

The following from a late number of the *Sydney Herald* may be well appended to the above :

* This included a small quantity of foreign wool. The proportion from Australia and Van Diemen's Land on this occasion was 26,134 bales. The early sales were held at Garraway's, and continued there from 1817 to 1843, when the locality was transferred to the Hall of Commerce, where they still continue The first bale at the first sale, from the novelty of the circumstance, realized 10s. 6d. per pound.
[Mark-Lane Express, 7th, 11th, and 21st Oct. 1844.

† There are five shipping ports in Australia Felix ; namely, Melbourne, or its port of Williamstown, Gee long, Portland, Belfast, and Port Albert, or Alberton, in Gippa's Land. The quantity for the present year (1847) may be estimated at about 25,000 bales, of which five-sixths are shipped at Williamstown and Geelong.

" Our two time-honored competitors in the production of fine wool, Spain and Germany have been fairly beaten out of the field. The climate and pasturage of these colonies, and of the congenial settlement at the Cape of Good Hope, backed by the energies of their Anglo-Saxon race of flock-masters, have more than compensated for our greater distance from British markets. We have been enabled to supply a good article—in vast and ever-increasing quantities—and at prices which, notwithstanding the cost of carriage, have, through our facilities of production, left us a remunerating profit, but which our ancient rivals have found to be insufficient to replace prime cost.

" But although Spain and Germany have ceased to vie with us as sellers of the raw material in England, they have done so only to renew the contest in another form. They have enlarged their manufacturing operations. Since they can no longer sell their fleece at a profit, they have resolved on working it up in their own looms. To that extent, therefore, they will cease to import wrought woolen fabrics; and in so far as their imports were from Great Britain, there will be a corresponding decrease in the British consumption of our wools. The woolen cloths imported into those two countries from Great Britain, in the year 1841, amounted, in declared value, to £1,026,481 sterling; and if we add the quantities imported in the same year into Holland and Belgium, the amount would have been about a million and a half. We must therefore be cautious, as prudent men, not to allow our spirits to be too much exhilarated by the apparent victory we have gained over ' our hereditary enemies,' seeing that, though seemingly vanquished, they have but shifted their position and varied their tactics.

" A judicious writer says, in 1844 : ' Of late years cottons have, from their cheapness, in a great degree superseded the lower qualities of cloths—a circumstance which, joined to the increasing rivalry of France, Germany and Belgium, *renders it improbable*, unless new markets shall be opened in China or elsewhere, *that much extension will in future be given to our manufacture of woolen cloths.*'*

" While, however, the Spanish and the German wool-growers have thus ceased (or are expected very shortly to cease) to compete with us as exporters to England, another competitor has sprung up in a new and quite unexpected quarter. In addition to corn, bread stuffs, rice, tobacco, cotton, sugar, and an endless catalogue of ' notions,' in which Brother Jonathan has hitherto prided himself as a mighty producer, he has now taken it into his head that he can breed sheep and export wool on a large scale. And it would seem that in England his whim has by no means been thought whimsical. For, say certain Liverpool brokers to him, under date of 3d September, 1846 : ' The arrivals of wool from the United States last year, for the first time to any extent, *made quite a sensation in this country*, as it was generally considered that you required to import these qualities, and there was no knowledge that your growth of wool was of such importance. *We have seen it estimated at sixty-five million pounds;*† and from your vast (and to us almost incredible) means of production, we believe *it will cause a kind of revolution in the wool trade.*'

" Jonathan's own opinion of the matter is thus expressed through the medium of the New-Orleans *Commercial Times : ' * Wool can be grown as cheaply, and to as great advantage, in the cotton-growing States as in any part of the world. There is nothing in the climate to prevent it. If it may be found desirable to grow that of the finest grades, it can be done without fear of the animals becoming covered with hair in a few years.' He has evidently some misgivings, however, as to the policy of his attempting the finest grades, for he immediately subjoins, ' However, we are inclined to think that wool *of a coarser quality will be found most profitable*, mutton being also an object with us.'

" If the United States already produce four times the quantity of wool that we do, and if there is a reasonable chance of their producing it of a quality equal to ours, and at no greater cost, then have we indeed much to fear from their formidable rivalry. The vast extent of their territory, the almost illimitable resources of their soil and climate, the indomitable spirit of their citizens, combined with their proximity to the British market, will render their competition, if successful at all, successful in no ordinary degree.

" ' Wool,' says another Liverpool correspondent, addressing an American, ' requires in its production *great attention in crossing the breed*, otherwise the quality degenerates very quickly. The maintenance of its fineness depends also very much on the nature of the pasturage on which the sheep graze. And we may remark that *your own samples are of a particularly good kind.*' "

Here is a word of encouragement for the Americans, with a word of caution for the Australians. Of the two requisites for the production and preservation of a superior staple, one, suitable pasturage, is bountifully supplied to the Australian grower by Nature, while the other depends upon his own industry and skill. In this, it is to be feared, he has scarcely been just to himself. He has possibly presumed too much upon the natural advantages of the fine sheep-sustaining country in which his capital is staked. It will be well if this note of warning from the land of Stars and Stripes shall rouse him to a more vigilant attention.

[Simmonds's Colonial Magazine

* Waterton's Cyclopædia of Commerce, p. 672.
† The quantity of wool exported from New South Wales, including the district of Port Philip, in the year 1843, was 17,564,734 lbs.

APPENDIX.

ON SHEEP HUSBANDRY IN SOUTH CAROLINA.

Report on the Value of Sheep Husbandry. Read to the Agricultural Society, Pendleton, South Carolina.

In obedience to your resolution, requiring your committee to " report on Sheep Husbandry in the South," they beg leave to say that the resolution would seem to require a more extended examination than could be embraced in a report of an ordinary length.

They will therefore confine themselves to that part of the subject which, in their estimation, will best show the applicability and value of sheep husbandry to our neighborhood and section.

Although but little attention is given by any of us, to raising sheep, and by none to preparing wool for a foreign market, yet it will be admitted, that our native stock are healthy, growing to a fair size and produce a fair fleece, from two to five pounds, even under the great neglect with which they are treated.

There is, however, one question necessary to examine, and that is, whether the quality and quantity of the fleece deteriorate in our climate. The question has been very fully examined by Mr. H. S. Randall, a very intelligent and experienced wool grower in Cortland, New York. From his excellent letters, published in the Farmers' Library, (the perusal of which I take pleasure in recommending to the members of this society,) I draw the following statement :

" It is known that from Spain (north latitude 36 to 44 degrees) all the fine wooled flocks have sprung. And that in Saxony (north latitude 50 to 51 degrees 30 minutes) the Spanish Merino wool has been improved in fineness of fibre but lessened in quantity. In New York (north latitude 42 to 44 degrees) the fineness of the Spanish Merino is preserved and quantity increased. In Vermont (north latitude 43 to 45 degrees) the fineness and quantity of the Saxony wool are preserved."

South of us, in Madison county, Mississippi, (north latitude 32 degrees, 41 minutes,) the wool of the Saxony sheep has been found to maintain its original fineness, and increased in quantity. Recent experiments in Australia (south latitude 33 degrees 55 minutes) show that fine wooled sheep (the Merino) preserve the quantity and improve in quality of fleece.

The exports of wool from there in 1810 was only - - - - - 167 lbs.
"   "   "   in 1833   "   - - - - 3,516,869 "
"   "   "   in 1843   "   - - - - 16,226,400 "

In 1834, London price for best Spanish Merino, was - - - 67 cts
  Australian Merino,   - - - - - - - 100 "
  English wool,   - - - - - - - - 48 "

2 P

297

In England, (north latitude 50 to 56 degrees,) from some cause not yet settled, fine wool cannot be grown.

Near the Cape of Good Hope, (south latitude 34 minutes,) Merino Sheep do well, maintaining both quality and quantity of fleece with Spain.

The latitude of Pendleton is 34 degrees 40 minutes, but counting a degree for every 200 feet altitude, would throw us some twelve degrees farther north, and between the latitudes of Spain and Saxony.

So far then as latitude is concerned, experiments have been made both north and south of us, which show, that here, Spanish Merino wool neither degenerates in quality nor quantity of fleece.

It is also known that rich succulent green food generally adds length to fibre, but does not always make fine wool coarser, at least between latitude 28 and 50 degrees north, nor does it make coarse wool finer. Humidity and dryness of atmosphere seem to have, as well as climate, some influence on the fibre. But as yet, the laws which govern the fleece have not been satisfactorily ascertained. It is sufficient for the present inquiry that experiments verify the operation, that in this latitude Merino wool will not become coarser, nor the coarse long wool become finer, if each stock be kept pure. And as the climate is well adapted to both, the choice of stocks may very well be left to the fancy of those who try either.

We come now to test the value of sheep husbandry to this section of country. There are many methods by which this can be done. I shall, however, take the simple one of comparing the profits of this with the other pursuits of the country.

It is acknowledged in the Northern States that growing wool is a good business there, and I will first give a table from Mr. Randall's letter, showing the profits of growing wool in New York:

A. buys 100 ewes at $2,	$200 00
" 33½ acres of land at $20,	666 66
Cutting and curing 11 acres of the above for hay,	13 65
Pay for shearing,	4 00
For salt, tar, and summer care,	4 00
For labor of winter feeding,	5 00
Loss by death 2 per cent. above pulled wool from those that die,	4 00
	$837 31

RECEIPTS.

300 lbs. wool at 39¢,	$118 71	
60 lambs at $1,	80 00	
Summer manure equal to winter care,	5 00	
		$203 71

This is equal to 24 per cent. on the amount invested, and makes the cost of the wool to the farmer 27 cents per pound. A calculation founded on the same data for Pendleton makes the result more favorable:

A. buys 100 ewes at $1,	$100 00
Pays for shearing,	4 00
For salt, tar,	2 00
Loss 2 per cent. above skins and wool of those that die,	2 00
I make no charge for summer pasture, because it costs nothing, nor should rye or barley pastures for winter be charged; the crop is reaped afterwards. But charge it at 20 cents per head,	20 00
One hand's attention an hour in the morning to turn to pasture, and an hour in the evening to pen; this is one-sixth part of his time. Say his whole time is worth $72; one-sixth is	12 00
Total outlay and expense for feeding one year,	$140 00

RECEIPTS.

2 lbs. wool per head is 200 lbs., at 20 cents,	- - -	$40 00	
40 lambs at $1 when one year old,	- - - - - -	80 00	
		$120 00	

This is 85 per cent. In this instance the wool costs the farmer nothing. Deduct $90, the value of the original stock of ewes at the commencement of the next year, from $140, the total outlay, and you have $50, which the value of the lambs more than equal.

Compare it with farming or planting:

A. buys a negro for	- - - - - -	$700 00
Furnishes him with fifteen acres of land at $5,	- - - - -	75 00
Half the expenses of a horse and plough,	- - - - - -	50 00
For his board and clothing,	- - - - - -	20 00
		$845 00

RECEIPTS.

His labor, 160 barrels corn at 40 cents,	- - - -	$64 00
5 bags cotton at $30 a bag,	- - - - - -	150 00
		$214 00

This is equal to 25 per cent., certainly as much as any man in this neighborhood makes. I have purposely made this large estimate that no one can say it is under the truth.

B. buys 500 ewes and 20 bucks, common stock, at $1,	- - -	$520 00
Employs a shepherd,	- - - - - - - -	175 00
Pays 20 cents for winter feed per head,	- - - - -	104 00
Pays for tar and salt,	- - - - - - - -	20 00
B has $73 less than A. in the outlay,	- - - -	$819 00

RECEIPTS.

3 lbs. wool per head is 1560 lbs. at 20 cents,	-	$312 00	
80 lambs to the 100 ewes is 400 lambs at $1,	- -	400 00	
		712 00	
Deduct for loss over skins and wool of those that die, 2 per cent.,		18 40	
		$693 60	

The outlay of A. ($890) brings him $210, equal to 25 per cent. The outlay of B. ($819) brings him $693 60, equal to 85 per cent.

This calculation will do for the neighborhood of Pendleton or lower down, where sheep have to be fed during the winter. But for all that part of Pickens and Greenville district, extending south for twenty-six or thirty miles from the foot of the mountains, the profits would be larger. For in that belt of country, I am informed by many residents—General Garvin among them—that the *range* affords sufficient food for sheep the entire year. Even when snow is on the ground, they paw the snow away and get sustenance from the winter grass.

For that section I would alter the calculation thus:

B. buys 500 ewes and 20 bucks,	- - - - - - -	$520 00
Pays for a shepherd,	- - - - - - - - -	175 00
Pays for salt and tar,	- - - - - - - -	20 00
		$715 00

RECEIPTS.

1560 lbs. wool at 20 cents,	- - - - -	$312 00
90 lambs to the 100 ewes is 450 at $1,	- - -	450 00
		762 00
Deduct for loss 2 per cent. over skins and wool of those that die,	19 00	
Making $28 more than 100 per cent.		$743 00

But try it on a scale that every one can compare with his own experience.

B. buys 3 ewes and 1 buck for		$4 00
He shears 12 lbs. of wool at 20 cents,	2 40	
2 lambs at $1,	2 00	
Over 100 per cent.		4 40

These are suppositions. Take what has actually occurred in Pickens district. Mr. Stribling, as I am informed by himself, bought one ewe for $1:

In 1846 she had 3 lambs,		$3 00
Sheared 2 lbs. wool at 20 cents,		40
		$3 40
In 1847, same ewe had 2 lambs,		2 00
Sheared again 2 lbs. wool at 20 cents,		40
Each one of the last year's lambs had a lamb apiece,		3 00
And sheared from the 3, 6 lbs. wool at 20 cents,		1 20
		6 80

This is an increase of eight in two years from one ewe, and Mr. Stribling says at this time the whole are alive. I admit this is an extraordinary case, and it is only mentioned to show there are cases of actual increase far above any of those calculations made above.

If, then, the climate be not only adapted to fine wool, but also to the coarse—if the range, which is abundant and sufficient to feed a flock the entire year; and if the above calculations are founded on data anywhere near correct, what more can be desired to show the applicability and value of sheep husbandry to this section.

From three very respectable wool growers, one from each of the states of Pennsylvania, New York, and Ohio, I was told the way to make a calculation on the profits of sheep husbandry in a reasonably safe way, was to put down every year one-ninth less of lambs than you have ewes, and then deduct one-tenth from the whole for deaths in that year.

Thus, say you have		100 ewes.
From 100 deduct one-ninth, and you have		89 lambs.
		189
Then deduct one-tenth for deaths,		18
And you have		171

To start with the next year. On this basis I have taken one hundred ewes and run the calculation on for eight years. The result was:

Wool sheared in eight years,		17,190 lbs.

The gentlemen above alluded to, say that half the wool will pay all expenses, even when the winters require five months' feed:

Deduct, then, one-half, 8,595, at 20 cents,		$1719 00
The increase amounted to 2067 sheep, at $1,		2067 00
Total,		$3786 00

This result, if attained, would exceed fifty per cent. compounded. They say, too, it is a safe estimate to say that the lambs of each year will pay the expenses of the whole flock for that year in Pennsylvania. They more than

do it here.   Allow eighty lambs to one hundred ewes, and you have from
five hundred ewes, four hundred lambs ; deduct one-tenth for deaths, and

You have 360 at $1, - - - - - - - - - - - -	$360 00
Charge 20 cents per head for 900 sheep, makes - - - 180 00	
Charge for shepherd, - - - - - - - - - 150 00 — 330 00	
Lambs over-pay expenses by - - - - - - - -	$30 00

All these calculations are made on the supposition that sheep get a par'
of that regular attention which all farmers give to their other domestic ani
mals.   And to make sheep husbandry successful, it is not only necessary
that this attention should be given, but every one who attempts it shoulu
know something of their diseases and the cures, and also the summer and
winter management.   This can be acquired only by their own, or the experience of others.   Easy access can be had to the experience of northern
wool growers, who are proverbial for being close observers.   They say the
disease called the "scours" is the principal one to which sheep are liable,
produced by excess of rich green food, and cured easily by a change to
dry, but if allowed to continue, is fatal.

From the introduction into the United States, in 1808. of Merino sheep
from Spain, owners of this stock have considered it of the first importance
to preserve the quality and quantity of their fleece, and if possible, to im
prove both.   They ascertained that lambs from young and healthy sires
improved, while those from old ewes fell back both in quality and quantity
of fleece.

Among them it is now a settled practice not to breed from ewes over
seven or eight, nor from bucks over six years old.   It is very important to
resort to the evidence of age the teeth afford.   Their books have been par
ticular in describing these evidences, which I will copy : "During the first
year, lambs have eight small teeth in front, in the upper jaw, called nippers ;
at a year old, the centre two shed, and two larger teeth take their place.
At two years old, the next two are lost, and supplied by two larger ones.
Thus losing and being supplied by two larger ones annually, till five—then
they have a full set.   At eight or nine they begin to lose their nippers—two
every year—and by thirteen or fourteen years old, they have lost their
entire set."

It is evident that during the time ewes are losing their teeth, they become
less and less able to supply themselves with food, consequently afford less
and less milk for their young.   Thus the degeneracy is accounted for.   In
Vermont, where wool is as much their staple as cotton in South Carolina,
so important do they consider it not to breed from ewes after they begin to
lose teeth, that although mutton is not used by the inhabitants for the table,
they sell their old stock to be fed to hogs.

In most of the other northern states, their ewes at that age are kept from
the bucks, and fattened for market.   From their known skill in managing
well what they undertake, we may safely take their usage as a guide, when
it is applicable to our situation.   With them grass is the entire food of their
flocks—green meadows for summer pasture and hay for winter.   Their winters require five months' constant feeding, during which they estimate each
sheep to consume fifty cents worth of hay.   All stock is then kept enclosed,
and the attention to turning sheep to pasture in summer, and feeding sheep
in winter, requires but little labor in addition to their other stock.

Flocks require close attention but at three times in the year—the tupping,
the lambing, and shearing seasons.   Ewes go with lamb one hundred and
twenty-five days, or five months, and they so manage as to have the lambs

brought in April and May; (here February is esteemed by many a better month.) To do this, the bucks must be kept impounded, except at the time desired. From the great superiority of early lambs, this part of sheep husbandry is esteemed very important.

During lambing season, close attention is required to guard both ewe and lamb from storms and cold winds—to see that the ewe acknowledges her lamb, and to keep up the marking and altering as the lambs get between five and ten days old.

Ten days previous to shearing time they prepare for it by washing the wool on the sheep's back, that the natural oil of the wool, which is destroyed by the washing, may have time to be renewed, without which the wool feels and works harsh. The washing is done at spouts or pools, prepared for the purpose, and the other flocks are kept on clean pastures till sheared. As the proper preparation of an article for market is justly considered of the next importance to a good article, I will describe this process minutely, as practised for the northern market. The shearing is done on a clean floor, and each fleece is kept unbroken. When sheared, it is rolled outside in, until it is reduced to a lump about ten inches each way, and then tied with strings in two places. For packing the wool, a bag is used nearly as large as our cotton bags, which, after sewing to a hoop at top, they hang through a hole in an upper floor, a little less in diameter than the hoop; here the fleeces are packed down by the foot, the weight of an ordinary sized man being a sufficient pressure. Thus finished, the wool is ready for a northern or European market. Manufacturers will not give full price for wool unless the fleeces are unbroken; because, before manufacturing, they consider it necessary to have each fleece divided into five qualities, which they cannot do if the wool of different fleeces is emptied together.

Different modes of summer management are followed in different countries. The one followed in Australia, as described by Mr. Randall, is not only less troublesome, but would be better suited to our mountain region. There they roam over the plains, under charge of a shepherd, in flocks of three hundred to one thousand. Every night, some two or more of these flocks are penned together, during the entire year. Breeding promiscuously from the bucks that run with the flocks, allowing three or four to the hundred ewes. At the Cape of Good Hope, he says the same practice prevails; and from both of these places the wool exported is equal, and in some instances superior in quality to the Merino.

The statements here given, Mr. President, we have carefully considered, and believe to be true—and we submit them under the belief that if the importance of sheep husbandry was duly considered, especially by the districts lying north and contiguous to the mountains, much additional comfort and wealth might be added to that already delightful region.

All which is respectfully submitted.

R. F. SIMPSON.

Note.—Since this report was read, I have been informed by Mr. Thomas M. Sloan, that his wool does not cost him more than five cents a pound—and by Mr. Morris, at Pickens C. H., that for care, shearing, &c., of a small flock of eight ewes, worth eight dollars, kept on Ocone mountain, he paid one-half of the spring clipping, equal to three-quarters of a pound per head, and that from them he had, after paying all expenses, $3 60, and eight lambs worth eight dollars—$11 60. And by Mr. Shepherd a tenant on

Mr. J. O. Lewis's Tamosa estate, that in 1846 he took with him from Green-ville a few sheep,

Among which were 4 ewes, worth - - - - - - -	$4 00	
They had 4 lambs, worth - - - - - - -	4 00	
Thinks he sheared 3 lbs. per head, but say 2, which is 8 lbs. at 20 cts.	1 60 — 5 60	
In 1848, from 8 sheared 14 lbs. wool at 20 cents, - - -	2 80	
And had 8 lambs, worth - - - - -	7 00 — 9 80	
		$15 40

Salt and shearing was the only expense. He had only offered to feed them once in the two years, and then they refused to taste corn shelled to them.

H. ANCRUM ON WOOL MATTRESSES.

ASHLEY, Pike county, Mo.

A sound man in one night of seven hours' sleep, generally perspires fifty ounces avoirdupois or four pounds troy weight; we cannot wonder at that, since there are above three hundred thousand millions of pores in the body of a middle-sized man, and that in the last hours of sleep one perspires most; hence the impropriety and the weakness of lying too long in a soft bed, and the necessity of lying on a comparatively hard elastic bed, such as a wool mattress. In France, wool mattresses are generally adopted, conse-quently you never meet with a bad bed there. I have travelled all over France, and never met with a bad bed, and a very recent intelligent Ameri-can traveller of great observation, mentions, on his removal from England to France, that he found the French beds delicious, because the beds are wool mattresses.

Mode of Making a Wool Mattress.—The first thing to constitute a good healthy bed is, that it must be absolutely flat, therefore all bedsteads should have wooden laths instead of sacking, which always gives and forms a hol-low; the wool is carded by hand, and all knots and extraneous matter taken out; the great point is to make it thick enough. The best bed I slept in, in my life, had sixty pounds of wool in it, but the bed was a very large extra size; half that quantity will make a small bed, but if you wish to lie luxuriously, yet hard, do not stint the wool, that makes all this difference; it lasts for ever—the covering is washed once a year; the wool is carded, and a few pounds of wool added, and the bed is sweet and new. However luxuriously he may be, let any gentleman have a good wool mattress made and let him ride forty or fifty miles and thoroughly fatigue himself, he will then know the value of such a bed. My object is also to increase the home consumption of our wool. There are twenty odd millions in the United States; say five to each family, four millions; say three beds to each family, taking the whole population, twelve millions; say thirty pounds of wool to each bed, three hundred and sixty millions of pounds of wool; say thirty-four millions of sheep in the United States, say eighty million pounds of wool; this will consume more than four years' clip of our wool. This ought to be promulgated to increase the consumption of our wool, and such wool as cannot be sold abroad. Independent of the benefit to all in their health, who adopt wool mattresses on account of their cleanliness and dura-bility, in the end, they are cheaper than any other bedding.

Every thing that increases the home consumption of our wool is of na-tional importance, as is every thing that will promote the general health of our people. In this changeable and rigorous climate in winter, if all were to wear flannel, particularly narrow-chested and delicate females, it would be of the greatest benefit to their health, and save them many a fit of sickness. When we consider how cheap the English sell their Welsh flannel, it ought to stimulate our manufacturers; I must, however, observe, that I never purchased any flannel in the United States equal to the real Welsh flannel, or that did not shrink, or that wore near so long as the English flannel. The English flannel has a nap on both sides, which renders it warm and soft, and it washes soft to the last. The United States flannel that I have used washes harsh, and the wear is not near so agreeable as real Welsh flannel, but surely all these difficulties can be overcome by our people, and they can make as good flannel as the best Welsh flannel.

It is well known that woollen clothes, such as flannels, worn next the skin, promote insensible perspiration. May not this arise principally from the strong attraction which subsists between wool and the watery vapor which is continually issuing from the human body? That it does not depend entirely on the warmth of that covering is clear, because one degree of warmth produced by wearing more clothing of a different kind does not produce the same effect. The perspiration of the human body being absorbed by a covering of flannel, it is immediately distributed through the whole thickness of that substance, and by this means exposed by a very large surface to be carried off by the atmosphere, and the loss of the watery vapor which the flannel sustains on the one side by evaporation, being immediately restored from the other in consequence of the strong attraction between the flannel and this vapor, the pores of the skin are disencumbered, and they are continually surrounded with a dry and salubrious atmosphere. It is astonishing that the custom of wearing flannel next the skin should not have prevailed more universally; it is certain it would prevent a number of diseases, and there certainly is no greater luxury than the comfortable sensation which arises from wearing it, after one is accustomed to it. It is a mistaken notion that it is too warm clothing for summer; it may be worn in the hottest climates, at all seasons of the year, without the least inconvenience arising from wearing it.* It is the warm-bath of a perspiration confined by a linen shirt, wet with sweat, which renders the summer heats of southern climates so insupportable; but flannel promotes perspiration and favors its evaporation, and evaporation, as it is well known, produces positive cold. I can vouch for the truth of every word of this. I wear the same kind of flannel all summer as I do in winter with sleeves; when I take extra exercise and perspire freely, my body and flesh is always cool and comfortable, and in part I owe it to wearing flannel that I have never had either fever or ague in this western country, which is full of it. All this may appear trivial, and sanitary rules are disregarded, but it is all of the utmost importance and to all. Say fifteen millions of our people wear flannel next their skin, and three flannel waistcoats to each, that is forty-five millions of waistcoats, at two yards each, (not enough with sleeves as they ought to be made,) ninety millions of yards of flannels in waistcoats only. Old people, delicate women and children, and above all, consumptive people, ought all to wear flannel drawers as well as a flannel waistcoat; if this was adopted, the great sickness that prevails in the United States would be much diminished. Men drink spirituous liquors to increase the animal heat, and feel that glow that is called comfortable. Let them wear flannel next their skin instead, and keep the body warm and the head cool.

* The firemen in steamboats could not exist if they wore linen instead of flannel shirts. The pleasant wt. of all mattresses is one made of a mixture of wool and hair. [EDS PLOUGH, LOOM, AND ANVIL.]

SHEEP HUSBANDRY IN TEXAS.

BY H. S. RANDALL, LL.D.,

Author of " Life of Thomas Jefferson ;" Editor of Randall's " Youatt on the Horse," etc., etc.

EDITORS OF TEXAS ALMANAC: In pursuance of your request, I proceed to give you some of the results of my experience and investigations in regard to wool-growing, and my views of the adaptation of this husbandry to the climate, soil, and other existing conditions of Texas.

CLIMATE.—The best climate for the *cheap* production of wool, other things being equal, obviously is that which furnishes the most abundant and suitable pasturage during the greatest portion of the year. This, speaking generally, is to be found in the Northern hemisphere, between latitudes 30° and 40° on the eastern margin of each continent, and between about 38° and 48° on the western. The south half of this wool-growing zone, where the other conditions besides climate are favorable, excels the northern.

North of the wool-growing zone, the growth of vegetation is suspended, and the nutritiveness of grasses destroyed by cold, during considerable portions of the year; and then sheep require more expensive dry feed, suitable winter shelter, etc. South of the wool-growing zone, vegetation, where it flourishes, is too rank and tropical for the smaller ruminating animals, and the heat too intense for those carrying so dense a pelage as the fine-woolled sheep.

Local exceptions exist to the above classification, owing to a variety of causes; most prominent of which are altitude, the shelter of mountains from northern and southern winds, the contiguity of large bodies of water, etc.

I have said, "other things being equal," the question will be immediately asked whether wool of the same variety of sheep grown in latitude 30° is as fine as that grown in latitude 40° or 45° I doubt whether it is. Sheep transported from a climate of long winters to one of perennial, or nearly perennial, pasturage, increase visibly in size, and their descendants permanently become a larger variety. The constant supply of succulent food produces more copious and uniform animal secretions than an interrupted supply, or than an alternating supply of green and dry food. In theory, we should expect the same causes to affect the fleece as well as the carcass. They do visibly increase the length of the staple. The increase of its diameter (admitting that it does increase) during over twenty years of breeding—about as far as my personal observations have extended on that point—is not, I think, perceptible to the naked eye. But be it greater or smaller, it is more than compensated for by the increased softness and evenness of wools grown in warm and more uniform climates, and on more uniformly succulent nutriment. I must be content to state this as a well-established practical fact. I have not room to array authorities on every point.

SOIL.—It would present a very tangible, and by no means a bad test of the proper soils for wool-growing, to say that they are those which produce, or which can be made to produce, the most continuous supply of fine sweet grasses. A marshy soil, a soil containing so great an excess of clay as to poach into mud, and remain long wet after rains, a low, rankly, rich river bottom alluvion, and especially such an alluvion, if annually replenished by slimy deposits of decaying vegetable matter, all inju riously affect the health of sheep. None of these soils produce the grasses I have mentioned. The best lands for sheep are those which are dry and "sound ;" which admit of the rapid percolation or drainage of water ; and an admixture of sand or gravel in them, is a favorable, though by no means an indispensable condition. Enormous flocks of sheep flourish, in perfect health, on the plains of Illinois, which are "sticky" after every shower. But water does not stand on them as on a stiff clay, nor does it constantly saturate them as it does boggy lands.

It is sometimes claimed—particularly by that class of tyros who are ready to jump at conclusions on a very limited experience—that sheep on their farms, or in their localities, defy all the preceding conditions. They flourish, if we may believe these gentlemen, in stagnant fens, in "hog-wallows," and on river bottoms, where the malaria is almost visible, as it steams up from the decomposing mass. It is true that fifty sheep, like a small family of human beings, will occasionally, and for a limited period, appear unaffected by such unpropitious circumstances. But, by and by comes the destroyer—the pestilence that walketh by noon-day—and the increase of years is suddenly swept away. Cholera, yellow fever, or bilious fever depopulates the human settlement—rot, or some other epidemic, passes, like a tornado, over the sheep-ranch. The causes of disease do not bear fruit every year, but the laws of nature are never abrogated.

Sheep tolerate almost all chemical varieties of soil. With the proper conditions in other respects, they are seemingly equally healthy on the sterile, pulverized granite of New England, and the rich, calcareous wheat lands of Ohio, or the Valley of Virginia—on the tertiary sands of our Atlantic border, and among the rocky cliffs of the Alleghanies. No animal is so necessary to man, and therefore none has been adapted to the circumstances of so large a portion of the earth's surface.

ELEVATION.—Elevation is, I rather think, a pleasing condition to an animal, which, like the goat, the ibex, etc., zoölogists consider the natural denizen of mountain regions ; and mountains and hills often present the other condition which are specially adapted to sheep—firm dry soils, short sweet grasses, pure air, and clear water. But elevation is of no consequence *per se ;* and if the same favorable conditions are found on plains, they are as healthy localities for sheep as mountains.

GRASSES.—Sheep will thrive on almost all varieties of grass, when they are first springing up tender and succulent from the earth. No grass is suitable for them, when its stems have become dry and woody. Tough, aquatic grasses are always unfavorable. We are to give the preference, then, to those varieties which do not send up coarse seed stems—those which are constantly supplying a fine verdure from the root. No variety is preferable to the small, spontaneous, white clover of the north, or the finest spontaneous musquite grasses of Texas. Red clover, Timothy, June, or Blue grass ;[*] indeed, all the grasses cultivated in the north, will do very well if kept fed down, and this might be the case with many of the coarser varieties in Texas. Some small flock-masters have fancied that sheep would thrive on the dry stems of tall, coarse grasses—because they thrive *among* them. But a few sheep will find tender, nutritious plants, which are screened from casual observation among these taller ones. When the former are gone, sheep will promptly and visibly fall off in condition.

The fact that the *natural* grass is too coarse for sheep, by no means

---

[*] I am not sure that the pure grass of New York and the Blue grass of Kentucky are the same, never having specially investigated the subject : but the late Mr. Clay wrote me that they were the same.

proves that proper "artificial" varieties would not flourish on the same soil, particularly if the first flush of its virgin fertility was a little reduced by cropping.

WATER.—Water is not indispensable for sheep, when at pasture. The juices of the grass and the dew and rains supply their wants. Thousands and thousands of good sheep-pastures in the north are wholly without other sources of supply. But, of choice, I would prefer water in sheep-pastures—clear springs, or rapidly running brooks. Sheep will visit these as regularly as cows or horses.

ADAPTATION OF TEXAS.—As I remarked in my answers to your interrogatories on this same subject, last fall, (published by you in the *Galveston News*,) I have never set my foot in your State. But, after the experience of many years in sheep-breeding, and after a close and diligent investigation, extending to all available sources of information—many of these the minute and careful statements of your own most intelligent and candid citizens—I do not entertain a particle of doubt, first, that you have vast regions in Texas admirably adapted to sheep husbandry; and secondly, that wool can now be raised more cheaply in those regions than in any other portion of the globe, where sufficiently good government prevails to make life tolerable and secure, and such property as sheep safe from frequent and extensive depredations. In no such portion of the earth, are lands, furnishing perennial pasturage, (or the use of such lands,) so-cheap. In none are the general circumstances more favorable, the accidental and occasional disadvantages fewer. In nearly every particular, Texas possesses decided advantages over our other Southern States, and enormous ones over the Northern and Eastern States.

As between it and the latter, a brief statement disposes of all controversy. The sheep-lands of the Northern and Eastern States cost, on an average, thirty dollars an acre ; and sheep are frequently kept on these worth from forty to sixty dollars an acre. On these high-priced lands, sheep must be fed on dry feed—hay and grain—about five months of each year. Expensive shelters must be erected, or the sheep-farmer will lose the cost of them in the loss of life and condition in his flocks.

In Texas, prime and desirable pasture-lands can be bought at two dollars an acre—frequently for considerably less. He who owns a homestead of a few acres, can pasture thousands of acres of unoccupied land. The pasturage of much of Texas is perennial. Large and small flock-masters have proved this to be a practical fact. Mr. Kendall has wintered a large and constantly increasing number of sheep, for three years, without, he writes me, giving "an ounce" of dry feed, or providing any artificial shelter, though he agrees with me that a little of both would be desirable for emergencies. He has encountered wet winters and dry winters with equal success. His sheep are perfectly healthy. His testimony is fully confirmed by that of some twenty other candid and intelligent gentlemen, scattered over various parts of the State, who have favored me with minute accounts of their experience in sheep-raising. Theory would anticipate these facts when the natural conditions of Texas are known; but it is always satisfactory to have the suggestions of theory established and made certain by actual experiment.

Your country cannot always enjoy this entire priority in the conditions for cheap wool production. The success of this husbandry of itself will aid in reducing its profits. Your sheep-lands skirt noble and navigable rivers. Unlike our rough sheep-ranges of the North, they are topographically adapted to the construction of those railroads which the business of your State will soon demand. An enterprising population is pouring in upon you from the other States of the Union, and from Europe. The boy is now born who will see, not only the good soils in all the counties at present organized in Texas, but in its regions where now roves the wild Camanche, worth twenty or thirty dollars an acre. Then the sunny but unarable slopes of the Alleghanies, in Virginia, the Carolinas, etc., may successfully compete with you in wool-growing, owing to their greater cheapness.

PROFITS OF WOOL-GROWING.—Northern flock-masters usually estimate the consumption of eight American Merino sheep equivalent to that of a cow. All prime American Merino flocks should average as high as about five pounds of washed wool per head, or seven and a-half pounds of unwashed wool. Choice breeding flocks should do better still. To show you that I speak from actual knowledge instead of conjecture, let me say that I have two flocks of pure Spanish (American Merino) yearling ewes, which averaged, this year, five pounds and six ounces of well-washed wool per head—equivalent to eight pounds and one ounce unwashed; and the yearling never produces as much wool as the adult sheep. A choice small flock of my grown ewes averaged six pounds and a fraction of washed wool per head, equivalent to nine pounds unwashed. In none of these flocks were there any rams or wethers to increase the yield of wool. None of them received any pampering, or were sheared at an unusual time. I have cross-breeds between the French and Spanish which averaged still higher; but they are larger, and will consume more. For many years my entire number of full-blood sheep of all ages, exceeded five pounds of washed wool per head.

The price of American Merino wool (washed) has been as follows, on the first day of August, in the years indicated :—

1851	42 to 44 cents.
1852	40 to 43 "
185?	49 to 53 "
1854	38 to 40 "
1855	37 to 38 "
1856	44 to 46 "
1857	45 to 48 "
1858	37 to 41 "
1859	44 to 46 "

Assuming five pounds to be the weight of fleece, and eight ewe sheep to be the equivalent of a cow in consumption, it follows that the feed of a cow would have returned this year eighteen dollars in wool, and as many lambs as eight ewes would raise, which would be at least seven. What these seven lambs would additionally be worth to the Texas grower, I leave you to estimate. Even among common sheep, the lamb is always considered to be worth as much as its dam's fleece. If rams and wethers raise no lambs, they produce greatly more wool than ewes. Now say what are the profits of a cow in Texas, and deduct the difference in the trouble of looking after her and the eight sheep, and you have a comparative view of the profits between the animals, which will prove instructive! Do you obtain five dollars per head of net annual profits on cows, on the average? And yet you raise cattle on a scale which conveys the idea that you find it as profitable as any other of your branches of husbandry on the pasture-lands of Texas. In my former letters to you I placed the entire cost of keeping sheep, including interest on land, at fifty cents a head per annum. In this estimate, I included the cost of shelters, of a month's winter feed, and some other contingencies, all of which I am assured by intelligent Texians are unnecessary. I also proceeded on the supposition that no lands were to be pastured but those which had been bought and paid for by the flock-master. If these items be struck out, the cost of keeping large flocks ought not to exceed one half of my former estimate. I confess, however, that this sounds almost too favorable to be true. Mr. Jefferson deeply lamented the dismemberment of that Texas from the United States which he had bought with Louisiana —considering it the very garden of our Southern country. But the "Sage of Monticello" hardly expected to find El Dorado in your sheep-pastures, or Aladdin's lamp on the bank of the Colorado! I repeat it, the story *must* be too good to be *all* true.

The first cost of embarking in breeding full-blood sheep is considerable. But the sale of surplus ones at extra prices to newer breeders will soon offset this; and, at all events, it is so soon repaid by the enormous profits of the husbandry, that it is not to be kept in view as an annual part of the account. Interest ceases to run after the principal is paid!

Another important fact in favor of sheep is always to be taken into view. If the steer or colt dies before it is sold or used—if the cow dies before she has produced young—the loss is nearly a total one. At best, the colt keeps you waiting on him, say three years, and the steer and heifer at least two, before they commence making returns. The sheep is a prompter paymaster. He pays you annually. And he never dies in your debt. If he dies before he is six months old, he has cost you nothing that is appreciable. If he dies afterwards, before his first shearing, his wool will more than pay for what he has consumed; and this is true of him at whatever age his death occurs, taking the aggregate of his life together.

BEST BREED OF SHEEP.—When wool is the main object, and mutton is only an incidental one—as always must be the case in a large and thinly inhabited country like Texas, not yet containing populous cities— there is but one breed of sheep worth consideration, so far as comparative intrinsic value is concerned. I can declare on a pretty extensive experience—but it really needs no experience to arrive at that conclusion—that no other breed makes a remote approach to the value of the Merino for the production of wool; and its mutton is good and palatable. Half and three-quarter breed Merino mutton is especially so, and five Americans out of six would prefer it, on the table, to the tallowy meat of the large long-woolled English mutton varieties.

Well-bred Merinos yield about as much wool per head as the largest English long-woolled breeds—yield farmore than English middle-woolled breeds—yield about twice as much value of wool for the amount of feed consumed as any English breed—and are hardier, and herd (that is, thrive when kept together in large numbers) better than any of the more valuable English varieties. Their length of life is much greater. A Merino is not, to use a common expression, "older at eight" than a Bakewell or Southdown sheep is at five. And, what may not be quite as well understood by those who have not experimented with both races, (as I have,) the Merino is decidedly hardier than the high-bred English sheep. It is less addicted to colds or snuffles, bears extremes of weather better, is capable of travelling farther for its food, and will endure a scarcity of food with far greater impunity. The English sheep has the advantage of arriving earlier at maturity—a matter of much importance in a mutton breed, but of comparatively little in a wool-growing one.

Of the unimproved English and Scotch varieties, it is unnecessary to speak. Several of them are much superior to the improved varieties of those countries in hardiness and herding properties; but they all produce coarse, and, compared with the Merino, light fleeces. None of them are, in my opinion, really essentially superior for wool-growing to what is termed the "native stock" in New-York and New-England.

SPANISH MERINOS.—I shall not here consume space with the past history of any breed. The ancient Spanish sheep, as imported into this country by Humphreys, Jarvis, and other breeders, whose character gave full warrant of supposed purity of blood, produced of washed wool less than four pounds in the ewes, and seven in the rams. The flocks of Spain, taken as a whole, are declared by that most intelligent observer and investigator, Chancellor Livingston, to have averaged of washed wool, only two pounds and a half in the ewe, and four and a quarter in the ram. (See Livingston's Essay on Sheep, p. 39.) They resembled the present American Merino in form, but were smaller.

SAXON MERINOS.—The Saxon was the first great offshoot from the parent Merino stem. Partly from the principle of selection first adopted, and partly from the system of management, and the special objects pursued in breeding, this variety materially dwindled from the size, constitution, and weight of fleece of the Spanish sheep, but improved in the quality of the wool. The first considerable importation into the United States took place in 1824, and the mania for these feeble little things raged to such an extent, for the three succeeding years, that the most miserable mongrels and grade sheep were introduced and sold under their name. (See the authoritative and undenied statements of the distinguished German importer and shepherd, Henry D. Grove, on this subject, made to me as the chairman of a Committee appointed by the N. Y. State Agricultural Society, in 1837, to report on "The condition and Comparative Value of the several Breeds of Sheep in the United States." Society's Transactions, 1841, p. 313.)

The American Saxon of the present day is a much larger and stronger sheep than its imported ancestor, with, I think, about the same quality and a greater quantity of wool. Some pure flocks now produce an average of nearly three and a half pounds of washed wool, and others dashed with a strain of fine American Merino blood, not materially changing the quality of the fleece, average very near four pounds. It continues, however, to require considerably more care than the American Merino, and does not rear so large a per centage of lambs.

SILESIAN MERINOS.—These appear to be something between the Saxon and American Merino—heavier fleece than the former, finer fleece than the latter—and between the two in size. Some specimens I have seen had fine plump forms, and their wool the oil and external black gum of the gummiest family of Merinos. They might, I should say, prove a desirable variety under certain circumstances, and I think a cross with them would improve the Saxon type of sheep. They are the only Merino family which I have not bred.

FRENCH MERINOS.—The selection and management of the Spanish Merinos in France, first carried into that country a little more than seventy years ago, produced precisely the contrary effects in several important particulars from those produced by the German system. The carcass was made larger, the fleece heavier and coarser. But a portion

of those introduced from France into the United States, within a few
years, have not been characteristic specimens of the variety in either of
these particulars. They have been exaggerated specimens or caricatures
of the breed. They were those exceptional animals to be found in all
flocks, larger and heavier-fleeced than the great body of those flocks. I
apprehend also that some of them were exceptional in certain other
particulars, as for example, in that enormous "throatiness" which renders
them such a marvel to the multitude.

There is no doubt, however, that as a variety, the French Merinos are
larger and heavier-fleeced than any other Merino family. But in respect
to fleece, the best of them do not excel the best American Merino as
much as in weight of carcass; in other words, they do not produce as
much wool in proportion to size, and, accordingly, to consumption.
There are two causes for this. The first is, their wool has, for reasons
which I can not explain, been bred proportionably short. As some very
remarkable misconceptions appear to prevail on this particular point, let
me state a positive and tangible fact for which I am willing to be held
responsible. I recently (to test the accuracy of previous impressions and
experiences) carefully measured, lying unstretched on a table, fifteen
specimens of French Merino wool, taken indiscriminately (with the aid
of the owner) from the sheared fleeces of *one year's growth* of a pure
blood flock, descended from prime sheep of Mr. Taintor's importation,
and considered by French breeders prime sheep of the variety. The
fifteen specimens averaged a little less than two inches in length. I have
owned French sheep from the best imported flocks, have seen hundreds,
if not thousands, of others, and they have been almost uniformly rather
short-woolled sheep—shorter woolled than the longest stapled American
Merino flocks.

The second reason for the comparative lightness of French fleeces, is
their dryness—their freedom from oil and gum. Though there are exceptions, there is a constant tendency in this direction among them, and
unless care be taken in breeding to prevent it, the wool becomes almost as
dry as cotton, and then the fleece has no proportionate weight for its bulk.
The difference in appearance and handling between such and good
American wool is very analogous to that between American and choice
Italian sewing-silk. The first, in both cases, even when the actual fineness (diameter) is the same, is comparatively light, unelastic, unglossy,
and "cheap looking." I have sometimes fancied that the pile or fibre of
the American Merino wool is actually denser and heavier of its size than
the French, independently of all extraneous substances. This, however,
may be but a fancy.

French wool washes much cleaner than the oily and gummy American
Merino wools; and I think more slowly recovers its maximum of oiliness,
after being washed on the back.

The overgrown animals of a variety, are rarely of good form. Hence
not a few of the imported French sheep, and their immediate descendants, when denuded of their fleeces, were most unsightly, scraggy,
"lathy' animals, excessively bony, crooked in the back, bad in the cross,
(that is, hollow behind the shoulders,) and so thin in the chest, that both
forelegs seemed, in horseman's phrase, to "come out at the same hole."
The last defect is perhaps rather characteristic of the variety; but I have
seen not only individuals, but flocks of French sheep, of moderate size, as
well formed in every other particular, and perhaps even in that, as any
other family of Merinos.

Another difficulty followed the selection of these huge sheep. Over-grown parents do not always produce overgrown offspring; but the *inored* must be kept up, and to do this, a concealed, or at least an un-avowed course of pampering was resorted to in some cases. The lambs were dropped two months before the usual time of having lambs dropped in the North—the ewes were stuffed with unusual and succulent food during the winter, regardless of cost; they were kept in close, warm stables at yeaning time; the lamb was often given the aid of a "sucking-bottle," or a foster-dam, in addition to its natural parent; it was taught as soon as possible to eat roots and grain; it was kept housed from every storm, and even from the dews of the night, during the *entire year*. This added greatly to the beauty and weight of the wool—to its weight, (when sheared unwashed,) because none of its natural oil was washed out by rains. When at length it was exhibited, without any explanations of the preceding facts, at some State fair, in autumn, it presented about twice the size of carcass, and twice the length of wool, that it would have done if dropped at the common time, and treated in the common way. If not sold at a year old, it was not sheared, and the entire fleece was left growing to increase the *marvellous* product at two years old; or if it was feared that this fraud would be too apparent, (beyond the gulli bility of the particular market in view,) the yearling was "stubble-sheared," that is, shared a half-inch or inch from the skin, leaving three or four months' growth of wool thereon, to go into the next fleece. I do not say, that the breeder is not authorized to conduct his business to suit himself—men clearly have the right to pamper, and to manufacture "marvels." But he who does so, is bound to give warning, "fair and true," to the *buyer*, whether questioned or unquestioned.

Helas! what was so soon the matter with those gigantic French rams, which first scattered like wild-fire over the North? There came a chilly rain-storm, and they sneezed and coughed. Soon they began to mope, and fall off from their feed. They grew thin, and then weak. Their heads drooped; yellow waxy matter collected about their dim, half-closed eyes; a sticky discharge clung about their nostrils; at length the faint but rapid heaving of the flanks began to indicate a low fever. Then an-other heavy cold shower, and the farmer's boys presently ran into the house, crying: "Father, father, the great ram is dead!" The farmer had not known that he had set a hothouse plant *out of doors!* Thus "departed this life," a majority—ay, a *majority*—of the first inundation of great French rams—many of them without getting a lamb. When they lived, it often proved a greater disaster to their owners. They spoiled the carcass and constitution of his flock, lowered the quality of his wool, and not unfrequently actually diminished its quantity.

These circumstances created a violent reaction against French sheep, and I should say, between eighty and ninety per cent. of our best North-ern and Eastern wool-growers now thoroughly detest them. I believe they have jumped off the bridge "on the other side!" Circumstances led me into an extensive course of fresh investigations on this subject last winter. I found French, like other sheep-raisers, divided in about the usual proportion, between quacks and legitimate breeders. I found French Merino flocks, and especially very high-bred grade French flocks, based on an American Merino foundation on the maternal side, which exhibited fine forms, sufficiently rugged constitutions, a good quality and large quantity of wool. If the wool lacked a little of the gloss and style of the choice American Merinos, it nevertheless was a desirable article.

and especially so for the manufacturer, on account of its cleanness. Perhaps, after my preceding remarks, justice requires me to add that I found breeders raising the very *largest* French sheep with undeniably legitimate objects. They considered that great size desirable, and were therefore (erroneously and unavailingly, in my opinion) attempting to perpetuate it without perpetuating its usual accompanying defects.

The careful and certainly disinterested examination of many flocks, which had been judiciously bred for a number of years, and down to the present epoch, somewhat modified, I confess, my own previous views. I believe indeed, I found hardier and every way better sheep than the French stock first brought into our country. I made up my mind, that the prejudice against them was violent and excessive, and that by and by another reaction will set in their favor, and that they will be extensively used for an object which I shall treat under another head.

THE AMERICAN MERINO.—About the same amount of fraud and deception attended the introduction of the Spanish Merino into the United States, (after Livingston, Humphreys, Jarvis, and a few other elevated men had done their part,) that heralded the advent of the Saxon and French varieties. Like the French, the former sunk into contempt, before it received the general approbation of the country. And it encountered a far more dangerous foe than contempt, in an almost universal admixture with the puny Saxons. But a remnant was fortunately kept pure, and many flock-masters, after a Saxon cross, bred back to their previous Merino standard.

The American Merino of the present day is a considerably heavier and stronger animal than his Spanish progenitor. He has been kept in smaller flocks than in Spain, better fed, (or more uniformly fed,) and subjected to a more careful and intelligent system of breeding. As long ago as 1841, the celebrated early importer and subsequent breeder, Hon. William Jarvis, of Vermont, wrote me that "twenty-five years' experience satisfied him that the wool of the Spanish Merino had rather improved" in this country; that his own wool was better than the samples received from Spain, when he purchased his imported flock. (The whole of this admirable letter will be found in the N. Y. Agricultural Society's Transactions, 1841, pages 320–328.) The same kind of improvement has continued down to the present time, in many flocks.

The different Spanish varieties were, as a general thing, soon intermingled with each other in this country, as they had been in France, so that the names of Paular, Negretti, Gaudeloupe, etc., now have no meaning, unless in a very few instances, when applied to American sheep. But in point of fact, the same varieties, or somewhat analogous ones, have been reproduced in our country by the systems of breeding pursued by particular persons. Some men, for example, have carefully shunned " oil " and " gum," and made fineness of wool the primary consideration. These have substantially reproduced the Spanish Escurial, a sheep closely resembling the Saxon, except in its larger size. Others have made weight of fleece the primary consideration, at some sacrifice of fineness ; and to this end they have bred as much oil into, and external gum upon the fleece, as practicable. The extreme of these sheep become coated over a few months after shearing, with a natural covering of gum of the color of tar, extending about an eighth of an inch into the wool, which in warm weather sticks to the hand, and in cold becomes a hard rigid crust. The interior of the fleece looks as if oil had been poured into it, as it exists there not merely as a coating of each filament of wool, but rather wears the appearance of a mass of oil, with filaments of wool growing out through it.

Some breeders take a middle course, cultivating the oil, but avoiding the gum. Some cultivate a medium share of both; and so on.

The well-bred American Merino is probably now the densest and heaviest-fleeced sheep of its size in the world, without the help of any extraneous circumstances; but when you read of the *enormous* fleeces you sometimes do, (whether they belong to H. S. R. or A. B. C.,) it is time that you, and all other intelligent men, understand that this enormous extra weight is made up of *oil* and *gum.* In the first place, wool of this character can not, at best, be *well* washed on the back of the sheep. In the second, many breeders do not desire to so wash it; they choose to leave these heavy animal excretions in the wool, and they let their flocks run long enough between washing and shearing, to restore all that has been washed away. In fact, washing with them, is little better than a name, a pretence, to prevent the buyer from deducting the usual one-third from the gross weight, as on unwashed wool. Then, further to mislead the purchaser, they do up each fleece in two parts—claiming that if that personage sees fit to judge the wool solely by weight of fleece, instead of quality and condition, (as he often does,) it is but a fair retaliation, a warrantable "spoiling of the Philistines," to take a course which will compel him to judge the article by legitimate tests, or to suffer the consequences. (That is to say, they assume that if the buyer is a blockhead, or screw, it is right to cheat him, if it can be done by silence.)

These excessively oily and gummy sheep are rather "the rage" at present in the North. There are two reasons for it. The wool-buyer has obstinately refused to make any proportionate difference in the price paid for their wool and that paid for cleaner wools. He will usually pay within three or four cents per pound, as much for the first as for the last, when the "greasy" wool weighs two pounds most to the fleece, when it costs no more to raise it, and when it will lose twenty-five per cent. more in cleansing. The manufacturers could have corrected this evil, if they had chosen to do so; and a class of sham-hating men have continued to breed clean wools, expecting them ultimately to do justice in the matter. But indifference, or the temptation to force these breeders to sell (or sacrifice) their beautiful clips at two or three pennies above the price of "greasy" wool, has generally triumphed over all more manly considerations, though in regions where clean wools are extensively grown, and where the breeders can and will stand by each other, they have fared better.

The other reason for the popularity of excessively oily and gummy sheep, exists in the fact that they generally sell better to those *beginners* who are willing to pay breeders' prices. The first thing in a variety or breed, which attracts the eye of a novice, is its salient peculiarities— whether they involve valuable characteristics or the contrary; and they are very apt to become his standards of purity of blood and individual excellence, until experience has taught him better. The Merino, compared with others, is an oily and gummy sheep, and "argal," the more oil and gum he possesses, the "more Merino" is he to the novice. The same remarks apply to "throatiness"—large corrugations or folds of pendulous skin about the neck or throat, and similar folds on other parts of the body.

Breeders defer more or less to the tastes of buyers, and thus more "grease and wrinkles" are produced than would otherwise be. A pettier personage—your nomadic ram peddler—carries his complaisance still further. He *manufactures* traits or peculiarities to please purchasers! He buys up half or three-quarter bred Merinos, which chance to have abundance of "wrinkles," (the mongrel get of a very "wrinkly" ram often show this peculiarity quite as strongly as his full-blood descendants,) and if the natural gum is wanting, he puts it on by daubing them over, immediately after shearing, with a pigment of linseed oil and burnt amber—a composition known in the North as the "Cornwall finish," from the fact that it was first used (as a winter protection to sheep I presume) in Cornwall, Vermont. It soon makes a nearly black external coating, so similar to the natural gum as to be entirely undistinguishable from it, except to a very practised eye. I should say, however, that it was usually a little more evenly put on, and a shade *handsomer*, than the natural article! A second good oiling, with clear oil, towards fall, helps along. Armed with these painted mongrels, a demure face, and a certificate of pedigree, purporting to be signed by a "Deacon," and a "Judge of Probate," your ram-peddler sallies forth, Macedonian-like, conquering and to conquer—greenhorns!

THE BEST BREED FOR TEXAS.—In the views I shall express under this head, I desire neither to advance nor to injure the interests of any individual—certainly to wound no man's feelings. But if I speak at all, of course I am called upon to express those candid convictions, for which I am willing to be held responsible.

I am free to say, on the start, that I believe there is altogether too much one-sidedness in the views entertained by individuals, and the public, in regard to this and other analogous questions. The current sets in some particular direction, and then all influenced by conviction, or the desire to take advantage of other people's convictions, jump into it irrespective of circumstances.

I have no doubt that every variety of the true wool-growing sheep, the Merino, has an appropriate and profitable place. As long as costly woollen fabrics are demanded by the wealthy and luxurious, the delicate Saxon sheep is a want in agriculture. The Silesian supplies the next want, and so on down. If the production is accurately proportioned to the consumption, the laws of trade declare that all these breeds must be profitable, (and something like equally profitable,) under the best circumstances, for their respective cultivation.

I desire to make another statement. In spite of all the pretences and quackeries of rival breeders, I have no doubt that the *best* animals of each of these varieties, produce about the same value (not amount) of wool for the amount of food consumed. Some, however, require more human labor and supervision than others, some demand milder climate than others, and so on.

The time may very probably come when each of the Merino families will be profitably grown almost side by side, in Texas. Your climate is as mild as the feeblest-constitutioned ones can elsewhere find. The extra labor demanded for the supervision of such, ought to be as cheap with you as in other portions of the United States. You have abundant sustenance for the strongest fine-wool breeds. And who can say that when you have railroads covered with cattle and sheep-cars, that you can not supply the mutton eaten in our Atlantic cities, more cheaply than it can be grown nearer to them, and that it may not thus be made profitable to you to grow coarse as well as fine wools?

At present, fine wools pay best in the United States; and among these medium qualities find the most extensive demand and the most remunerating prices. This class of wool is borne both by the American and French Merino.

As a pioneer and experimental sheep, you want the hardiest variety—one capable of resisting a change of climate and circumstances, general or local difficulties in the way of acclimation, and the effects of inexperienced management. As a distinct variety, there can be no reasonable doubt, I think, that the pure American Merino is better able to "rough it" in a new country than the pure French Merino, though I apprehend the latter will ultimately do well enough in your mild climate. No person acquainted with both breeds will hesitate to believe, that in a summer drouth, or during a severe winter scarcity, the former will suffer less than the latter—as much less as black cattle would suffer, under like circumstances, than the larger Shorthorns. The impression is also universal, and certainly every appearance and analogy would seem to favor it, that the American Merino will herd best in *very* large numbers.

But it is not necessary that you limit yourselves exclusively to one variety. He who wants to grow very large wethers for plantation use, or for sale, or he who wishes to grow a larger fine-woolled sheep as a matter of taste, and is prepared to take care of them, will find his wishes met by the French sheep; or if wethers weighing from 150 lbs. to 200 lbs., and ewes weighing from 120 lbs. to 140 lbs., are large enough for him, he can get them by a cross between the French and American varieties.

These last, if well bred, compose a beautiful and hardy sub-variety. I bought a flock of them, last winter, mostly French, which were dropped in May, 1858. They were kept in the common way, without pampering. They were thoroughly washed and sheared at the common time, about the close of June, 1859. Their wool was destitute of gum, about as white at the outer as inner end, and seemingly almost as free from oil as cotton. They averaged six pounds and four ounces of wool per head. I believe that a cross between such ewes and a very heavy and "greasy" fleeced American Merino ram, would carry the average a pound higher in the produce. I am so confident of this, that I am availing myself of the results of such a cross on a comprehensive scale. The ewes are uncommonly fine-woolled of their kind. The ram used weighs but 150 lbs. in full fleece, and his washed fleece (as well as it could be washed) weighed 14 lbs. Of course he was excessively oily and gummy, and it was for that very reason he was selected for the experiment. His lambs, two months after birth, began to be colored very perceptibly by the oil inherited from their sire.

It is legitimate in breeding, to counteract one defect with another. I have noticed for twenty-five years that the oiliest and gummiest Merino rams cross best with the dry coarse-woolled varieties. I have recently noticed that the cross between the French ram and such varieties rarely results very satisfactorily. There is another reason for this. Where the size of the male is greatly disproportioned to that of the female, the unborn lamb has not room to expand in the womb, and it is born crooked and unshapely—generally thin-chested and flat-ribbed. Hence I entirely prefer the cross between the French ewe and American Merino ram to the one made conversely.

Am I·asked at this point, if it is legitimate to breed extraneous substances, like oil and gum, and sell them for wool? Certainly not. On the other hand, is it legitimate for the wool buyer and manufacturer to make no fair distinction between clean and dirty wool? He who sells "greasy" wools in broad daylight, without splitting his fleeces or resorting to any other trick, at leasts commits no fraud! On the whole, I would push towards neither extreme. In your climate, I think you will have to obtain pretty dark-colored and oily rams to keep up the proper medium in that particular in your full-blood American Merino flocks, and still oilier and darker ones, to produce that medium in a cross with coarse sheep.

The wool of the French crosses I have described was a shade coarser and a shade lest "stylish" than fair American Merino wool. But the difference in cleanness was in its favor, and the difference in fineness was so little against it, that every lot I mentioned sold, or could have been sold separately, in July, for forty-four or forty-five cents a pound. An intelligent breeder of these, and of French full-bloods, candidily admitted to me last winter, that he considered the former worth quite as much as the latter for wool-growing. Under common care, and exposed to any disadvantages, I think they would prove most profitable. And such a cross would promptly give additional weight of fleece, and especially size, to American Merino flocks deficient in either particular. But the cross must always be well made not to result in failure.

SELECTION OF SHEEP.—The points of a good Merino's carcass are, good but not extraordinary size for the variety, the barrel well rounded, the chest deep, the cross full, the back level forward of the hips, the loin and buttocks comparatively wide, the flank and twist well let down, the neck round and set on level with the shoulders, the head fine but broad between the ears, the eye lively and mild, the legs straight and moderately long, the whole figure wearing a marked appearance of compactness and solidity.

The degree of throatiness is rather a matter of taste. It is a great impediment to smooth and rapid shearing; but as a badge of blood, and as an indication of that loose, large skin which is a characteristic mark and valuable property of the highbred Merino, (and which is often found without throatiness,) it is liked to a reasonable extent by most breeders. The skin should be of a fresh pink color—not dead white, and especially not tawny.

The wool of the Merino should be compact at all hazards, and of as great length as can be found united with compactness. It should open with some appreciable resistance to the hand, not drop apart at the touch, like the fur of furbearing animals. The pile, in addition to its fineness, should be finely and regularly crimped from one extremity to the other. This is an important indication of quality, and in the case of the American Merino, of blood. The pure French sheep does not so perfectly or so uniformly exhibit it. The interior of the wool (after it has gained length subsequently to shearing) should be brilliantly glossy, and when properly opened by the hand, every spire of its crimped filaments should seem to be moving, as if instinct with life. This last appearance (of which I can give no definite idea on paper) is the highest possible indication of good breeding. A dry, lustreless appearance, especially a dead appearance, is very objectionable. If, in addition to this, the wool is destitute of crimp, it is wholly inferior. Except near the outer end, wool should be white, or of a faint golden tinge. If saffron-colored near the skin, it is "yellowed," (by some abnormal secretion,) and injured for sale. Slightly brownish or nankeen-colored wools, unless so stained by earths, indicate defective breeding. French wools are oftener of this color than those of any other family of the Merino.

The gum whi:h is permitted to exist, should be on the outer extremity of the fleece, not scattered through it in small yellow particles resembling bee-bread, or in occasional white waxy concretions. The former defect is commonest in the American, the latter in the French Merino. Neither of them appertain to the Saxon. The oil of the fleece should appear like a delicate white perfectly transparent varnish, or some thinner fluid, barely coating over every fibre to give it lustre. As already said, it is objectionable to have it fill up the interstices of the wool, as if it had been poured in, and doubly so if its color is yellowish. If quite yellow and viscid, it is called " yolk."

The wool of the Merino should closely cover every wool-bearing part. It should be thick and long on the belly as well as on the back, and the bare spots for the movements of the legs, etc., should occupy only the surface absolutely necessary for that purpose. It should look, when its pelage is out at full length, like a bundle of wool on legs. But wool below the knees and hocks, and on the point of the nose, is like throatiness, one of those " fancy points" which is highly valued by some, and objected to by others. The wool on these parts is inferior, and trifling in weight. It does not, as novices often imagine, specially indicate a heavy fleece. That on the legs gets foul with mud or dug, when it comes in contact with it, and that on the nose often so impedes the sight, that unless it is sheared away two or three times a year, the animal can see neither forward nor backward, nor scarcely sideways, without awkwardly twisting about its head. I confess I rather like the peculiarity ; but there can be no doubt it would be undesirable in sheep which must travel and " look out for themselves" on extensive plains, and particularly so, if there was any chance of their being attacked by dogs or beasts of prey.

PRICE OF MERINOS.—I shall recur to this subject, because the incidental discussion which has taken place on it, in your paper, renders me desirous to submit some definite and tangible statements. I therefore say, definitely and tangibly, that pure-blood American Merino flocks of good quality, including the usual admixture of all ages and sexes, up to four years old, can be bought for eight dollars a head, where one hundred are taken ; for ten dollars a head, where fifty taken ; for twelve dollars a head, where twenty-five are taken ; for twenty-five dollars a head, where a half-dozen are taken. The pure-bred French sheep are comparatively few, and though unpopular with the mass of wool-growers, are highly prized by their breeders on account of their salableness in new regions. I can give no approach to a uniform price on them. Good high-bred French, grades, (a cross with the American Merino) resembling full-blood French can be bought at from fifteen to twenty-five per cent. advance on the price of American Merinos.

Mongrel American Merinos—not unfrequently denominated " full-bloods," by sheep-growers who have no record of pedigree, oftentimes no distinct conception of what constitutes a pedigree—can be purchased in nearly all the Northern and Middle States, at from two to four dollars a head, according to the prevailing market prices of sheep at the time. Most of them are a cross between the Saxon and " Native" sheep, with a later infusion of Merino blood. Where the Saxon admixture was strong, these sheep are often as fine as pure-blood Merinos. But their fleeces are lighter ; their constitutions much less vigorous; and like all mongrels made up between distinct races, they are lacking in uniformity.

COST OF IMPORTATION.—There are three ways of getting sheep from the Northern States to Texas—by the Ocean and Mississippi River routes, and by the land route. Where time is no object, and the number of sheep to be taken large, the latter is by far the cheapest. Freights from New-York City to Galveston, in ship-houses, (water found,) will average about three dollars per head at proper seasons of the year. When enough are sent to fill a ship-house, the usual cost is two dollars a head. The cost of arranging ship-house, keep, and attendance on the passage is then to be added. It should not exceed two dollars per head. Under proper arrangements, the passage is as safe as that of the human passenger of the vessel.

CROSSING WITH COARSE SHEEP.—It may be laid down as a settled rule, that the Merino can be improved, as wool-producing sheep by a cross with no other breed whatever. All legitimate crossing, for that object, is confined to the several varieties of its own breed. Secondly, there is no other breed the quality and quantity of whose wool is not improved by a Merino cross. It is a matter of economy first to stock an extensive wool estancia with coarse, cheap breeds of sheep. Any thing, from English long-wools down to the puny, miserable Mexican sheep, can be used; and with well-selected rams, (medium-sized, compact, oily, gummy, and heavy-fleeced American Merinos,) the rapidity of the improvement will appear almost miraculous to inexperienced persons. In selecting the coarse sheep, the carcass is of vastly more importance than the fleece, and hence the Mexicans are the least valuable. But even they are preferable to nothing.

None but the full-blood Merino ram should be used under any circumstances. A different course would, at best, lead to a retardation of the desired improvement, of more amount than many times the cost of the necessary full-blood rams; and the degree and kind of improvement would become wholly a matter of uncertainty.

Every breeder whose means admit of it, will do well also to start with a more limited flock of full-blood ewes. They constitute the foundation of a future pure flock, and are the nursery to draw rams from, without the expense of resorting to new purchases every two or three years. To meet this latter object, the ewes and rams originally imported should be of different strains of blood, and so marked as to be readily distinguishable from each other. All extensive breeders should keep two or three separate strains of blood, for the convenience of purchasers.

MISCELLANEOUS SUGGESTIONS.—Every new breeder should start with an established system of marks which will at once point out to him the blood of the particular animal. The brands may be cut out of wood, or constructed of iron, and they are dipped in some pigment and applied to the sheep (to prevent mistakes) as soon as it is sheared. On one side stamp the owner's initials, on the other a cross, a circle, a triangle, or the like, (or a combination of these marks,) to indicate the precise family.

Every sheep of inferior carcass of fleece, should receive a mark at shearing, which indicates that it is to be killed or sold.

On the subject of winter shelter and keep, I shall here offer nothing. In this particular, experience is the only guide.

But I repeat my former adjuration, to *keep down the dogs*—that curse of sheep-raising in Virginia, the Carolinas, etc., which is more fatal than all others, and which it is next to impossible to get rid of, where it has once got a firm footing.

Yours truly and sincerely,   HENRY S. RANDALL.

Cortland Village, New-York, Aug. 12, 1859.

SHEEP RAISING IN TEXAS.

BY GEORGE W. KENDALL, ESQ.

On commencing a third article on "Sheep Raising in Texas," giving my experience for another year, or since September, 1858, I find that I have but a repetition of the same old story to offer the many readers of your valuable almanac; my good fortune has continued without an interruption, and my losses amount to next to nothing. My flocks have all been extremely healthy, and in the best possible condition, while the few losses I have sustained, (not one per cent. probably,) have mostly resulted from casualty or accident of some kind; no disease has visited my folds. You may recollect, that I last year said that I could not hope for a continuance of such good luck or fortune as had followed me through the years '56–7 and a part of '58; it has continued up to this 1st of August, 1859, and my sheep are now in finer order than I have ever before seen them.

In the fall and early winter of 1858, or during the months of October, November, and December, I felt not a little uneasy about the effect of the acorns, of which we had a most abundant crop in the mountains. I had read in one book that they were hurtful to sheep; I had been told, by those who pretended to know, that their effect would certainly be injurious. To keep my flocks away from them was entirely out of the question; within a hundred yards of the pens where they were nightly kept, and in almost every direction, they must enter an oak range when turned out in the morning, every tree loaded heavily with acorns. The sheep devoured them with avidity, would run from tree to tree in the morning searching for such as had dropped fresh during the night, and this continued until the heaviest mast we have had in many years was exhausted. What with the grass they cropped meanwhile, (and it was noticed that each day the sheep would graze for hours,) they would come home to their pens at night well filled. And all this time the flocks were bright, healthy, and never in better condition, thus proving past all doubt, that acorns, instead of being injurious to sheep, are a positive benefit, and hereafter the heavier the mast may be at my place, the better I shall like it.

In April last, at shearing time, I smeared the noses of my sheep, and especially the lambs, bountifully with tar, and so far they have not been troubled at all with grub in the head. Last year, it may be remembered, I lost a few lambs from this cause; the tar certainly can do no harm, costs but a trifle, and I believe is beneficial during the spring and early summer months.

About the 15th of August, 1858, I weaned my lambs, over eleven hundred in number, all on the same morning: as nearly all were dropped in the month of April preceding, they were then about four months old. The Merino buck lambs I turned into my regular wether flock, where I keep all my bucks save during the short tupping season in the fall; the ewe and wether lambs I have kept in a flock by themselves up to this time, and all have grown and thriven remarkably well—far better than when the old ewes with their lambs ran together, and from one end of the year to the other.

Every one who knows anything about sheep must be well aware that long after a ewe has nearly dried up—when she gives but a drop of milk—the lamb will hang on and worry her, forty times a day, for that drop. It does the latter no good—it pulls down and pesters the old ewes—in short, injures both. For a day or two after they are separated, of course there will be a terrible outcry and clamor, lambs bleating for their mothers, and mothers calling for their lambs. But this is soon over; both soon set to work in earnest cropping their food, they have the entire day to fill themselves, and my experience has proved that both commence fattening within a week after the weaning is over. The old ewes have a chance to recruit and strengthen themselves before frost sets in, are in finer condition for the bucks in November, and pass through the winter in far better order.

I know that where a person has but a single flock of sheep, and that flock small, it creates an additional expense to separate and wean the lambs. But I hold that anything that is worth doing at all is worth doing well, and the additional expense will be more than repaid by the increased size, strength, condition, and constitution of the flock.

My last year's ewe-lambs, (those dropped in the spring of 1858,) I shall put to buck on the 1st of the coming November, or when they are some nineteen months old. They will then be two years of age when they have lambs; and I am confident this plan is much better than the one so often practised in Texas, of allowing yearling ewes to run with bucks and have lambs before they have attained their growth, and before they are well able to sustain their offspring. I do not increase my stock so fast by following this system; but I materially improve it, both in size and constitution, and that is what I am constantly striving after. We can all afford to be patient in Texas.

I shall have some two thousand ewes to put to buck this fall. Of these, about one hundred and thirty are full-blood Merinoes, which I shall turn into a pasture with two of the best bucks I can find, on 20th of October. On the 25th of the same month I shall put half of my grade ewes to buck, and on the 1st of November the balance. For *six weeks only* will the bucks be allowed to run with the ewes; I never wish to see a lamb come in one of my flocks later than the 15th of May. I have proved to my own satisfaction, that a lamb dropped on 1st of April, when the grass is young and fresh and the days comparatively cool, will be larger and better formed the day it is three months old, than will a lamb dropped on the 1st of July, when the grass is apt to be coarse and dry, and the days scorching hot, when it is six months old; and the former will turn out the best sheep in every respect. Many persons, anxious to increase the number of their flocks, may be loath to believe all this, but let them try both or all systems. The custom of allowing bucks to run with the ewes the year round, and having lambs come twice a year, or during every month in the year, I cannot but believe ruinous. It would worry me more to see a buck among my ewes in July, August, or September, or in February, March, or April, than a wolf: the latter might kill half a dozen, and there end; the former would cost me more real loss in the long run.

I am induced to give this statement in relation to my system because I am continually receiving letters from persons just starting in the sheep business, making inquiries on the subject. I do not say that I am right; I ask no one to follow my general plan of management. I shall change it the moment I hear of any one who has had better success than has befallen me, but not until then.

In the Texas almanac for 1859, I see that Thos. Decrow, Esq., after an

interesting account of his own great success in sheep-raising on Mata-
gorda Bay, sees fit to disagree with me in my estimate of the necessity
of breeding from no other than pure Merino bucks. Now, Mr. Decrow
may be right, and I altogether in the wrong; yet his argument does not
convince me that a grade buck, which is perhaps just as apt to breed
back as ahead, is as useful in a flock of Mexican ewes as a square-built,
compact, stout, vigorous, well-woolled, thorough-bred Merino, an animal
perfect in all those parts where the Mexican is naturally defective. I this
year sheared many grade sheep, three and four removes from common
Mexican ewes, which yielded 8, 8½, and some of them 9 pounds of wool,
and wool so fine that it would require a sharp sampler to distinguish it
from pure Merino, while the animals were perfect in form, lusty, and of
most vigorous constitution. I could not have got along so fast with grade
bucks, and I think Mr. Decrow was wrong when he says that he had
made up his mind to "sell or exchange his thirteen pure Merino bucks,
even at half-price, and furnish his newly purchased Mexican ewes (600 in
number) with rams of his own raising, from his own flock in preference."
The flock master who breeds altogether from pure Merino bucks, knows
always where he is, and where he will come out at the expiration of a
certain time ; but if he uses no other than grade rams, he is ever living
in uncertainty, and will never reach any particular end. My great object
is to breed up until every sheep I may own, may be safely marked a
thorough, full-blood Merino ; and in the course of a few years, should I
live, I shall achieve this result. But not in a century could I attain an
end I deem so desirable, were I to breed continually from grade rams.

I do not wash my sheep at all, and for what I deem good reasons.
About the middle of April, or at the time when one half my ewes have
young lambs at their sides, and the balance are about to drop, would be
the only time I could wash in this region. At this period I would not
race or worry my ewes at all on any account ; they should be pestered as
little as possible, and any advantage to the fleece from washing cannot
make up for the injury to the animal. I might wash my bucks and
wethers without injury, and my yearling lamb flock, made up of ewes and
wethers, and I may possibly try the experiment ; but my old ewes never.
Could my lambs come the latter part of February, as Mr. Decrow deems
best, I might then wash all ; but in this high mountain region yeaning
time cannot prudently come before the latter part of March or April, the
the very period when we must commence washing and shearing. We
are apt to have bad weather in February in this section, and even up to
the 15th or 20th of March. Now, as my lambs come at the outset at the
rate of over one hundred a day, a single cold, rainy or sleety norther
would carry off one half of those dropped during its continuance ; and
hence I say that in this parallel of latitude, and north of it, our yeaning
time cannot commence before the latter part of March, without running
great risk of loss.

I will not trespass farther upon your valuable space at this time :
another year, should you wish it, I will give you a fourth article upon my
experience in sheep raising in Texas.

Respectfully, your friend,

GEO. WILKINS KENDALL.

NEW BRAUNFELS, *August* 1, 1859.

INDEX.

2 Q

Barns for sheep, cut of. Page 205.
  ground-plan of, with sheds and yards, 209.
Barrack for hay, description and cut of, 209.
Bavaria, advantages of, for sheep husbandry, 114, 115.
Beans, value of, as a fodder, 213.
  straw of, fed to sheep in Germany, ...
Beet field, value of, as a fodder, 213.
  white Silesian, value of, as a fodder, 213.
Belgium, exports of wool from, 110.
  exports of wool from, to U. S. in 1846, 124.
  late increase of manufactures in, 294.
Beloochistan, advantages of, for sheep husbandry, 118.
Bermuda grass in the South, 38.
  its enormous product, 38.
  its adaptation to meadow or pasture, 38.
  its adaptation to barren sands, 38.
Bichloride of mercury, use of, in sheep medicine, 275.
Biflex canal, description of, 238.
  disease of, 261.
Bile, account of the, 231.
Biliary duct, description of the, 231.
Bladder, the, 233.
Blain, unusual in U. S., 222.
Blankets for slaves, description of, 87, 90.
  cost of manufacturing, 87, 90—92.
Bleeding, place for, 273, 274.
  rules for, 274.
  the quantity of blood to be abstracted in, 274.
Blood, the circulation of the, 235.
  the importance of purity of, in breeding, 168, 171, 172.
Blue grass, as the food of sheep, 212.
  in the North, 33.
  in the South, 37.
  on the Southern mountains, 44, 47, 48.
Blue Ridge of mountains, location of, 30.
  Also, see Apalachians.
  geology of, 30.
  soils and products of, 31, 44—47, 59.
  advantages of, for sheep husbandry, 44—47, 59.
Bone dust, as a manure in the South, 67.
Bot. See Grub in the Head.
Box for feeding grain to sheep, cut of, 203.
  for dipping lambs, cut of, 192.
Brain, description of the, 236.
Brazil, a portion of, in wool zone, 105.
  exports of wool from, 110.
  exports of wool from, to U. S. in 1846, 124.
Breeding, principles of, 168—172.
  importance of selection in, 168, 190.
  in and in, effects of, 169.
  in and in, how avoided, 170, 172.
  crossing, when admissible in, 170.
  crossing, how conducted, 172.
  crossing, method of starting flocks in the South by, 170.
  crossing, importance of selecting good rams for, 172.
  register, how kept, 180.
British America, exports of wool from, 110. to U. S. in 1846, 124.
British West Indies, exports of wool from, 110.
  to U. S. in 1846, 124.
Broad-tailed sheep introduced into the U. S., 151.
  wool and mutton of the, 151.

Bronchial tubes, the, 235.
Bronchitis, description and treatment of, 240.
Bronchocele. See Goitre.
Browse, feeding of, in winter, 217.
Buckwheat, value of, in producing live weight, wool and tallow, 214.
  per cent. of nitrogen in, 214.
  use of straw of, as a fodder, 213.
  straw of, fed to sheep in Germany, 213.
Buenos Ayres, advantages of, for sheep husbandry, 105, 106.
  advantages of, for sheep husbandry, compared with U. S., 106.
  exports of wool from, 105.
  exports of wool from, to U. S., in 1846 124.
  pampas of, 105.
  inhabitants of, 105.
Burdock, injurious to wool, 131.

C.

Cabbage, value of, as a fodder, 213.
Cabul, advantages of, for sheep husbandry. 118.
Cachectic diseases, 254, 255.
Camphor, use of, in sheep medicine, 275.
Cape of Good Hope, 65, 119.
  Merinos introduced in, 26.
  Merinos, their increase in, 26.
  exports of wool from, 110.
  exports of, to U. S. in 1846, 124.
  wool of, compared with Australia, 26.
  advantages of, for sheep husbandry, 65, 119.
  climate of, 26, 119.
  climate of, effect of, on quality of wool, 26.
  face of the country in, 119.
  remarkable drouths in, 119.
  prevalence of wild beasts in, 119.
Capillaries, functions of the, 234.
Caraway seeds, the use of, in sheep medicine, 275.
Cardiac opening, the, 231.
Carrots, value of, as a fodder, 213.
Castration of rams, 180.
Cataract, the, 239.
Catarrh, common, description and treatment of, 240.
  malignant epizootic, description of, 240—247.
  malignant epizootic, ravages of, in U. S., 240.
  malignant epizootic, treatment of, 245, 246.
Catechu, use of, in sheep medicine, 275.
Cattle doctor, the most dangerous of maladies, 226.
Caul. See Omentum.
Census of U. S., inaccurate in its wool returns, 18.
Cerebellum, the, 236.
Chalk, the use of, in sheep medicine, 275.
Chelmsford plains, for slave cloths, 86, 90.
  quality and cost of manufacturing, 90—93
Cheviot sheep, introduction of, into U. S. 149.
  description of, 149, 150, 154.
  low quality of their wool, 151.
Chili, portion of, in the wool zone, 105.
  exports of wool from, 110.
  exports of wool from, to U. S. in 1846 124.

APPENDIX.